Annual Reports in Organic Synthesis—1976

Annual Reports in Organic Synthesis

ANNUAL REPORTS IN ORGANIC SYNTHESIS—1970
John McMurry and R. Bryan Miller, Eds.

ANNUAL REPORTS IN ORGANIC SYNTHESIS—1971
John McMurry and R. Bryan Miller, Eds.

ANNUAL REPORTS IN ORGANIC SYNTHESIS—1972
John McMurry and R. Bryan Miller, Eds.

ANNUAL REPORTS IN ORGANIC SYNTHESIS—1973
R. Bryan Miller and Louis S. Hegedus, Eds.
John McMurry, Series Editor

ANNUAL REPORTS IN ORGANIC SYNTHESIS—1974
Louis S. Hegedus and Stephen R. Wilson, Eds.
R. Bryan Miller, Series Editor

ANNUAL REPORTS IN ORGANIC SYNTHESIS—1975
R. Bryan Miller and L. G. Wade, Jr., Eds.

ANNUAL REPORTS IN ORGANIC SYNTHESIS—1976
R. Bryan Miller and L. G. Wade, Jr., Eds.

Annual Reports in Organic Synthesis—1976

edited by

R. Bryan Miller
University of California, Davis, California

L. G. Wade, Jr.
Colorado State University, Fort Collins, Colorado

ACADEMIC PRESS New York San Francisco London 1977
A Subsidiary of Harcourt Brace Jovanovich, Publishers

Academic Press Rapid Manuscript Reproduction

ACADEMIC PRESS, INC.
111 Fifth Avenue, New York, New York 10003

United Kingdom Edition published by
ACADEMIC PRESS, INC. (LONDON) LTD.
24/28 Oval Road, London NW1

LIBRARY OF CONGRESS CATALOG CARD NUMBER:

ISBN 0-12-040807-4

PRINTED IN THE UNITED STATES OF AMERICA

CONTENTS

CONTENTS

PREFACE

One of the most difficult problems facing chemists today is that of "keeping up with the literature." For several reasons, the problem is particularly severe for the synthetic organic chemist. Bits of information of potential use to the synthetic chemist are scattered throughout common chemistry journals and can be found in any paper, not just those dealing strictly with synthesis. Thus a synthetic chemist must read a large number of journals. He must organize and index what he reads to make the information available for future reference. All snythetic chemists do this; but the task is becoming more difficult each year as the flow of information increases.

The problem however is shared to some extent by all. Most organic chemists are at some time faced with the problem of synthesizing a desired material and, for many, the problems are formidable. Nonspecialists faced with a synthetic problem are most likely not to have kept pace with the developments in synthetic chemistry that may well solve their problems and will not have the necessary information in their files.

Thus, we felt that an organized annual review of snythetically useful information would prove beneficial to nearly all organic chemists, both specialist and nonspecialist in synthesis. It should help relieve some of the information-storage burden of the specialist and should aid the nonspecialist who is seeking help with a specific problem to become rapidly aware of recent synthetic advances. Ideally also, such a review should be minimally priced to be within the means of potential users including graduate students, and it should appear as promptly as possible after the close of the abstracting period.

In producing *Annual Reports in Organic Synthesis—1976* we have abstracted 47 primary chemistry journals, selecting useful synthetic advances. We have tried to present the information in an organized manner, emphasizing rapid visual retrieval. Only the common journals received by our libraries have been extracted. Any journal received after March 1, 1977 will be covered in the next volume. We have also exercised selectivity in choosing which papers to abstract. Our general guidelines have been to include all reactions and methods that are new, synthetically useful, and reasonably general. Each entry comprises primarily of structures, accompanied by very few comments. The purpose of this is to aid the reader in scanning the book. The mind is capable of absorbing a whole picture in an instant, but is considerably slowed by having to read sentences. If the pictures presented catch the reader's interest, he should then seek details from the original paper.

For the first time we have included an author index to aid the user, but we still have no subject index. To include one would have greatly increased both the cost of the book and the delay time before publication. Instead, we have chosen to use an extensive table of contents. Chapters I–III are organized by reaction type and constitute the major part of the book. The organization of these sections is self-explanatory, and there should be no difficulty in locating a new method of oxidation or a new cyclopropanation procedure. Chapter IV deals with methods of synthesizing heterocyclic systems. Chapter V covers the use of new protecting groups and is also self-explanatory. Chapter VI is divided into three main parts and covers those synthetically useful transformations that do not fit easily into the first three chapters. The first part deals only with functional group synthesis. The second and third parts of Chapter VI are self-explanatory. The third part involves useful multistep sequences, the individual steps of which may be well known. Future volumes of this series will maintain the present table of contents as much as possible. If no entry is found for a particular section, the last volume in which one appears will be cited in the table of contents—see I.A.7.a.1b.

Any undertaking of this type involves a series of compromises. We have chosen to emphasize reasonable cost, rapid publication, and rapid visual retrieval of information at the admitted expense of detail and beauty. This volume is the seventh in an annual series. We welcome suggestions for improvement of future volumes and would like to thank Professor Samuel Danishefsky for his useful comments concerning the author index.

The arduous task of drawing the multitude of structures appearing in this review was carried out by Ms. Linda Benedict and Ms. Sandi Hanson. We thank them very much for their efforts.

R. Bryan Miller
L. G. Wade, Jr.

JOURNALS ABSTRACTED

Accounts of Chemical Research
Acta Chemica Scandinavica
Angewandte Chemie International Edition in English
Annales de Chimie
Australian Journal of Chemistry
Bulletin of the Chemical Society of Japan
Bulletin de Societes Chimiques Belges
Bulletin de la Societe Chimique de France
Canadian Journal of Chemistry
Chemical Communications
Chemical and Pharmaceutical Bulletin
Chemical Reviews
Chemical Society Reviews
Chemische Berichte
Chemistry and Industry
Chemistry Letters
Collection of Czechoslovakian Chemical Communications
Comptes Rendus Hebdomadaires de Seances de l'Academie des Sciences (C)
Doklady Chemistry
Experientia
Fortschritte der Chemischen Forschung
Gazzetta Chimica Italiana
Helvetica Chimica Acta
Indian Journal of Chemistry
Israel Journal of Chemistry
Journal of the American Chemical Society
Journal of the Chemical Society (Perkin I)
Journal of the Chemical Society (Perkin II)
Journal of General Chemistry (USSR)
Journal of Heterocyclic Chemistry
Journal of the Indian Chemical Society

JOURNALS ABSTRACTED

Journal of Organic Chemistry
Journal of Organic Chemistry (USSR)
Journal of Organometallic Chemistry
Journal fur Praktische Chemie
Liebig's Annalen der Chemie
Monatschefte fur Chemie
Pure and Applied Chemistry
Recueil des Travaux Chimiques des Pays-bas
Russian Chemical Reviews
Steroids
Synthesis
Synthetic Communications
Tetrahedron
Tetrahedron Letters
Zeitschrift fur Chemie
Zeitschrift fuer Naturforschung, Teil B

I.A.1-1 J. d'Angelo, Tetrahedron, 32, 2979 (1976).

Review: Ketone Enolates: Regiospecific Preparation and Synthetic Uses.

I.A.1-2 T. Tanaka and T. Mukaiyama, Chem. Lett., 1976, 1259.

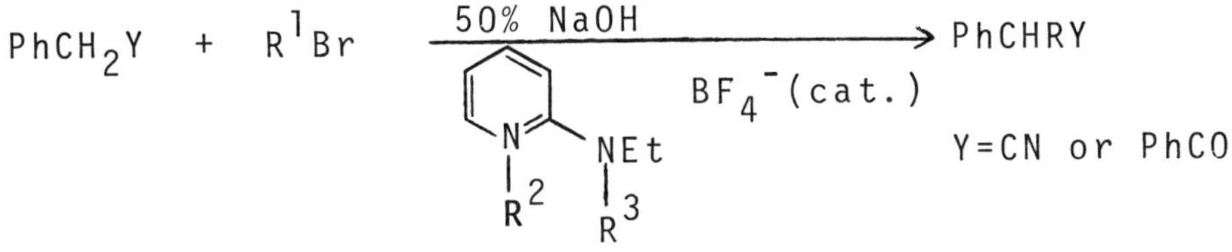

A study of the two-phase catalyst, 2-dialkylamino-pyridinium salts, in alkylation reactions was carried out.

I.A.1-3 W. P. Reeves and R. G. Hilbrich, Tetrahedron, 32, 2235 (1976).

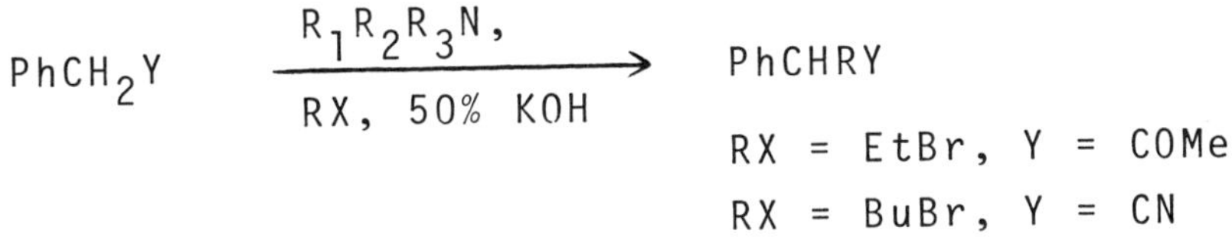

A study of the phase transfer catalysis of various 1°, 2°, and 3° amines was carried out.

I.A.1-4 J. F. Bunnett, X. Creary, and J. E. Sundberg, J. Org. Chem., 41, 1707 (1976).

$$PhCH{=}CHBr \xrightarrow[h\nu]{CH_2{=}C(OK)Me} PhCH{=}CHCH_2COMe \; (48\%) + PhCH_2CH{=}CHCOMe \; (34\%)$$

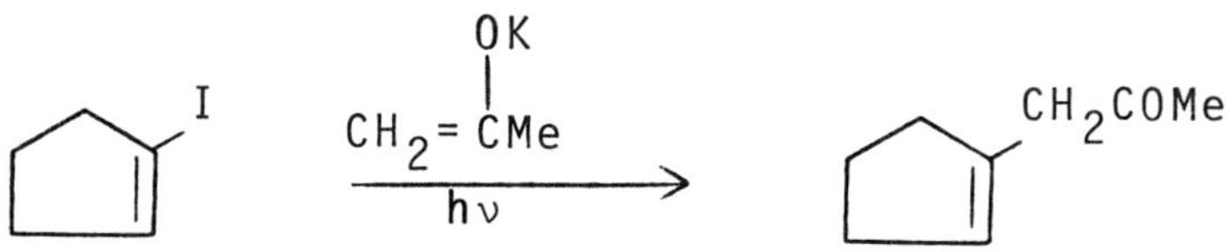

I.A.1-5 J. M. Fortunato and B. Ganem, J. Org. Chem., 41, 2194 (1976).

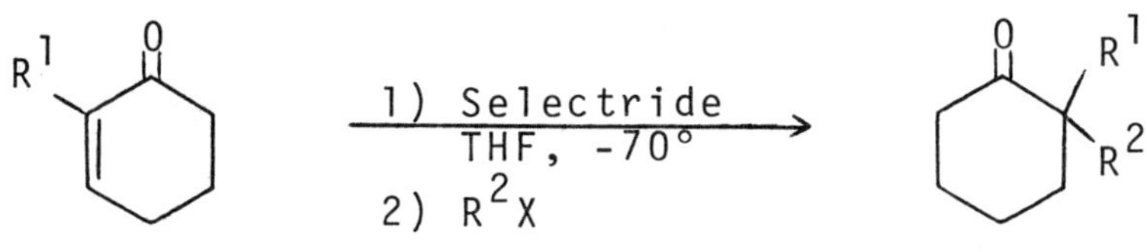

Selectride	R^1	R^2X	% Yield
K	H	$CH_2{=}CHCH_2Br$	55
L	H	$CH_2{=}CHCH_2Br$	85
L	Me	$MeO_2CCH{=}CHCH_2I$	60

α,β-Unsaturated esters can also be reductively alkylated.

I.A.1-6 P. A. Chaloner and A. B. Holmes, J. Chem. Soc., Perkin Trans. I, 1976, 1838.

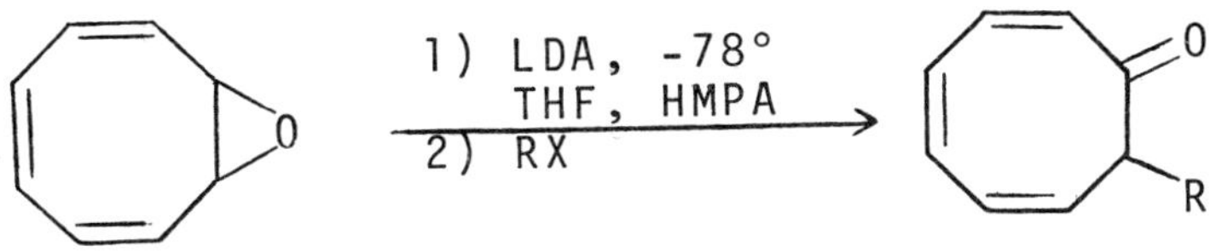

RX	% Yield
MeI	78
CH_2=$CHCH_2Br$	74
EtI	81
$PhCH_2I$	37 (51% of R=H)

I.A.1-7 M. Hajek and J. Malek, Coll. Czech. Chem. Commun., 41, 746 (1976).

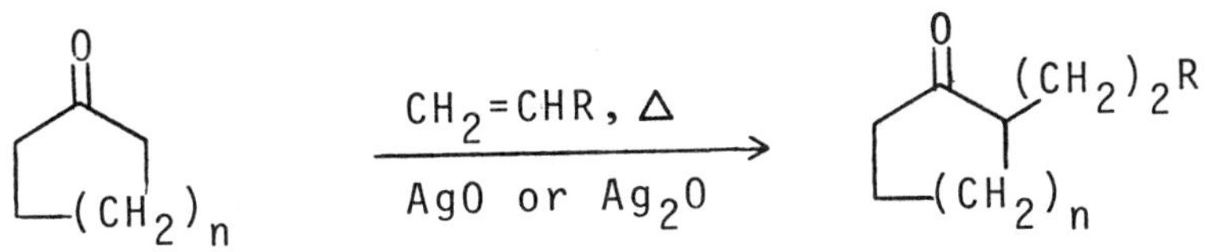

The procedure uses a ten-fold excess of ketone.

I.A.1-8 H.-J. Bestmann and R. W. Saalfrank, *Chem. Ber.*, **109**, 403 (1976).

$$R^1CH_2COR^2 + Ph_3P{=}C{=}C(OEt)_2 \longrightarrow Ph_3P{=}CHC(OEt){=}CR^1COR^2$$

R^1	R^2	% Yield
H	Ph	65
H	Me	55
H	OEt	74
$-(CH_2)_4-$		69

I.A.1-9 Y. Leroux and C. Jacquelin, *Synth. Commun.*, **6**, 597 (1976).

$$PhC{\equiv}CCH(OEt)_2 \xrightarrow[\text{2) RX; 3) } H_2O]{\text{1) 2 eq. n-BuLi-TMEDA, THF, }-78°} PhCR{=}CHCO_2Et$$

RX	% Yield
Me_2SO_4	79
EtI	85
i-PrBr	86
$EtOCH_2Cl$	57

I.A.2-1 Y. Masuyama, Y. Ueno, and M. Okawara, Tetrahedron Lett., 1976, 2967.

$$NCCHR^1SC(S)NMe_2 \xrightarrow[\text{aq. NaOH}]{R^2X} NCCR^1R^2SC(S)NMe_2$$

R^1	R^2X	% Yield
H	MeI	82
H	$CH_2{=}CHCH_2Cl$	86
Me	$n\text{-}C_5H_{11}Br$	94
Et	$n\text{-}C_6H_{13}Br$	88

The products are readily converted to the corresponding ketones or aldehydes.

I.A.2-2 A. J. Birch and J. Slobbe, Aust. J. Chem., 29, 2737 (1976).

$$MeCH{=}CHCH{=}CHCO_2H \xrightarrow[\text{2) RX; 3) esterification}]{\text{1) Li, } NH_3(\ell)} EtC(H){=}C(H)CHRCO_2Me$$

RX	% Yield
MeI	∿ 60
i-PrBr	∿ 70
$CH_2{=}CHCH_2Br$	∿ 60

Protonated material (R=H) was a significant by-product (25-40%).

I.A.2-3 P. Hullot, T. Cuvigny, M. Larcheveque, and H. Normant, Can. J. Chem., 54, 1098 (1976).

$$R^1R^2CHCONMe_2 \xrightarrow[\text{2) } R_3X\text{, } Et_2O \text{ or THF}]{\text{1) } Et_2NLi\text{, PhH, HMPA}} R^1R^2R^3CCONMe_2$$

R^1	R^2	R^3X	% Yield
H	H	$C_6H_{13}Br$	82
H	i-Pr	MeBr	70
Me	Me	n-BuBr	86
Me	Me	$CH_2=CHCH_2Br$	80
H	i-Pr	$CH_2=C(Br)CH_2Br$	69

I.A.2-4 R. A. Gorski, G. J. Wolber, and J. Wemple, Tetrahedron Lett., 1976, 2577.

$$R^1CH_2\overset{O}{\overset{\|}{C}}SR^2 \xrightarrow[\text{2) } R^3X\text{, -78°}]{\text{1) LDA, THF, -78°}} R^1R^3CH\overset{O}{\overset{\|}{C}}SR^2$$

R^1	R^2	R^3X	% Yield
H	$PhCH_2$	MeI	75
H	t-Bu	$CH_2=CHCH_2Br$	71
$\text{-(}CH_2\text{)}_2\text{-}$		MeI	74
$\text{-(}CH_2\text{)}_2\text{-}$		n-BuI	35

I.A.2-5 B. M. Trost and L. S. Melvin, Jr. J. Am. Chem. Soc., 98, 1204 (1976).

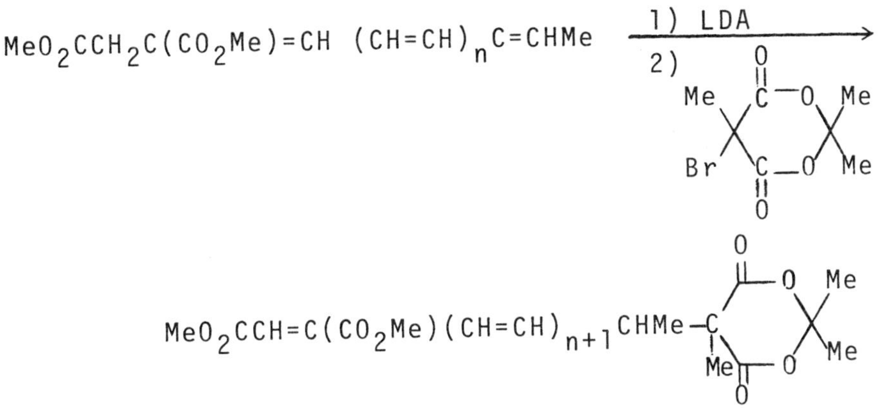

n = 0, 66%

n = 1, 38%

I.A.2-6 K. Saigo, M. Osaki, and T. Mukaiyama, Chem. Lett., 1976, 769.

$R^1R^2C(OMe)_2 + R^3R^4C{=}C(OSiMe_3)(OMe) \xrightarrow[CH_2Cl_2,\ -78°]{TiCl_4} R^1R^2C(OMe)CR^3R^4CO_2Me$

R^1	R^2	R^3	R^4	% Yield
$PhCH_2$	H	Ph	H	100
$PhCH_2$	H	$-(CH_2)_5-$		92
Me	Me	$PhCH_2$	Me	94
H	MeO	$PhCH_2$	H	77

I.A.2-7 E. Nakamura, M. Shimizu, and I. Kuwajima, <u>Tetrahedron Lett</u>., <u>1976</u>, 1699.

$$Me_3SiCH_2CO_2Et \xrightarrow[Bu_4NF]{R^1R^2CO} R^1R^2C(OSiMe_3)CH_2CO_2Et$$

R^1	R^2	% Yield
PhCH=CH	H	81
n-PrCH=CH	H	82
Ph	Ph	88
Ph	Me	0

If R^1R^2CO has acidic α-hydrogens the reaction fails.

I.A.2-8 L. M. Lerner, <u>J. Org. Chem</u>., <u>41</u>, 2228 (1976).

$$(EtS)_2CHCO_2Me \xrightarrow[2)\ RX]{1)\ NaH,\ DMF} (EtS)_2CRCO_2Me$$

RX = MeI, 100%

RX = i-PrI, 91%

I.A.2-9 W. Beck and M. Girnth, <u>Chem. Ber</u>., <u>109</u>, 965 (1976).

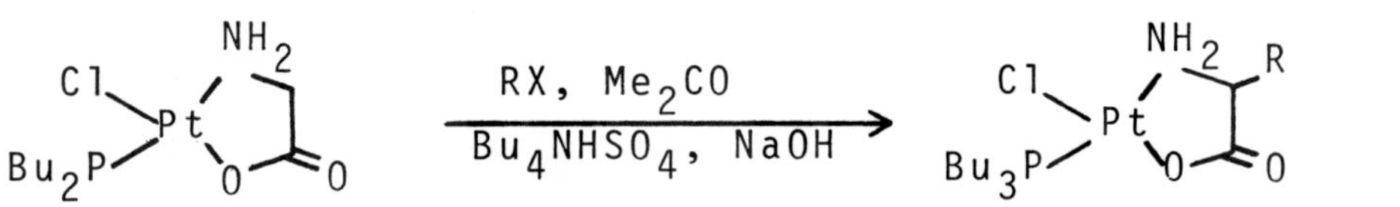

RX = $MeOSO_3Me$ or $PhCH_2Cl$

70-80%

I.A.2-10 A. P. Krapcho and E. A. Dundulis, Tetrahedron Lett., 1976, 2205.

$$PhCONHCH_2CO_2H \xrightarrow[\text{2) } RCH_2X]{\text{1) 3 eq. LDA, TMEDA, THF}} PhCONHCH(CH_2R)CO_2H \quad 23\%-60\%$$

$$PhCONHCH_2CO_2Et \xrightarrow[\text{2) } RCH_2X]{\text{1) 2 eq. LDA, TMEDA, THF}} PhCONHCH(CH_2R)CO_2Et \quad 48\%-71\%$$

The products can be hydrolyzed to α-amino acids.

I.A.2-11 G. Stork, A.Y.W. Leong, and A. M. Touzin, J. Org. Chem., 41, 3491 (1976).

$$PhCH{=}NCH_2CO_2Et \xrightarrow[\text{2) RX}]{\text{1) LDA or t-BuOK, THF}} PhCH{=}NCHRCO_2Et$$

R = n-C_8H_{17}, 90%

R = i-Pr, 75%

R = CH_2CO_2Et, 82%

A one-flask sequential dialkylation procedure is also described.

I.A.3-1 H. Oediger and F. Möller, Justus Liebigs Ann. Chem., 1976, 348.

$$R^1CH_2R^2 \; + \; 2\, R^3Br \xrightarrow[\text{DMF}]{\text{DBU}} R^1R^2CR^3_2$$

R^1	R^2	R^3	% Yield
CO_2Me	CN	$PhCH_2$	77
COMe	COMe	$PhCH_2$	84
CN	CN	$PhCH_2$	91
CN	SO_2Ph	n-Bu	75

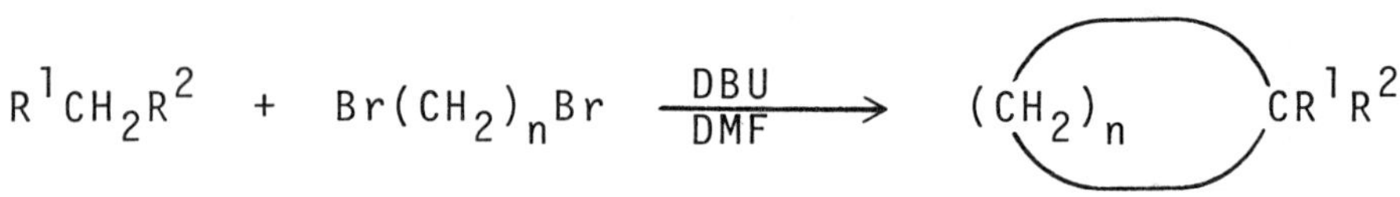

n = 4 or 5 ∿60%

I.A.3-2 T. Kurihara, Y. Nakajima, and O. Mitsunobu, Tetrahedron Lett., 1976, 2455.

$$NCCH_2CO_2Et \; + \; HO(CH_2)_nOH \xrightarrow[\text{Ph}_3\text{P, THF}]{EtO_2CN{=}NCO_2Et}$$

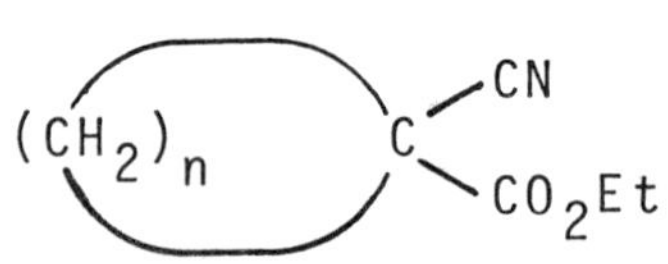

n = 4, 46%
n = 5, 40%
n = 6, 21%

I.A.3-3 J. Nakayama, J. Chem. Soc., Perkin Trans. I, 1976, 540.

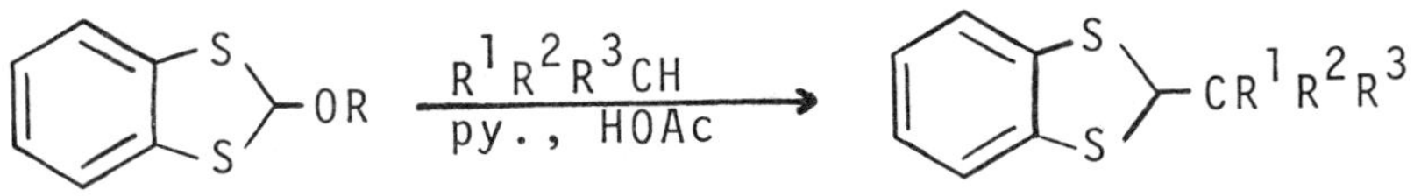

R^1	R^2	R^3	% Yield
H	MeCO	MeCO	90
Me	MeCO	MeCO	73
CHO	$-CO(CH_2)_3$		55

I.A.3-4 B. Eistert, A. Schmitt, and T. J. Arackal, Chem. Ber., 109, 1549 (1976).

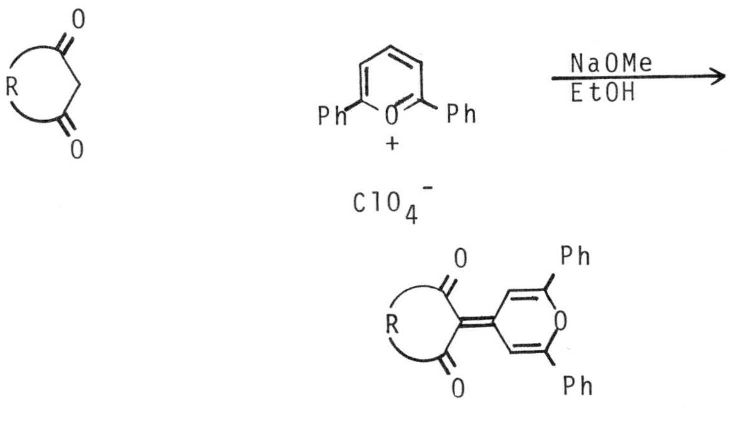

69-97%

Ring size of β-diketone is 6-membered.

I.A.3-5 J. Schnekenburger, D. Heber, and E. Heber-Brunschweiger, Justus Liebigs Ann. Chem., 1976, 1799.

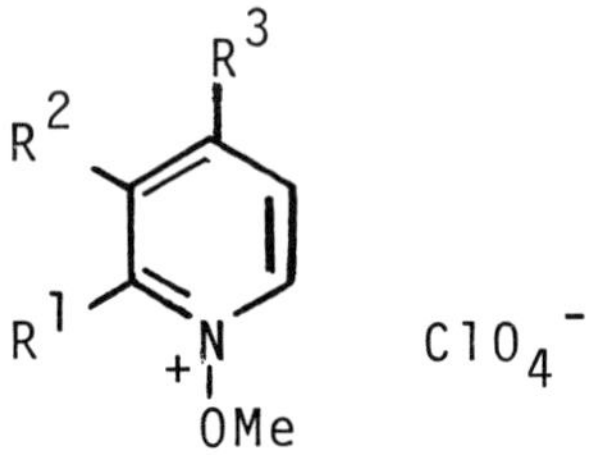

$\xrightarrow[\text{DMF or HMPA}]{CH_2R^4R^5,\ Et_3N}$

$R^4R^5CH{=}CHCR^3{=}CR^2CR^1{=}NCOMe$

R^1	R^2	R^3	R^4	R^5	% Yield
CN	H	H	$COCO_2Et$	CO_2Et	48
H	CN	H	COMe	CN	31
H	H	CN	COMe	COMe	37
CN	H	H	$COCH_2CMe_2CH_2CO$		38

I.A.3-6 N. Ono, H. Eto, R. Tamura, J. Hyami, and A. Kaji, Chem. Lett., 1976, 757.

$$NaCR^1(CN)(CO_2Et) \quad + \quad R^2R^3C(Br)NO_2 \quad \xrightarrow[120°]{HMPA}$$

$$R^2R^3C{=}CR^1CN$$

R^1	R^2	R^3	% Yield
i-Pr	Me	Me	70
i-Pr	$-(CH_2)_5-$		63
$PhCH_2$	Me	Et	75

I.A.3-7 F. Cooke and P. Magnus, J. Chem. Soc., Chem. Commun., 1976, 519.

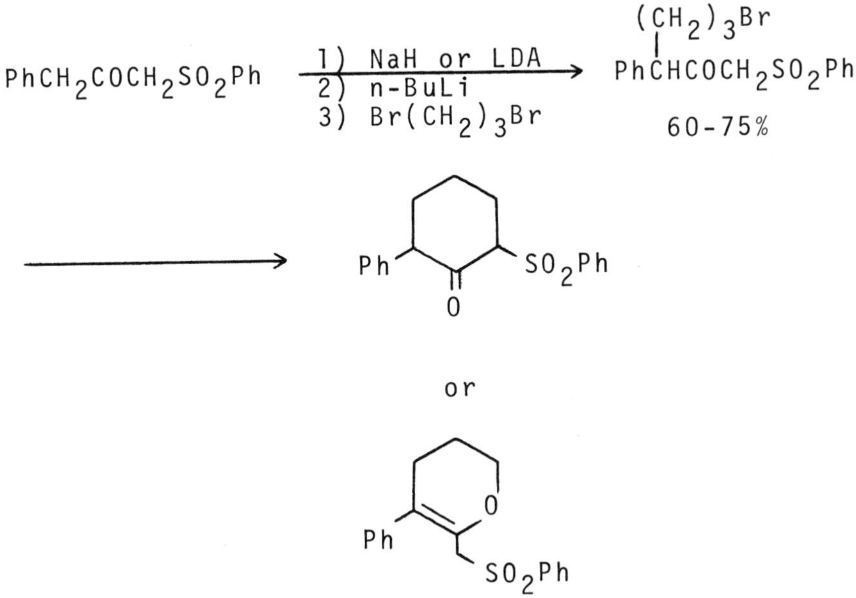

A study of conditions for C- vs. O-alkylation was carried out.

I.A.4-1 H. H. Wasserman, M. J. Hearn, B. Haveaux, and M. Thyes, J. Org. Chem., 41, 153 (1976).

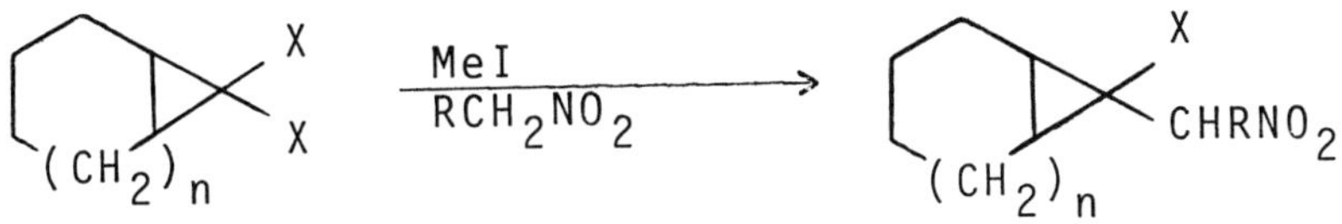

n	X	R	% Yield
1	N-pyrrolidino-	H	78
1	N-pyrrolidino-	Me	45
0	N-piperidino-	H	50
0	N-piperidino-	Me	26

I.A.4-2 R. Henning, F. Lehr, and D. Seebach, *Helv. Chim. Acta*, **59**, 2213 (1976).

$$ArCH_2CH_2NO_2 \xrightarrow[\text{2) RX}]{\text{1) 2 eq. n-BuLi, THF, HMPA, }-78^\circ} ArCHRCH_2NO_2$$

R = n-C_6H_{13}, Ar = p-ClC_6H_4, 76%

R = $PhCH_2$, Ar = Ph, 75%

$$PhCH_2CH_2NO_2 \xrightarrow[\text{2) } PhCH{=}CHNO_2]{\text{1) 2 eq. n-BuLi, THF, HMPA, }-78^\circ} PhCH(CH_2NO_2)CH(CH_2NO_2)Ph$$

80%

I.A.4-3 M. E. Jung, P. A. Blair, and J. A. Lowe, *Tetrahedron Lett.*, **1976**, 1439.

$$R^1C(=N{-}OH)CH_2R^2 \xrightarrow[\text{2) } R^3I,\ -78^\circ]{\text{1) 2 eq. BuLi, THF}} R^1C(=N{-}OH)CHR^2R^3$$

R^1	R^2	R^3	% Yield
Me	H	Me	73
Me	Me	Me	81
Me	Me	i-Pr	46
$(CH_2)_4$		Me	74

If the *syn* alkyl group is disubstituted the reaction does not proceed cleanly.

I.A.4-4 W. G. Kofron and M.-K. Yeh, J. Org. Chem., 41, 439 (1976).

$$R^1C(=NOH)CH_2R^2 \xrightarrow[\text{2) } R^3X \text{ 3) hydrolysis}]{\text{1) 2 eq. n-BuLi, THF}} R^1COCHR^2R^3$$

R^1	R^2	R^3X	% Yield
Me	H	$PhCH_2Cl$	72
H	Me	$PhCH_2Cl$	37
$-(CH_2)_4-$		$CH_2=CHCH_2Br$	65
$-(CH_2)_3-$		n-BuI	59

I.A.4-5 R. R. Fraser and K. L. Dhawau, J. Chem. Soc., Chem. Commun., 1976, 674.

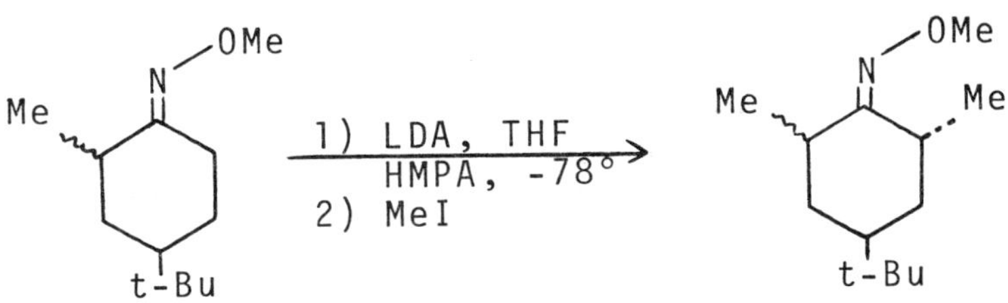

86-94%

Only *syn*-axial alkylation was observed independent of the relative stereochemistry of the starting material.

I.A.4-6 E. J. Corey and D. Enders, Tetrahedron Lett., 1976, 3.

$$R^1CH_2C(=NNMe_2)R^2 \xrightarrow[2)\ R^3X,\ -78^\circ]{1)\ LDA,\ THF,\ 0^\circ} R^1R^3CHC(=NNMe_2)R^2$$

R^1	R^2	R^3X	% Yield
H	$n\text{-}C_5H_{11}$	MeI	95
H	$n\text{-}C_5H_{11}$	i-PrI	83
Ph	H	MeI	95
$-(CH_2)_3CHMe-$		MeI	95

A mild oxidative hydrolysis procedure of the products to carbonyl compounds is also described.

I.A.4-7 T. Cuvigny, J. F. LeBorgne, M. Larcheveque, and H. Normant, Synthesis, 1976, 237.

$$R^1CH_2CH{=}NNMe_2 \xrightarrow[2)\ R^2Br]{1)\ Et_2NLi,\ PhH,\ HMPT} R^1R^2CHCH{=}NNMe_2$$

R^1	R^2	% Yield
Et	Et	75
Et	$CH_2{=}CHCH_2$	76
Bu	Bu	79

I.A.4-8 J. F. LeBorgne, T. Cuviguy, M. Larcheveque, and H. Normant, Synthesis, 1976, 238.

$$R^1R^2CHCH{=}NNMe_2 \xrightarrow[\text{2) } R^3Br]{\text{1) } Et_2NLi,\ PhH,\ HMPT} R^1R^2R^3CCN$$

R^1	R^2	R^3	% Yield
Me	Me	n-Bu	88
Et	Et	Et	93
Et	Et	$CH_2{=}CHCH_2$	87

I.A.4-9 G. R. Kieczykowski, R. H. Schlessinger, and R. B. Sulsky, Tetrahedron Lett., 1976, 597.

$$MeCH{=}CHCH{=}NC_6H_{11} \xrightarrow[\text{2) RX, } -78°]{\text{1) LDA, THF, HMPA, } 0°} CH_2{=}CHCHRCH{=}NC_6H_{11}$$

RX = MeI, 98%

RX = $CH_2{=}CHCH_2Br$, 100%

$$MeCH{=}CHCH{=}NC_6H_{11} \xrightarrow[\text{2) } BrCH_2CO_2Me]{\text{1) LDA, THF, HMPA, } 0°} MeCH{=}C(CH_2CO_2Me)CH{=}NC_6H_{11}\quad 100\%$$

Alkylation of the initial products leads to quaternary α-carbon atoms.

I.A.4-10 S. Brenner and H. G. Viehe, Tetrahedron Lett., 1976, 1617.

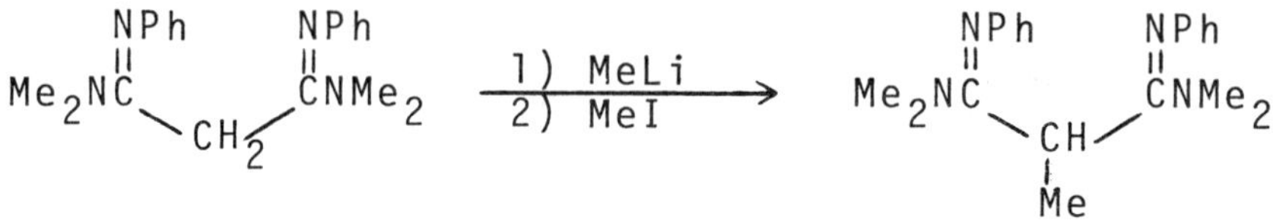

80%

Use of $NaNH_2$ as the base gave exclusive N-alkylation.

I.A.4-11 H. Hoberg and U. Griebsch, Synthesis, 1976, 830.

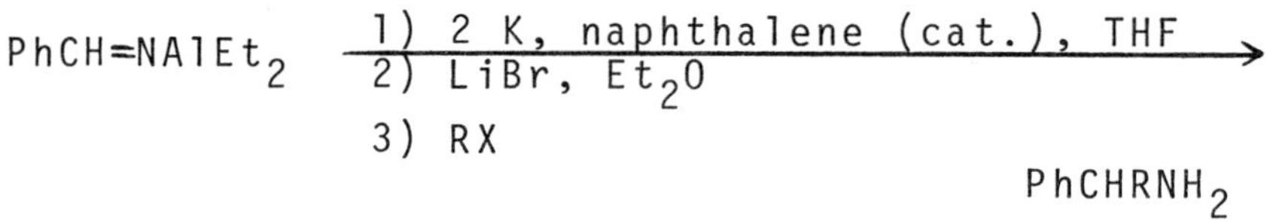

R = Me, 41%

R = CH_2=$CHCH_2$, 61%

I.A.4-12 R. R. Schmidt and J. Talbiersky, Angew. Chem. Int. Ed. Engl., 15, 171 (1976).

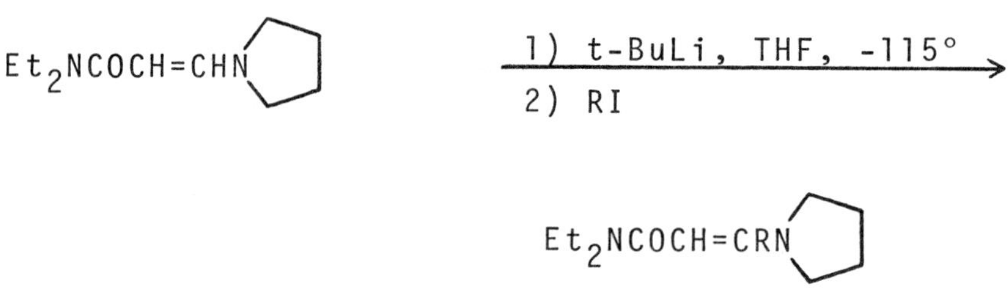

R = Me, 95%

R = Et, 60%

I.A.4-13 D. Seebach and W. Lubosch, Angew. Chem. Int. Ed. Engl., 15, 313 (1976).

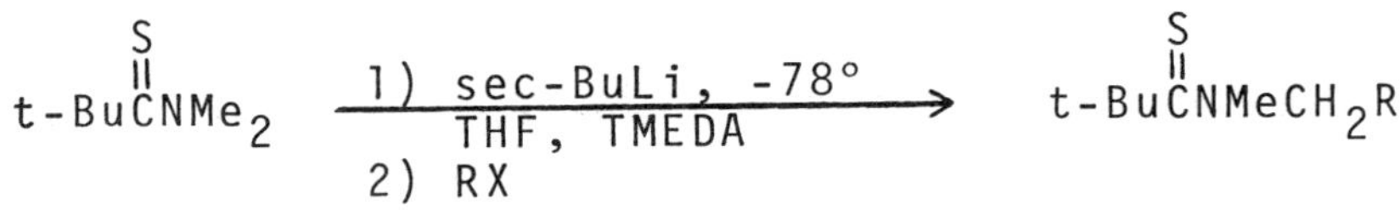

RX	% Yield
MeI	80
$n\text{-}C_5H_{11}I$	82
$PhCH_2Br$	44

I.A.4-14 C. F. Beam, C. W. Thomas, R. M. Sandifer, R. S. Foote, and C. H. Hauser, Chem. Ind. (London), 1976, 487.

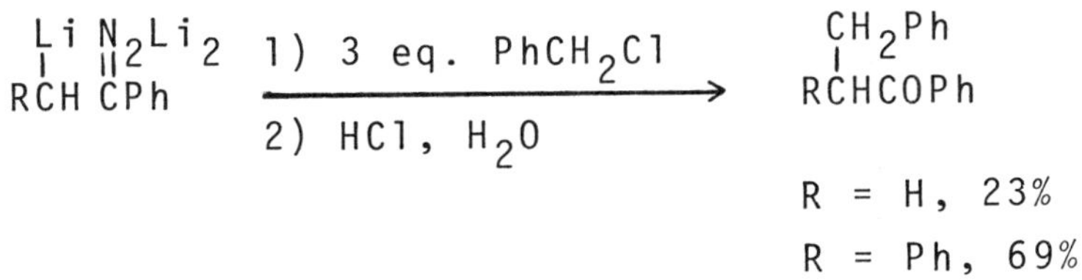

I.A.4-15 T. Cohen, D. A. Bennett, and A. J. Mura, Jr., J. Org. Chem., 41, 2506 (1976).

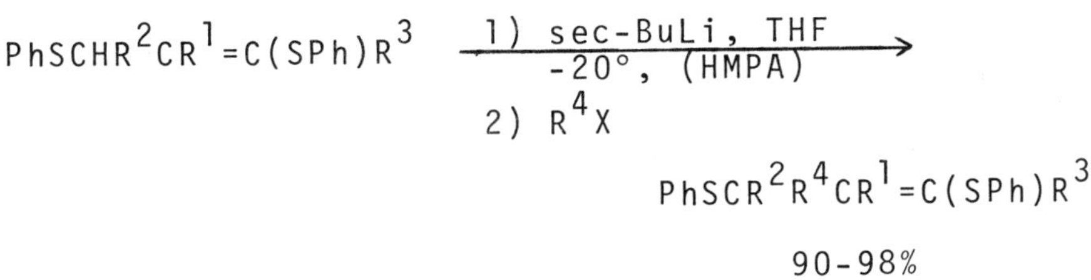

I.A.4-16 M. Julia and D. Uguen, Bull. Soc. Chim. Fr., 1976, 513.

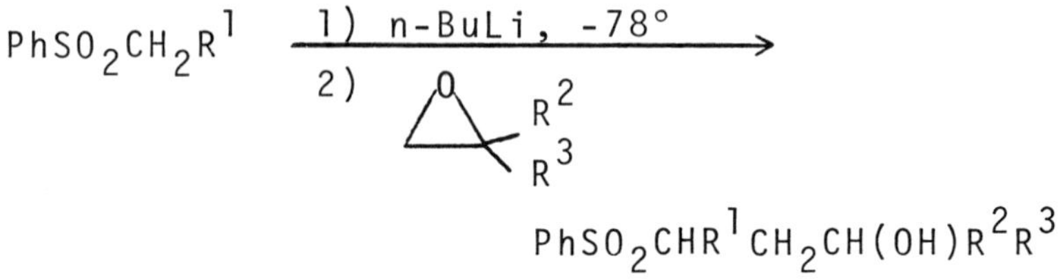

R^1	R^2	R^3	% Yield
H	Et	H	100
Ph	Me	H	98
$PhCH_2CH_2$	Et	H	94
$Me_2C=CH$	Me	$CH=CH_2$	97

I.A.4-17 Y. Gaoni, Tetrahedron Lett., 1976, 503.

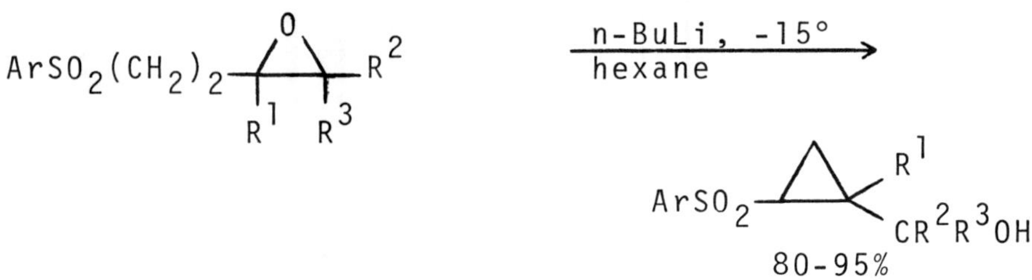

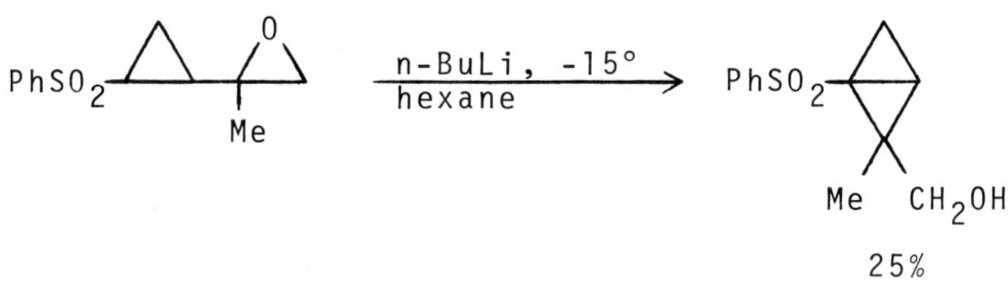

I.A.4-18 B. Corbel and T. Durst, J. Org. Chem., **41**, 3648 (1976).

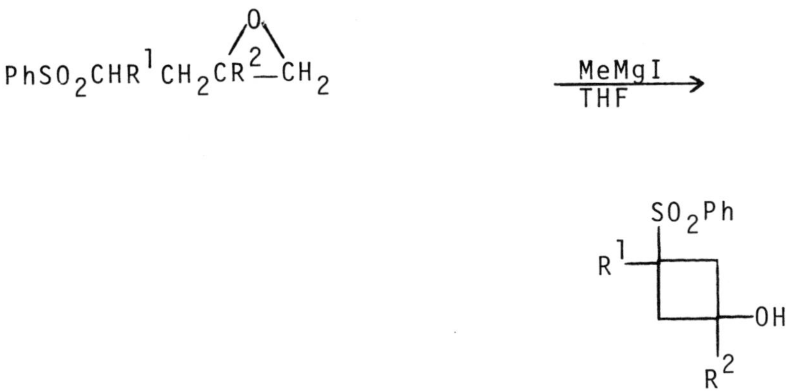

$R^1 = R^2 = H$, 96%
$R^1 = H$, $R^2 = Me$, 51%
$R^1 = Ph$, $R^2 = H$, 98%

I.A.4-19 K. Sachdev and H. S. Sachdev, Tetrahedron Lett., **1976**, 4223.

$$Me_3SiCH_2SePh \xrightarrow[\text{2) } RCH_2Br]{\text{1) LDA, THF, } -78°} Me_3SiCH(SePh)CH_2R$$

R = n-Bu, 94%
R = $Ph(CH_2)_2$, 88%

The products can be oxidized (30% H_2O_2) to aldehydes.

I.A.5-1 E. V. Ermilova, L. A. Remizova, I. A. Favorskaya, and N. L. Tregubova, J. Org. Chem. USSR, 11, 517 (1975).

$$HC{\equiv}CCH_2OH \xrightarrow[RCH_2Br]{LiNH_2,\ NH_3(\ell)} RCH_2C{\equiv}CCH_2OH$$

40-60%

Use of the lithium salt allows direct C-alkylation without protection of the hydroxyl group.

I.A.5-2 E. M. Kaiser and J. D. Petty, J. Organomet. Chem., 107, 219 (1976).

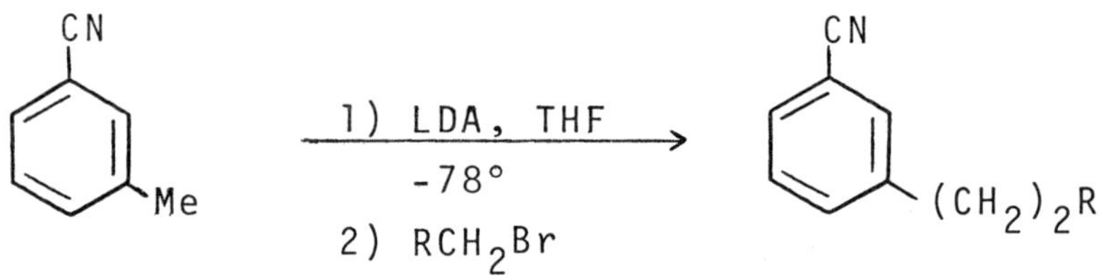

30-37%

I.A.5-3 J. E. Stemke, A. R. Chamberlin and F. T. Bond, Tetrahedron Lett., 1976, 2947.

$$n\text{-}C_6H_{13}\overset{\overset{NNHTs}{\|}}{C}Me \xrightarrow[2)\ RX]{1)\ 3\ eq.\ BuLi,\ TMEDA} n\text{-}C_6H_{13}CR{=}CH_2$$

R = Me, 71%
R = n-Bu, 65%

I.A.5-4 H. W. Thompson and B. S. Huegi, J. Chem. Soc., Perkin Trans. I, 1976, 1603.

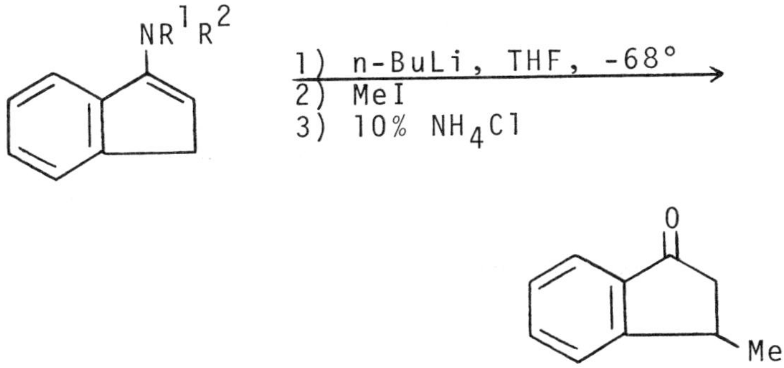

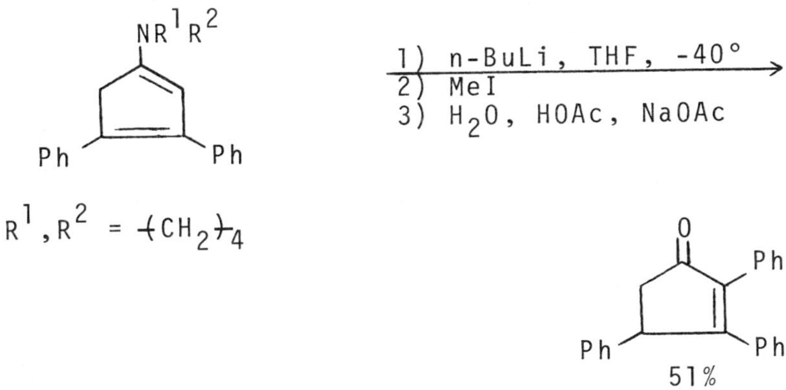

Also works well for 2-indanone.

I.A.5-5 D. Seebach and N. Meyer, Angew. Chem. Int. Ed. Engl., 15, 438 (1976).

$$n\text{-}Bu_3SnCH_2OH \xrightarrow[\text{2) RBr}]{\text{1) 2 eq. n-BuLi, n-}C_6H_{14}} RCH_2OH$$

45%

R = n-C_8H_{17} or $PhCH_2$

I.A.5-6 I. Vlattas, L. D. Vecchia, and A. O. Lee, J. Am. Chem. Soc., 98, 2008 (1976).

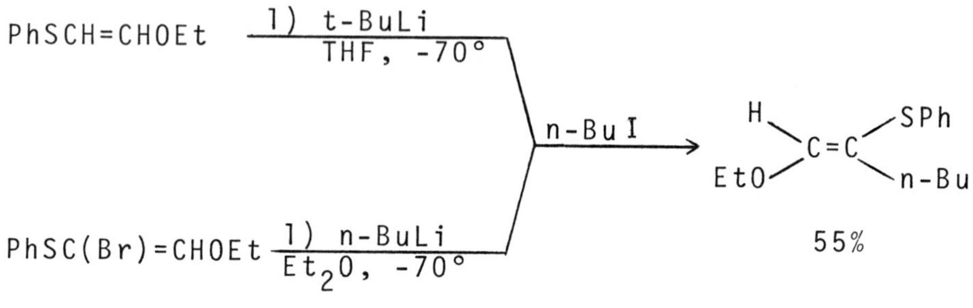

I.A.5-7 J. J. Eisch and J. E. Galle, J. Organomet. Chem., 121, C10 (1976); see also: *ibid*, J. Am. Chem. Soc., 98, 4646 (1976).

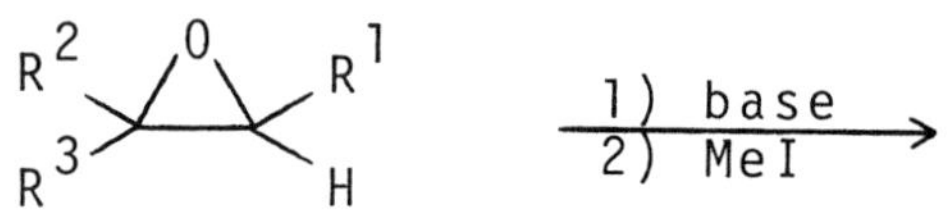

R^1	R^2	R^3	base	% Yield
Ph_3Si	H	H	n-BuLi	73
Me_3Si	$n\text{-}C_6H_{13}$	H	t-BuLi, TMEDA	80
$PhSO_2$	H	Ph	n-BuLi	46
$(EtO)_2P(O)$	H	H	LDA	37

I.A.5-8 T. H. Chan and B. S. Ong, Tetrahedron Lett., 1976, 319.

$$R^1R^2C(OSiMe_3)(SPh) \xrightarrow[\text{HMPA}]{R^3Li} R^1R^2R^3COSiMe_3$$

R^1	R^2	R^3	% Yield
Ph	H	Me	86
Ph	H	n-Bu	83
Me	Me	n-Bu	38
$(CH_2)_5$		Me	32

I.A.5-9 W. Oppolzer and R. L. Snowden, Tetrahedron Lett., 1976, 4187.

$$(CH_2{=}CH)_2CHOSiEt_3 \xrightarrow[\text{THF, } -78°\text{; 2) RX}]{\text{1) sec-BuLi}} CH_2{=}CHC(OSiEt_3){=}CHCH_2R \ (\mathbf{A}) + (CH_2{=}CH)_2CROSiEt_3 \ (\mathbf{B})$$

RX	% Yield	A:B
MeI	86	100:0
$CH_2{=}CH(CH_2)_2I$	77	66:33
$CH_2{=}CH(CH_2)_2OTs$	68	10:90
$PhCH_2Br$	84	80:20

I.A.5-10 J. A. Katzenellenbogen and A. I. Crumrine, J. Am. Chem. Soc., 98, 4925 (1976).

$$R^1CMe{=}CHCO_2H \xrightarrow[\text{3) } R^1R^3C{=}CR^4CH_2Br]{\text{1) LDA, THF; 2) CuI, THF}} R^1C(CH_2CH_2CR^4{=}CR^2R^3){=}CHCO_2H$$

Study of both regioselectivity and mode of substitution of allylic alkylating agent was carried out. The Cu enolates give high proportions of γ-alkylation.

I.A.5-11 W. Priester, R. West, and T. L. Chwang, J. Am. Chem. Soc., 98, 8413 (1976); see also: W. Priester and R. West, *ibid*, 8421 and 8426.

$$Me_nCH_{3-n}C{\equiv}CH \xrightarrow{\text{n-BuLi, hexane}} Me_nC_3Li_{4-n}$$

$$\xrightarrow{Et_2SO_4} Et_{3-n}Me_nCC{\equiv}CEt$$

Use of EtI as the alkylating agent caused oxidative coupling.

I.A.5-12 S. Bhanu, E. A. Khan, and F. Scheinmann, J. Chem. Soc., Perkin Trans. I, 1976, 1609.

$$BuCH_2C{\equiv}CI \xrightarrow[\text{hexane, } Et_2O \text{ 2) } n\text{-}C_5H_{11}Br]{\text{1) 3 eq. n-BuLi, } -60^\circ} BuCH(C_5H_{11})C{\equiv}CH \quad 52\%$$

A comparison of the reactions of 1-iodo and 1-boro-alkynes was carried out.

I.A.5-13 A.J.G. Sagar and F. Scheinmann, Synthesis, 1976, 321.

$$n\text{-}PrCH_2C{\equiv}CH \xrightarrow[\text{2) } n\text{-}C_6H_{13}Br \text{ 3) } Me_2SO_4]{\text{1) 2 eq. n-BuLi}} n\text{-}PrCH(C_6H_{13})C{\equiv}CMe \quad 50\%$$

I.A.5-14 J. Millon and G. Linstrumelle, Tetrahedron Lett., 1976, 1095.

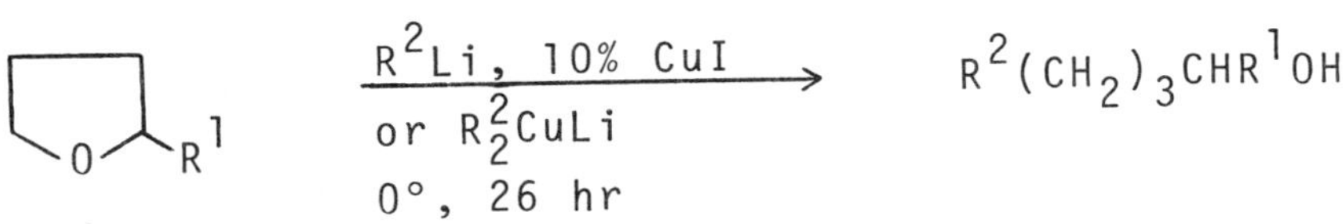

R^1	R^2	% Yield
H	n-Bu	66-68
H	n-Pr	60
Me	n-Bu	41

Similar reactions with tetrahydropyran gave an 8% yield of analogous product.

I.A.5-15 K. Kitatani, T. Hiyama, and H. Nozaki, J. Am. Chem. Soc., 98, 2362 (1976).

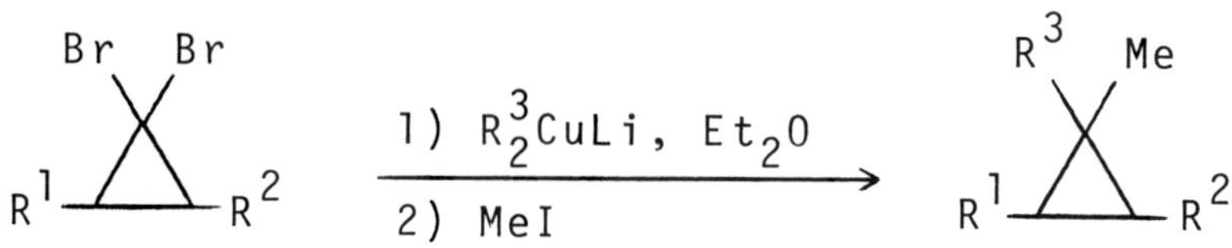

R^1	R^2	R^3	% Yield
Ph	H	n-Bu	100
Ph	H	sec-Bu	43
n-C_6H_{13}	H	n-Bu	50
$(CH_2)_4$		n-Bu	82

A stereoselective one-pot procedure is described.

I.A.5-16 P. A. Grieco, C.-L.J. Wang, and G. Majetich, J. Org. Chem., 41, 726 (1976).

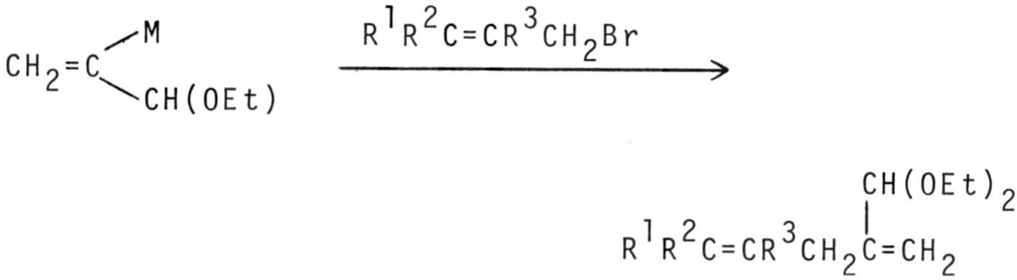

M = $PhSCu^-$, Li^+, 74-85%

M = Cu, 50-90%

I.A.5-17 L. Blaszczak, J. Winkler, and S. O'Kuhn, *Tetrahedron Lett.*, **1976**, 4405.

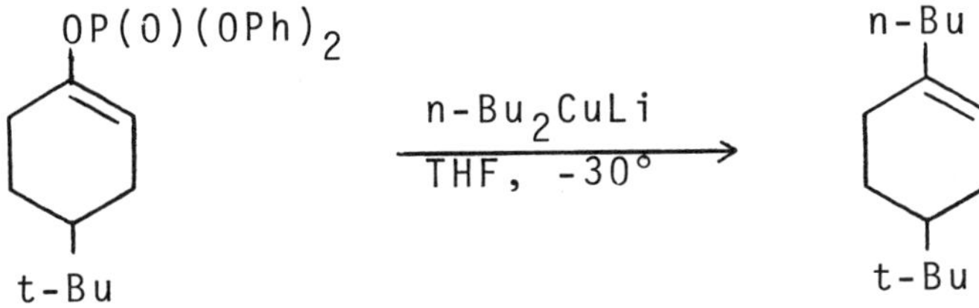

I.A.5-18 T. A. Baer and R. L. Carney, *Tetrahedron Lett.*, **1976**, 4697.

$$Br(CH_2)_nCO_2MgCl \xrightarrow[Li_2CuCl_4\ (cat.)]{RMgBr} R(CH_2)_nCO_2H$$

R = C_5H_{11}, n = 10, 94%

R = sec-Bu, n = 4, 88%

I.A.5-19 G. M. Whitesides and F. D. Gutowski, *J. Org. Chem.*, **41**, 2882 (1976).

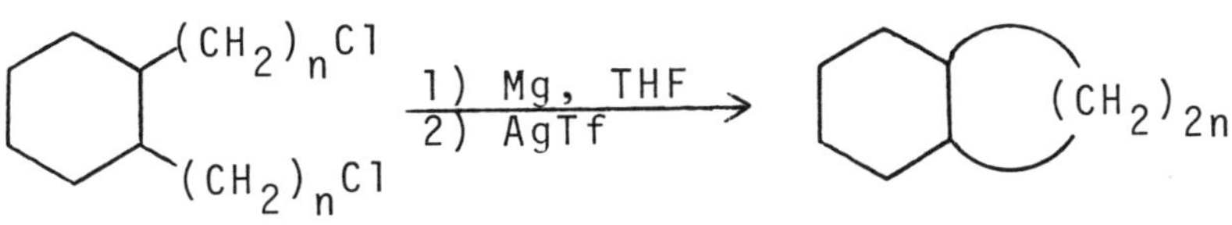

n = 1, 61%

n = 2, 57%

This procedure does not work for formation of 7-membered or higher ring systems.

I.A.5-20 G. N. Dorofeenko, L. V. Mezheritskaya, and E. S. Matskovskaya, J. Org. Chem. USSR, 11, 2476 (1975).

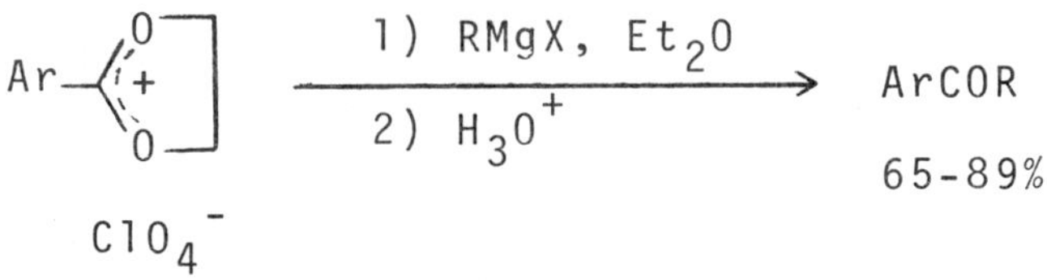

I.A.5-21 I. Degani and R. Fochi, J. Chem. Soc., Perkin Trans I, 1976, 1886.

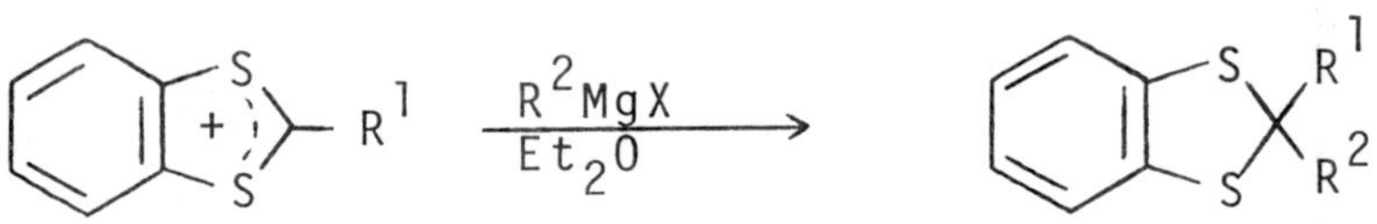

R^1	R^2	% Yield
H	Ph	87
H	n-Pr	88
Ph	Me	66
Ph	Ph	30

I.A.5-22 G. Zweifel and R. A. Lynd, Synthesis, 1976, 816.

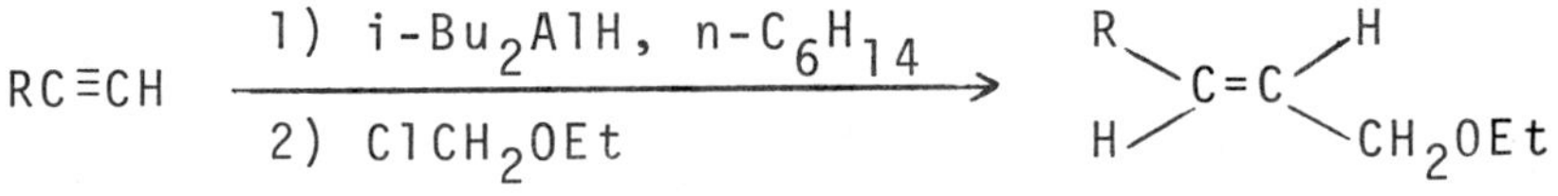

R = n-Bu, 80%
R = t-Bu, 75%

I.A.5-23 S. Baba, D. E. Van Horn, and E. Negishi, Tetrahedron Lett., 1976, 1927.

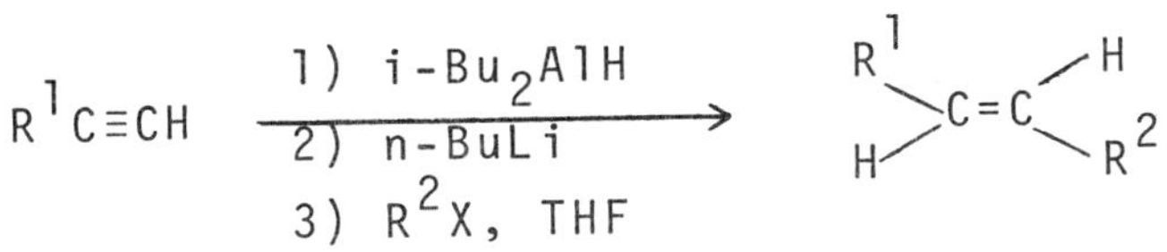

R^1	R^2X	% Yield
n-C_5H_{11}	MeI	65
n-Bu	n-$C_8H_{17}I$	49
n-Bu	CH_2=$CHCH_2Br$	73
n-Bu	$PhCH_2Br$	46

I.A.5-24 J. J. Eisch and G. A. Damasevitz, J. Org. Chem., 41, 2214 (1976); see also: K. Uchida, K. Utimoto, and H. Nozaki, *ibid*, 2215.

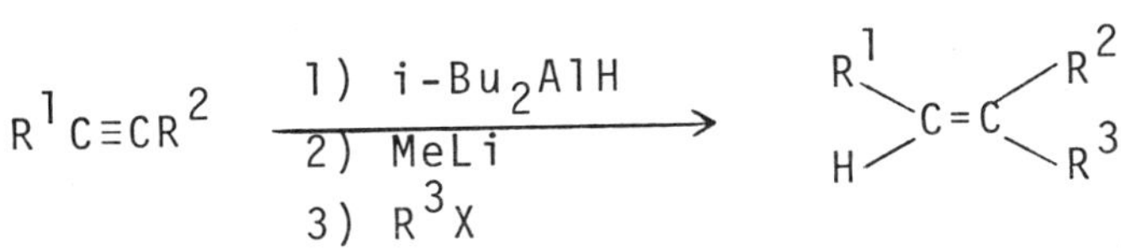

R^1	R^2	R^3	% Yield
Ph	Ph	CH_2=$CHCH_2$	50-70
n-C_6H_{13}	H	Me	60
Ph	$SiMe_3$	CH_2=$CHCH_2$	40*

*Depending on reaction conditions either the E or Z isomer can be synthesized stereoselectively.

I.A.5-25 K. Uchida, K. Utimoto, and H. Nozaki, J. Org. Chem., 41, 2941 (1976).

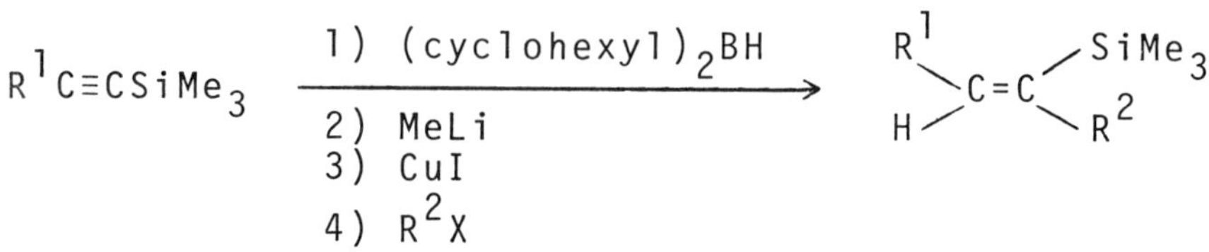

R^1	R^2X	% Yield
$n-C_6H_{13}$	MeI	94
$n-C_6H_{13}$	$CH_2=CHCH_2Cl$	93
$THPO(CH_2)_4$	EtI	85

I.A.6-1 M. Tokuda, T. Taguchi, O. Nishio, and M. Itoh, J. Chem. Soc., Chem. Commun., 1976, 606.

$$PhC{\equiv}CH \xrightarrow[n\text{-}Bu_4NI,\ HMPA;\ RX]{\text{electrolysis}} PhC{\equiv}CR$$

R = Me, Et, n-Bu

79-100% (39-81% conversion)

I.A.6-2 R. Alkabets and I. Granoth, J. Chem. Soc., Perkin Trans. I, 1976, 2380.

$$Ar_2CH_2 \xrightarrow[(MeOCH_2CH_2)_2O]{LiAlH_4,\ 150^\circ} Ar_2CMe_2$$

50-80%

I.A.6-3 N. M. Przhiyalgovskaya, E. N. Sidorenko, and A. T. Prudchenko, J. Org. Chem. USSR, 11, 1671 (1975).

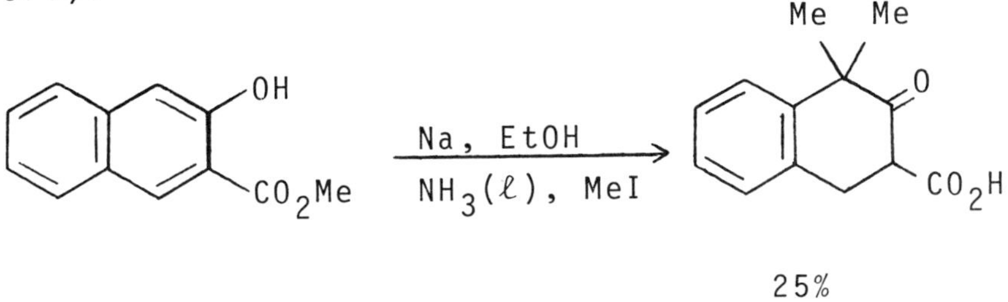

I.A.6-4 H. Kise, Y. Arase, S. Shiraishi, M. Senō, and T. Asahara, J. Chem. Soc., Chem. Commun., 1976, 299.

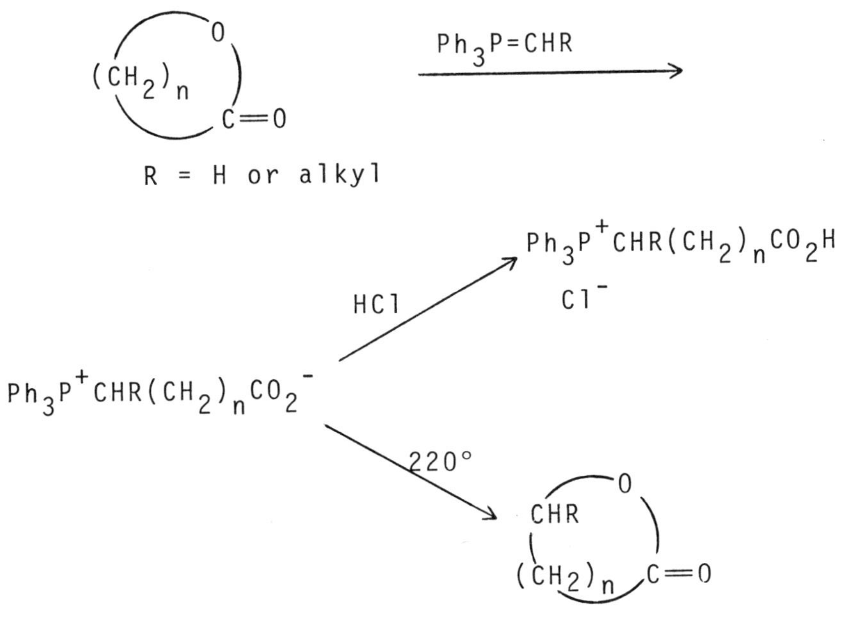

I.A.6-5 J. M. Midgley, J. S. Millership, and W. B. Whalley, J. Chem. Soc., Perkin Trans. I, 1976, 1384.

$$ArCH(OSiMe_3)C{\equiv}CH \xrightarrow[\text{reflux} \\ \text{2) } CO_2]{\text{1) EtMgBr, THF}} ArCHEtC{\equiv}CCO_2H$$

$$ArCH_2OSiMe_3 \xrightarrow{NaCH(CO_2Et)_2} ArCH_2CH(CO_2Et)_2$$

$Ar = 2,4,6\text{-}(MeO)_3C_6H_2$

I.A.7.a.1a-1 G. Wittig, Fortschr. Chem. Forsch., 67, 1 (1976).

Review: Old and New in the Field of Directed Aldol Condensations.

I.A.7.a.1a-2 G. W. Gokel, S. A. Di Biase, and B. A. Lipisko, Tetrahedron Lett., 1976, 3495.

$$R^1COR^2 \xrightarrow[MeCN]{\text{powdered KOH}} R^1R^2C{=}CHCN$$

$R^1 = Ph$, $R^2 = H$, 82%

$R^1,R^2 = -(CH_2)_5-$, 50%

I.A.7.a.1a-3 I. Kuwajima, T. Sato, M. Arai, and N. Minami, Tetrahedron Lett., 1976, 1817; see also: I. Kuwajima, N. Minami, and T. Sato, *ibid*, 2253.

$$R^1COMe \quad + \quad R^2CHO \xrightarrow[\text{THF, -40°}]{Me_2CHCH_2C(SiMe_3)_2OLi} R^1COCH_2CH(OH)R^2$$

71-86%

Use of the hindered base allows the directed aldol to proceed without prior formation of the ketone enolate.

I.A.7.a.1a-4 T. Mukaiyama and T. Inoue, Chem. Lett., 1976, 559.

$$R^1COCH_2R^2 \xrightarrow[\text{2) } R^3CHO]{\text{1) } Bu_2BOTf,\ R_3N,\ Et_2O} R^1COCHR^2CH(OH)R^3$$

R^1	R^2	R^3	% Yield
Ph	Et	Ph	80
Ph	H	$PhCH_2CH_2$	75
i-Bu	H	C_5H_{11}	70
$-(CH_2)_4-$		$PhCH_2CH_2$	72

I.A.7.a.1a-5 Ae. de Groot and B.J.M. Jansen, Recl. Trav. Chim. Pays-Bas, 95, 81 (1976).

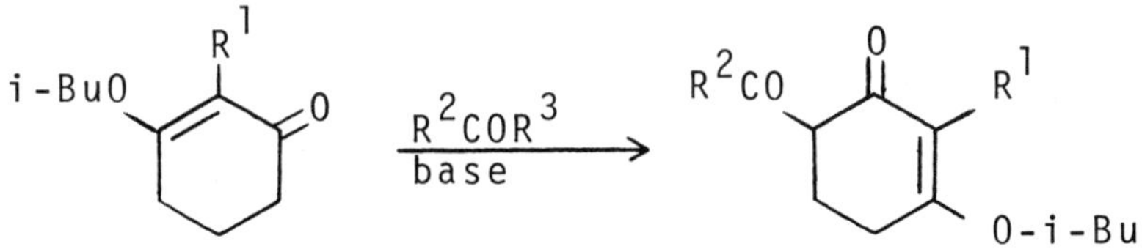

R^1	R^2	R^3	Base	% Yield
H	Me	OEt	$(Me_3Si)_2NLi$	78
H	H	O-i-Bu	NaH	95
Me	H	O-i-Bu	NaH	68

I.A.7.a.1a-6 G. Stork and G. A. Kraus, J. Am. Chem. Soc., 98, 2351 (1976).

MeCOCH=C(OMe)Me $\xrightarrow[\text{2) } R^1COR^2\text{, THF, -78°}]{\text{1) LDA, THF, -78°}}$

$R^1R^2C(OH)CH_2COCH{=}C(OMe)Me$

The product can be converted by reduction and acid hydrolysis to vinylogous aldols ($R^1R^2C(OH)CH_2CH{=}CHCOMe$).

I.A.7.a.1a-7 T. M. Harris, G. P. Murphy, and A. J. Poje, J. Am. Chem. Soc., 98, 7733 (1976).

$$RCOCH_2COCH_2COMe \xrightarrow[\text{THF, 0°; 2) } CO_2(g)]{\text{1) excess LDA}} RCOCH_2COCH_2COCH_2CO_2H$$

R = Ph, 82%
R = Me, 72%
R = Et, 42%

I.A.7.a.1a-8 G. P. Chiusoli and F. Gasparoni, Gazz. Chim. Ital., 106, 201 (1977).

$$RCH_2CO_2Li \xrightarrow[\text{2) } CH_2O]{\text{1) } Et_2NLi\text{, hexane}} HO_2CCHRCH_2CHRCO_2H$$

63%

I.A.7.a.1a-9 A. G. Schultz and M. H. Berger, J. Org. Chem., 41, 585 (1976).

$$Me_2CHCO_2CH_2OMe \xrightarrow{\text{LDA, THF}} Me_2C(CH_2OH)CO_2Me$$

69%

$$MeCH{=}CMeCO_2CH_2OMe \xrightarrow{\text{LDA, THF}} CH_2{=}CHCMe(CH_2OH)CO_2Me$$

61%

I.A.7.a.1a-10 D. A. Konen, P. E. Pfeffer, and L. S. Silbert, <u>Tetrahedron</u>, **32**, 2507 (1976).

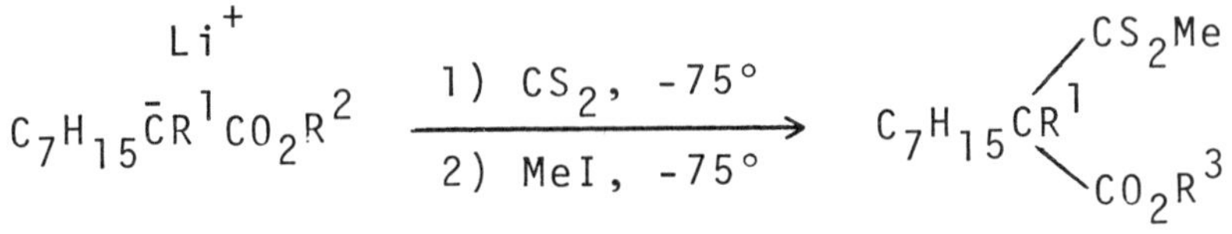

R^1	R^2	R^3	% Yield
H	Li	H	88
H	Me	Me	85
Me	Li	H	77
Me	Me	Me	75

I.A.7.a.1a-11 S. L. Hartzell and M. W. Rathke, <u>Tetrahedron Lett.</u>, **1976**, 2757.

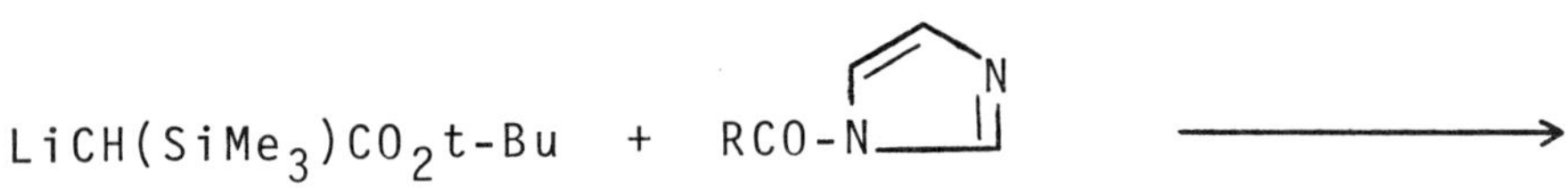

$RCOCH_2COt\text{-}Bu$

R = Me, 94%

R = PhCH=CH, 50%

I.A.7.a.1a-12 S. L. Hartzell and M. W. Rathke, Tetrahedron Lett., 1976, 2737.

$$(Me_3Si)_2CHCO_2t\text{-}Bu \xrightarrow[\text{2) RCHO}]{\text{1) LDA, THF, -78°}} RCH{=}C(SiMe_3)CO_2t\text{-}Bu$$

70-90%

I.A.7.a.1a-13 W. Beck and M. Girnth, Chem. Ber., 109, 965 (1976).

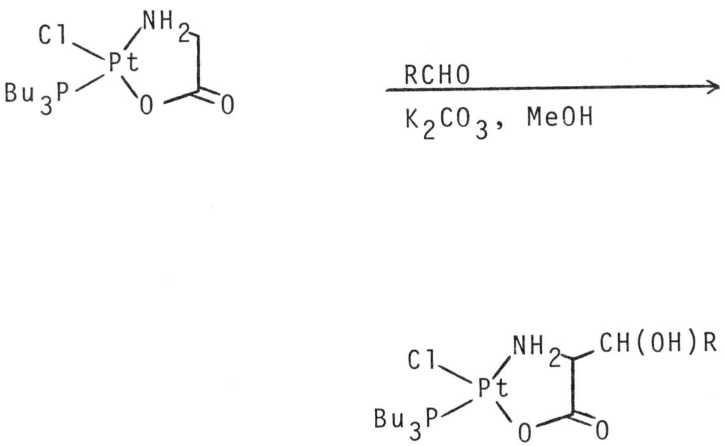

R = Me or Ph, 95%

I.A.7.a.1a-14 D. A. Evans, L. K. Truesdale, and K. G. Grimm, J. Org. Chem., 41, 3335 (1976).

$$\mathrm{RCHO} \xrightarrow[\text{KCN}\cdot\text{18-crown-6}]{\mathrm{M_3SiC(N_2)CO_2Et,\ CHCl_3}} \mathrm{RCH(OSiMe_3)C(N_2)CO_2Et}$$

R = Ph, 86%

R = n-C_5H_{11}, 86%

R = MeCH=CMe, 44%

I.A.7.a.1a-15 K. Smith and K. Swaminathan, J. Chem. Soc., Chem. Commun., 1976, 387.

$$\mathrm{HCON(i\text{-}Pr)_2} \xrightarrow[\text{Trapp mixture; 2) } R^1COR^2]{\text{1) t-BuLi, -95°}} \mathrm{R^1R^2C(OH)CON(i\text{-}Pr)_2}$$

R^1	R^2	% Yield
Ph	Ph	85
Ph	H	80
Et	H	62
PhCH=CH	H	68

I.A.7.a.1a-16 D. Seebach, W. Lubosch, and D. Enders, Chem. Ber., 109, 1309 (1976).

$$HC(=S)NMe_2 \xrightarrow[2)\ R^1COR^2]{1)\ LDA,\ THF,\ -100°} R^1R^2C(OH)C(=S)NMe_2$$

R^1	R^2	% Yield
Et	H	65
Ph	H	65
Me	Me	75
Ph	Ph	80
$-(CH_2)_5-$		50

I.A.7.a.1a-17 G. Kinast and L.-F. Tietze, Angew. Chem. Int. Ed. Engl., 15, 239 (1976).

$$R^1COCHR^2R^3 \xrightarrow[MeCN,\ 82°]{CH_2=\overset{+}{N}Me_2Cl^-} R^1COCR^2R^3CH_2\overset{+}{N}HMe_2Cl^-$$

R^1	R^2	R^3	% Yield
t-Bu	H	Me	58
H	Me	Me	81
$-(CH_2)_3-$		H	79
$-(CH_2)_3-$		Me	79

I.A.7.a.1a-18 S. Danishefsky, T. Kitahara, R. McKee, and P. F. Schuda, J. Am. Chem. Soc., 98, 6715 (1976).

$$R^1C(OSiMe_3)=CHR^2 \xrightarrow[\text{2) aq. NaOH}]{\text{1) } Me_2\overset{+}{N}=CH_2I^-\text{, } CH_2Cl_2} R^1COCHR^2CH_2NMe_2$$

$R^1,R^2 = -(CH_2)_4-$, 87%

$R^1 = CH{=}CHOMe$, $R^2 = H$, 95%

The dimethyl(methylene)ammonium iodide also reacts with enolates of lactones.

I.A.7.a.1a-19 T. Sato, J. Hanna, H. Nakamura, and T. Mukaiyama, Bull. Chem. Soc. Japan, 49, 1055 (1976).

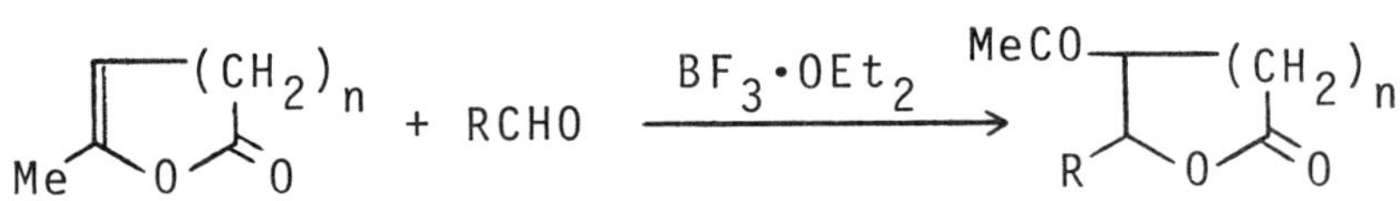

n	R	% Yield
1	Ph	62
2	Ph	73
1	H	93
2	Et	78
1	$ClCH_2$	90

I.A.7.a.1a-20 C. Bischoff and H. Herma, J. Prakt. Chem., 318, 773 (1976).

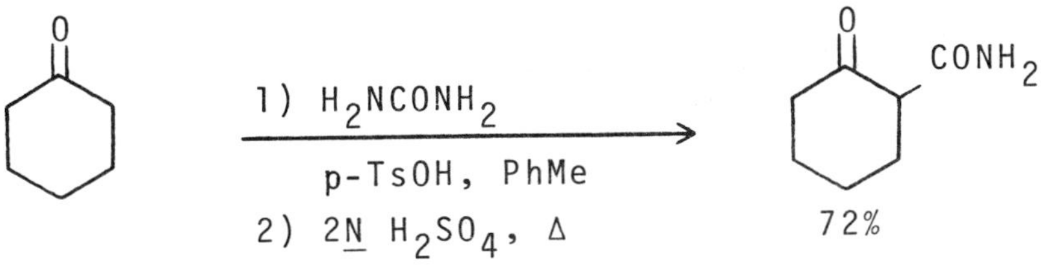

I.A.7.a.1a-21 T. Mukaiyama, T. Sato, S. Suzuki, T. Inoue, and H. Nakamura, Chem. Lett., 1976, 95.

$$R^1COCHR^2CO_2CH_2CCl_3 \xrightarrow[\text{Zn, DMSO}]{R^3CHO} R^1COCHR^2CH(OH)R^3$$

R^1	R^2	R^3	% Yield
Me	n-Bu	Ph	86
Me	n-Bu	Me	90
Et	Me	Ph	92
Et	Me	i-Pr	61

I.A.7.a.1a-22 H. Singh and A. S. Cheema, Chem. Ind. (London), 1976, 413.

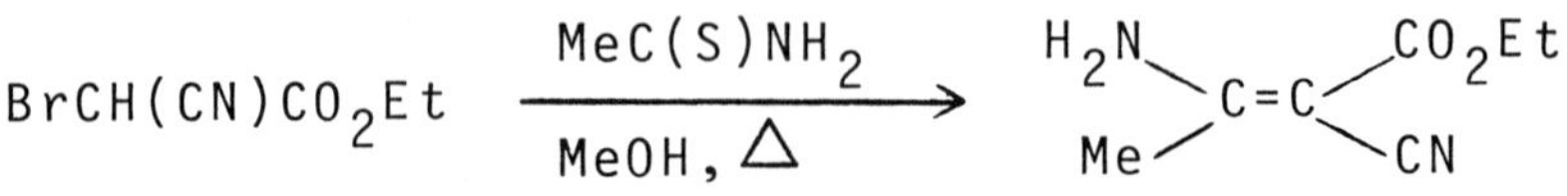

40-45%

I.A.7.a.1a-23 J. Liebscher and H. Hartmann, J. Prakt. Chem., 318, 705 (1976).

$$\mathrm{ArC(Cl)=CHCH=\overset{+}{N}Me_2\ ClO_4^-} \xrightarrow[\mathrm{HOAc,\ Et_3N}]{\mathrm{YCH_2CN}} \mathrm{ArC(CL)=CHCH=C(Y)CN}$$

Y = CO_2H, CO_2Et, $CONH_2$, CN, $CSNH_2$

50-80%

I.A.7.a.2-1 D. Seebach and F. Lehr, *Angew. Chem. Int. Ed. Eng.*, **15**, 505 (1976).

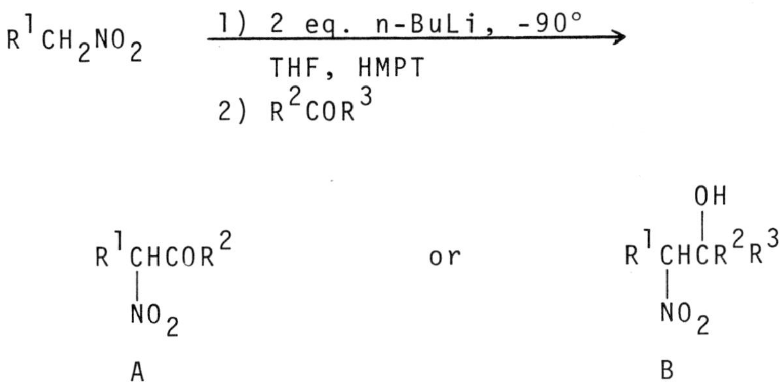

R^1	R^2	R^3	% Yield
Et	MeO	MeO	55(A)
i-Pr	MeO	Cl	65(A)
i-Pr	Me	OAc	30(A)
i-Pr	n-Pr	OMe	60(A)
Et	Ph	H	80(B)
Et	Me	Me	40(B)
Et	$-(CH_2)_5-$		65(B)

I.A.7.a.2-2 R. Henning, F. Lehr, and D. Seebach, Helv. Chim. Acta, 59, 2213 (1976).

$$R^1CH_2CHR^2NO_2 \xrightarrow[\text{THF, HMPA, -78°; 2) } R^3COR^4]{\text{1) 2 eq. n-BuLi}} R^2CH(NO_2)CHR^1C(OH)R^3R^4$$

R^1	R^2	R^3	R^4	% Yield
p-ClC_6H_4	H	Et	H	75
Ph	H	Ph	H	80
Ph	H	$-(CH_2)_4-$		69
H	Me	Ph	H	50

I.A.7.a.2-3 W. G. Kofron and M.-K. Yeh, J. Org. Chem., 41, 439 (1976).

$$R^1C(=NOH)CH_2R^2 \xrightarrow[\text{2) } R^3COR^4\text{; 3) hydrolysis}]{\text{1) 2 eq. n-BuLi, THF}} R^1COCR^2{=}CR^3R^4$$

R^1	R^2	R^3	R^4	% Yield
Me	H	Ph	Ph	55
Me	H	Me	$CH{=}CH_2$	17
$-(CH_2)_4-$		Me	Me	48
$-(CH_2)_4-$		Me	H	45

I.A.7.a.2-4 R. M. Sandifer, S. E. Davis, and C. F. Beam, Synth. Commun., 6, 339 (1976).

$$PhC(Me){=}NNHR^1 \xrightarrow{R^2R^3CO} PhC({=}NNHR^1)CH_2CR^2R^3OH$$

R^1	R^2	R^3	% Yield
PhCO	Ph	p-ClC_6H_4	84
$PhSO_2$	Ph	Ph	78

I.A.7.a.2-5 E. J. Corey, D. Enders, and M. G. Bock, Tetrahedron Lett., 1976, 7; see also: E. J. Corey and D. Enders, *ibid*, 11.

$$Me_3SiCHR^1CH{=}X \xrightarrow[\text{2) } R^2CHO \text{ 3) } H_3O^+]{\text{1) LDA, THF}} R^2CH{=}CR^1CHO$$

R^1	R^2	X	% Yield
H	n-C_5H_{11}	Nt-Bu	94
Me	i-Pr	Nt-Bu	88
H	n-C_5H_{11}	$NNMe_2$	90
H	Ph	$NNMe_2$	90

I.A.7.a.2-6 J. F. LeBorgne, T. Cuvigny, M. Larcheveque, and H. Normant, Synthesis, 1976, 238.

$$R^1R^2CHCH{=}NNMe_2 \xrightarrow[\text{2) } R^3COR^4]{\text{1) } Et_2NLi,\ PhH,\ HMPT} R^1R^2C(CN)CR^3R^4OH$$

R^1	R^2	R^3	R^4	% Yield
Me	Me	H	n-Pr	87
Et	Et	Me	Me	88
Me	Me	Ph	Ph	90

I.A.7.a.2-7 C. F. Beam, C. W. Thomas, R. M. Sandifer, R. S. Foote, and C. R. Hauser, Chem. Ind. (London), 1976, 487.

$$PhC(=N_2Li_2)CH_2Li \xrightarrow[\text{2) } HCl\text{-}H_2O]{\text{1) } Ph_2CO} PhCOCH_2C(OH)Ph_2$$

24%

I.A.7.a.2-8 U. Schöllkopf and H.-U. Scholz, Synthesis, 1976, 271.

$$H\text{-}\overset{N_2}{\overset{\|}{C}}\text{-}SiMe_3 \xrightarrow[\text{pentane, } -100°]{\text{1) n-BuLi, THF}} R^1R^2C(OH)\overset{N_2}{\overset{\|}{C}}SiMe_3$$
$$\text{2) } R^1COR^2$$

R^1	R^2	% Yield
Me	Me	65
H	Ph	80
$-(CH_2)_5-$		65

The product can be converted into epoxides and aldehydes by known procedures.

I.A.7.a.2-9 D. Seebach and W. Lubosch, Angew. Chem. Int. Ed. Engl., 15, 313 (1976).

$$t\text{-Bu}\overset{S}{\overset{\|}{C}}NMe_2 \xrightarrow[\text{THF, TMEDA}]{\text{1) sec-BuLi, } -78°} t\text{-Bu}\overset{S}{\overset{\|}{C}}NMeCH_2COHR^1R^2$$
$$\text{2) } R^1COR^2$$

R^1	R^2	% Yield
Ph	H	70
Ph	Ph	63
i-Pr	H	23
$-(CH_2)_5-$		17

I.A.7.a.2-10 R. R. Schmidt and J. Talbiersky, Angew. Chem. Int. Ed. Engl., 15, 171 (1976).

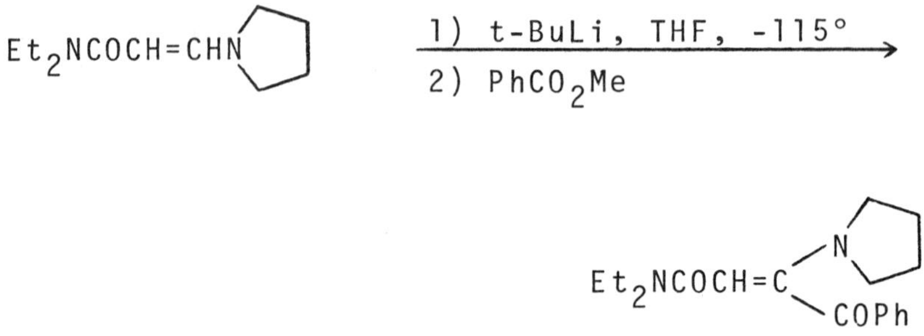

95%

I.A.7.a.2-11 H. Hoberg and U. Griebsch, Synthesis, 1976, 830.

$PhCH{=}NAlEt_2$ $\xrightarrow{\text{1) 2K, naphthalene, THF}}$
2) LiBr, Et_2O
3) PhCOR

$PhCHNH_2$
|
PhCROH

R = H, 58%

R = Ph, 75%

I.A.7.a.2-12 M. Schmidt and E. Weissflog, Z. Naturforsch., Teil B, 31b, 136 (1976).

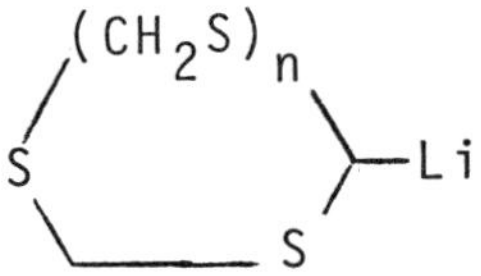

1) CO_2 or CS_2 ⟶
2) HCl

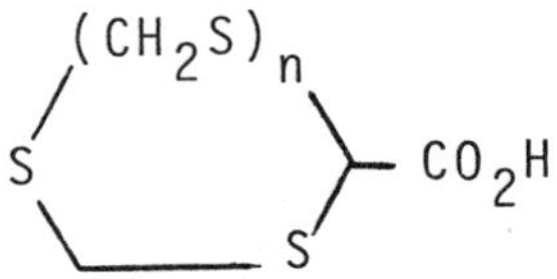

(or CS_2H)

n = 1,2, or 3
(70-95%)

$(MeS)_2CHLi$ $\xrightarrow[\text{2) HCl}]{\text{1) } CO_2}$ $(MeS)_2CHCO_2H$

75%

I.A.7.a.2-13 A. Bongini, D. Savoia, and A. Umani-Ronchi, J. Organomet. Chem., 112, 1 (1976).

RSO_2Me $\xrightarrow[\text{2) PhCHO}]{\text{1) 2 eq. n-BuLi, THF,}}$

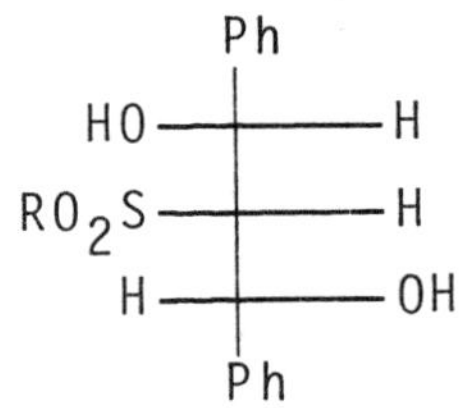

R = Ph, 60%
R = Me_2N, 50%

$PhSO_2Me$ $\xrightarrow[\text{2) } MgI_2 \text{ or } BF_3\ Et_2O \text{ 3) } R^1COR^2]{\text{1) 2 eq. n-BuLi, THF,}}$ $PhO_2SCH{=}CR^1R^2$ + $PhSO_2CH_2C(OH)R^1R^2$

∿35% each

I.A.7.a.2-14 K. Tanaka, R. Tanikaga, and A. Kaji, Chem. Lett., 1976, 917.

$$\mathrm{EtO\overset{S}{\overset{\|}{C}}SCHR^1CO_2Et} \xrightarrow[\text{2) } \mathrm{R^2COR^3}]{\text{1) LDA, THF, -78°}} \mathrm{R^2R^3C{=}CR^1CO_2Et}$$

R^1	R^2	R^3	% Yield
H	Et	H	52
H	Me	Me	85
H	$-(CH_2)_5-$		78
Me	Ph	H	82
Et	i-Pr	H	82

I.A.7.a.2-15 J. N. Denis, W. Dumont, and A. Krief, Tetrahedron Lett., 1976, 453.

$$\mathrm{(R^1Se)_2CR^2R^3} \xrightarrow[\text{2) } \mathrm{R^4COR^5}]{\text{1) n-BuLi}} \mathrm{R^1SeCR^2R^3COR^4}$$

R^1	R^2	R^3	R^4	R^5	% Yield
Ph	$-(CH_2)_2Ct\text{-}Bu(CH_2)_2-$		H	NMe_2	46
Me	H	C_6H_{13}	Ph	Cl	65
Ph	Me	Me	OMe	Cl	80
Me	H	C_6H_{13}	OMe	Cl	40

I.A.7.a.2-16 W. Dumont and A. Krief, Angew. Chem. Int. Ed. Engl., 15, 161 (1976).

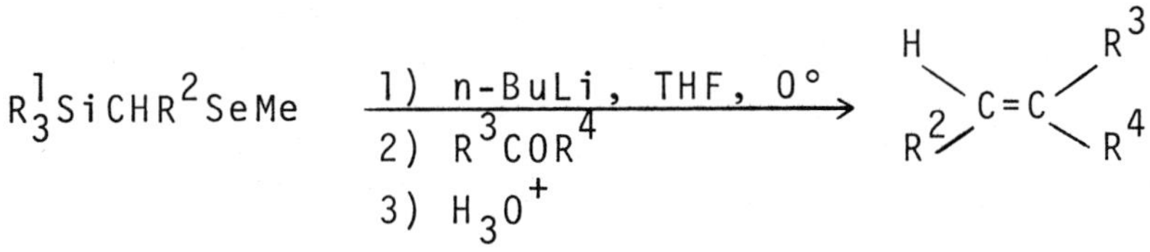

R^1	R^2	R^3	R^4	% Yield
Me	C_6H_{13}	H	C_6H_{13}	45
Me	C_6H_5	$-(CH_2)_5-$		40
Et	Me	$-(CH_2)_2CH(t\text{-}Bu)(CH_2)_2-$		40

I.A.7.a.3-1 E. Nakamura and I. Kuwajima, Angew. Chem. Int. Ed. Engl., 15, 498 (1976).

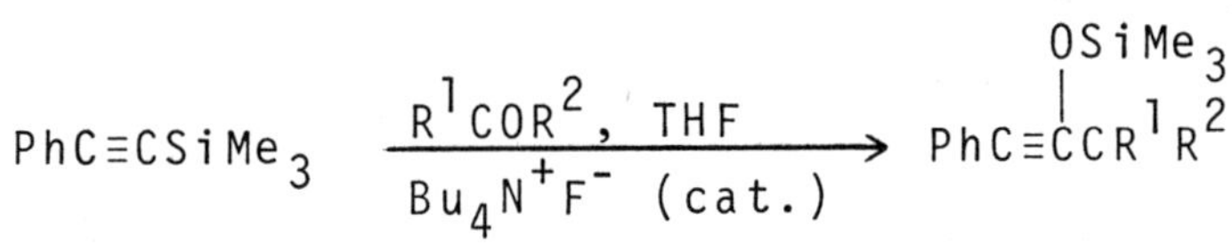

R^1	R^2	% Yield
Ph	H	76
$n\text{-}C_7H_{15}$	H	70
Ph	Ph	79
$-(CH_2)_5-$		87

I.A.7.a.3-2 D. Abenhaim, G. Boireau, C. Bernardon, A. Deberly, and C. Germain, Tetrahedron Lett., 1976, 993.

$$\text{RCOMe} \xrightarrow[\text{NiCl}_2\ (\text{cat.}),\ \text{C}_5\text{H}_{12}]{\text{NaAlEt}_4} \text{RC(OH)EtMe}$$

R = Ph, 84%

R = $PhCH_2$, 81%

I.A.7.a.3-3 J. F. Ruppert and J. D. White, J. Org. Chem., 41, 550 (1976).

$$R^1CH{=}CHCH_2Br + R^2COR^3 \xrightarrow[\text{flow system}]{\text{Zn}} R^2R^3C(OH)CHR^1CH{=}CH_2$$

R^1	R^2	R^3	% Yield
H	Ph	H	96
H	Ph	Me	97
Me	Et	Et	96
H	$-(CH_2)_5-$		97

I.A.7.a.3-4 A. A. Ponaras, Tetrahedron Lett., 1976, 3105.

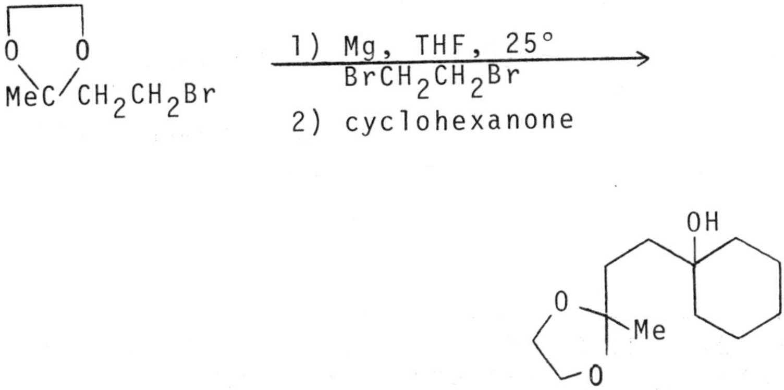

97%

I.A.7.a.3-5 D. Seebach, K.-H. Geiss, and M. Pohmakotr, Angew. Chem. Int. Ed. Engl., 15, 437 (1976).

$[CH_2{=}CH{-}CHS]^{=}\ 2Li^+ \xrightarrow[\text{2) } R^1COR^2 \text{ 3) MeI}]{\text{1) } MgBr_2} R^1R^2C(OH)CH(SMe)CH{=}CH_2$

R^1	R^2	% Yield
Et	H	91
Ph	H	73
Ph	Me	89
$(CH_2)_4$		94

I.A.7.a.3-6 N. L. Holy, Synth. Commun., 6, 539 (1976).

$$\overset{+}{CH_2=NMe_2}\ CF_3CO_2^- \xrightarrow[Et_2O]{RM} RCH_2NMe_2$$

RM = PhMgBr, 85%

RM = n-BuLi, 72%

I.A.7.a.3-7 H. G. Richey, Jr., R. C. McLane, and C. J. Phillips, Tetrahedron Lett., 1976, 233.

$$R^1R^2C{=}NOH \xrightarrow{R^3Li} R^1R^2R^3CNOH$$

R^1	R^2	R^3	% Yield
Ph	Ph	n-Bu	63
Ph	H	Ph	63
Me	Me	n-Bu	17
Pr	H	Me	46

I.A.7.a.3-8 D. S. Matteson and P. K. Jesthi, *J. Organomet. Chem.*, **110**, 25 (1976).

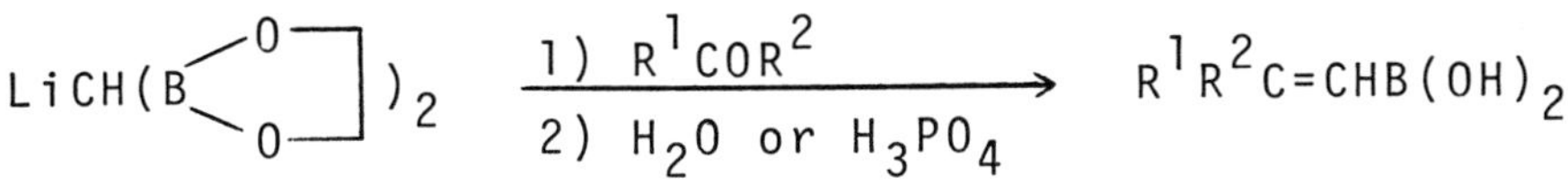

R^1	R^2	% Yield
Me	H	50-67
Ph	H	87
Me	Me	65
$\leftarrow CH_2 \rightarrow_6$		84

I.A.7.a.3-9 D. Seebach and N. Meyer, *Angew. Chem. Int. Ed. Engl.*, **15**, 438 (1976).

$$n\text{-}Bu_3SnCH_2OH \xrightarrow[\text{2) } R^1COR^2]{\text{1) 2 eq. n-BuLi, n-}C_6H_{14}} R^1R^2C(OH)CH_2OH$$

R^1 = Ph, R^2 = H, 62%

R^1,R^2 = $\leftarrow CH_2 \rightarrow_5$, 46%

I.A.7.a.3-10 W. C. Still and T. L. Macdonald, *J. Org. Chem.*, **41**, 3620 (1976).

$$CH_2{=}CHCH_2OSiMe_3 \xrightarrow[\text{2) } R^1COR^2]{\text{1) sec-BuLi, THF, HMPA, }-40°} R^1R^2C(OH)CH(OSiMe_3)CH{=}CH_2$$

$R^1,R^2 = (CH_2)_5$, 76%

$R^1 = Pr$, $R^2 = H$, 80%

I.A.7.a.3-11 W. Oppolzer and R. L. Snowden, *Tetrahedron Lett.*, **1976**, 4187.

$$(CH_2{=}CH)_2CHOSiEt_3 \xrightarrow[\text{2) } R^1COR_2]{\text{1) sec-BuLi, THF, }-78°} CH_2{=}CHC(OSiEt_3){=}CHCH_2CR^1R^2OH$$

$R^1 = Me$, $R^2 = H$, 92%

$R^1,R^2 = (CH_2)_5$, 73%

I.A.7.a.3-12 H. Neumann and D. Seebach, <u>Tetrahedron Lett.</u>, 1976, 4839.

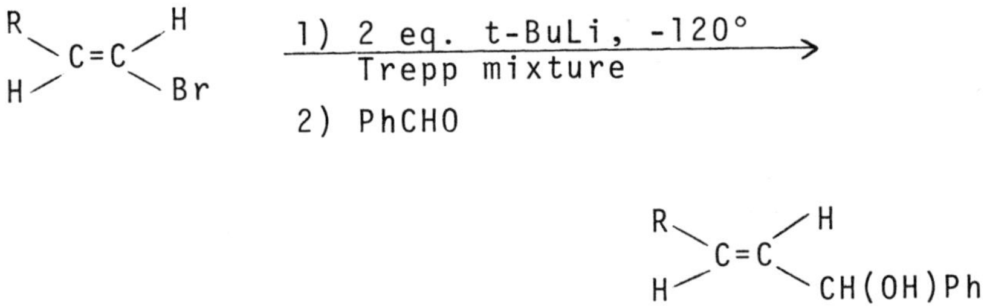

R = H, 76%
R = Me, 72%
R = Ph, 73%

I.A.7.a.3-13 J. E. Stemke, A. R. Chamberlin, and F. T. Bond, <u>Tetrahedron Lett.</u>, 1976, 2947.

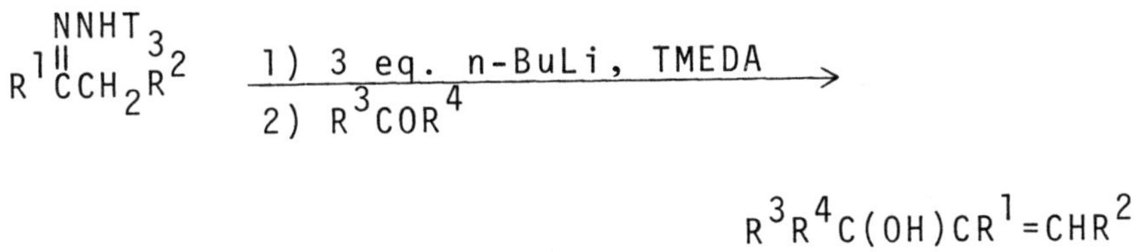

R^1	R^2	R^3	R^4	% Yield
C_6H_{13}	H	Ph	H	78
C_6H_{13}	H	$-(CH_2)_3CH=CH-$		67
$-CHMe(CH_2)_3-$		$-(CH_2)_5-$		48

α,β-Unsaturated carboxylic acids can be synthesized by reaction with CO_2.

I.A.7.a.3-14 C. N. Skold, Synth. Commun., 6, 119 (1976).

$$\mathrm{MeOCH{=}C(Br)OMe} \xrightarrow[\text{2) } R^1COR^2 \quad \text{3) } H_3O^+]{\text{1) BuLi, } Et_2O, -70°} R^1R^2C{=}C(OMe)CHO$$

R^1	R^2	% Yield
Ph	H	76
Ph	Me	51
$\text{-(}CH_2\text{)}_5\text{-}$		68

I.A.7.a.3-15 I. Vlattas, L. D. Vecchia, and A. O. Lee, J. Am. Chem. Soc., 98, 2008 (1976).

PhSCH=CHOEt $\xrightarrow[\text{THF, } -70°]{\text{1) t-BuLi}}$; PhSC(Br)=CHOEt $\xrightarrow[Et_2O, -70°]{\text{1) n-BuLi}}$; then $\xrightarrow{\text{2) } R^1COR^2}$

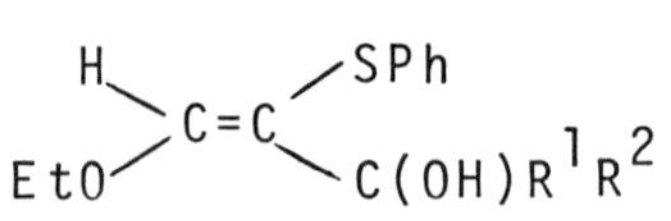

R^1 = Ph, R^2 = H, 75%

R^1,R^2 = $\text{-(}CH_2\text{)}_4\text{-}$, 78%

I.A.7.a.3-16 W. E. Parham and L. D. Jones, J. Org. Chem., 41, 2704 (1976).

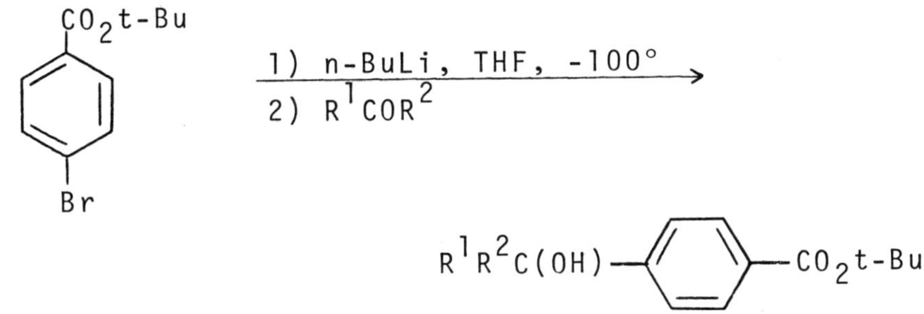

$R^1 = R^2 = Ph$, 76%

$R^1, R^2 = (CH_2)_5$, 75%

I.A.7.a.3-17 M. J. Manning, P. W. Raynolds, and J. S. Swenton, J. Am. Chem. Soc., 98, 5008 (1976).

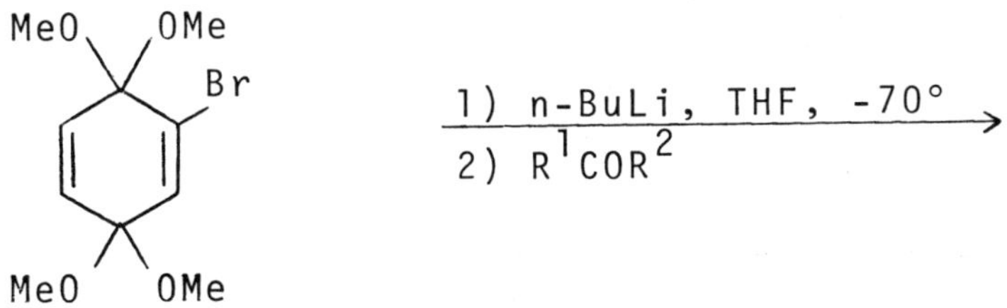

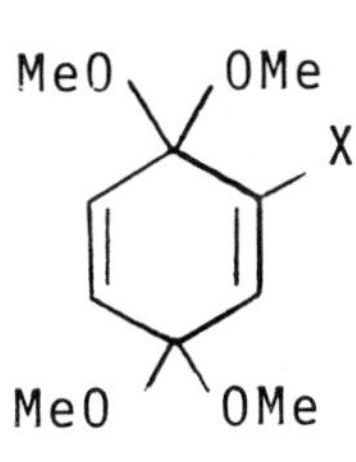

$R^1, R^2 = (CH_2)_5$, $X = C(OH)R^1R^2$, 81%

$R^1 = Ph$, $R^2 = OMe$, $X = PhCO$, 78%

$R^1 = Ph$, $R^2 = Cl$, $X = PhCO$, 67%

I.A.7.a.3-18 K. Okuhara, J. Org. Chem., 41, 1487 (1976).

$$\mathrm{ArLi} \xrightarrow[\substack{2)\ n\text{-BuLi},\ -50^\circ \\ 3)\ CO_2}]{1)\ CF_2{=}CCl_2} \mathrm{ArC{\equiv}CCO_2H}$$

72-84%

I.A.7.a.3-19 P. C. Traas, H. Boelens, and H. J. Takken, Tetrahedron Lett., 1976, 2287.

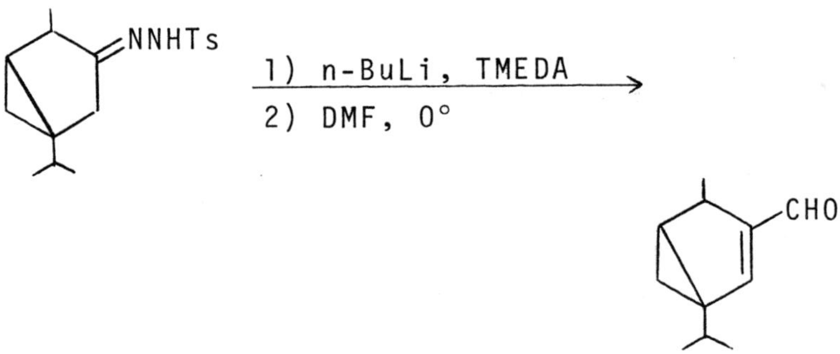

60%

I.A.7.a.3-20 J. J. Eisch and J. E. Galle, J. Am. Chem. Soc., 98, 4646 (1976).

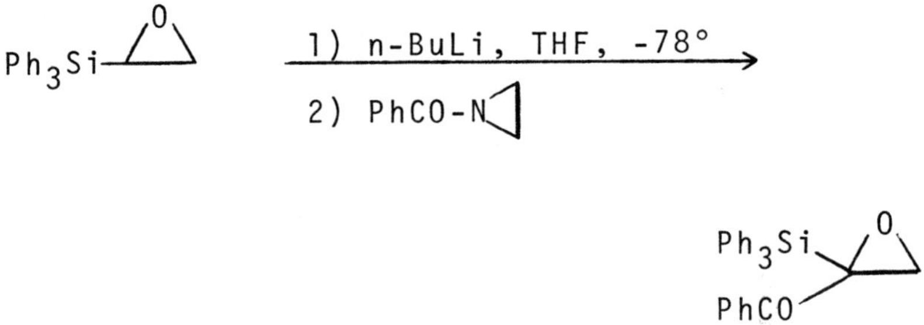

61%

I.A.7.a.3-21 P. Savignac and F. Mathey, Tetrahedron Lett., 1976, 2829.

$$(R^1O)_2P(Y)Me \xrightarrow[\substack{2)\ CuI,\ -35^\circ \\ 3)\ R^2COCl}]{1)\ n\text{-}BuLi,\ -70^\circ} (R^1O)_2P(Y)CH_2COR^2$$

Y	R^1	R^2	% Yield
O	Me	Me	80
O	Et	t-Bu	95
O	Et	$CMe{=}CH_2$	100
S	Me	Me	93

I.A.7.a.3-22 D. E. Bergbreiter and J. M. Killough, J. Org. Chem., 41, 2750 (1976).

$$R^1MgX \xrightarrow[2)\ R^2COCl]{1)\ MeCu} R^1COR^2$$

R^1	R^2	% Yield
n-Bu	Ph	85
PrCHMe	Me	90
t-Bu	Ph	93
Ph	t-Bu	100

I.A.7.a.3-23 C. Wiaux-Zamar, J.-P. Dejonghe, L. Ghosez, J. F. Normant, and J. Villieras, *Angew. Chem. Int. Ed. Engl.*, **15**, 371 (1976).

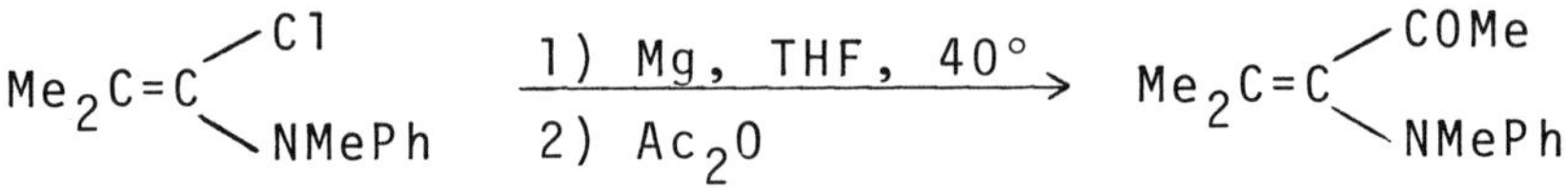

59%

I.A.7.a.3-24 H. Viola and R. Mayer, *Z. Chem.*, **16**, 335 (1976).

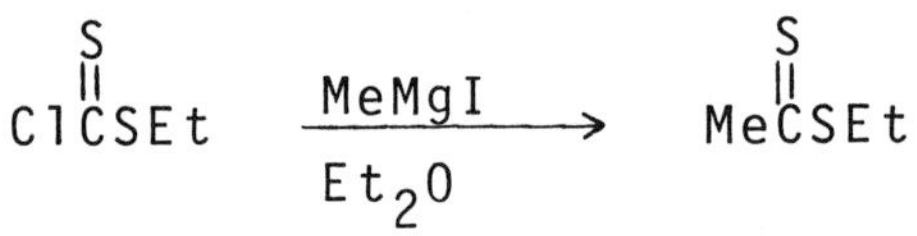

40%

I.A.7.a.3-25 G. Zweifel and R. A. Lynd, *Synthesis*, **1976**, 625.

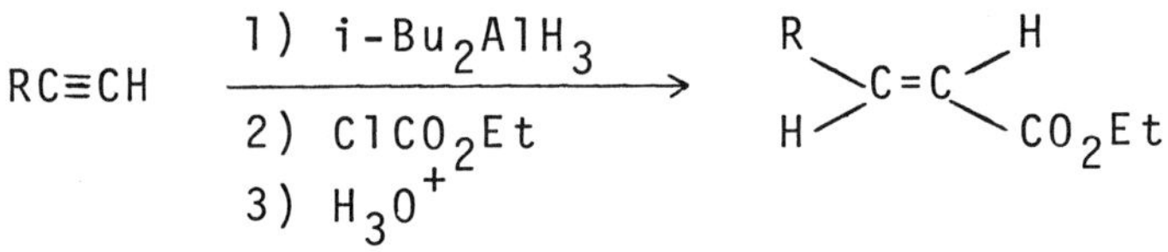

R^1	% Yield
n-Bu	64
i-Pr	64
t-Bu	72

I.A.7.a.3-26 G. Cahiez, A. Masuda, D. Bernard, and J. F. Normant, Tetrahedron Lett., 1975, 3155.

$$R^1MnI \xrightarrow[Et_2O]{R^2COCl} R^1COR^2$$

R^1,R^2 = alkyl

60-91%

I.A.7.a.3-27 K. Saigo, K. Kawata, and T. Mukaiyama, Chem. Lett., 1976, 771.

$$R^1CHO \xrightarrow[\text{2) t-BuOLi, NCS}]{\text{1) } R^2MgBr} R^1COR^2$$

R^1	R^2	% Yield
Ph	n-Bu	82
C_6H_{13}	Ph	78
PhCH=CH	n-Bu	67

I.A.7.b.1-1 R. E. Gawley, Synthesis, 1976, 777.

Review: The Robinson Annelation and Related Reactions.

I.A.7.b.1-2 H. Stetter, Angew. Chem. Int. Ed. Engl., **15**, 639 (1976).

Review: Catalyzed Addition of Aldehydes to Activated Double Bonds — A New Synthetic Approach.

I.A.7.b.1-3 K. Saigo, M. Osaki, and T. Mukaiyama, Chem. Lett., **1976**, 163.

$$R^1COCH{=}CR^2R^3 + R^4R^5C{=}C(OMe)OSiMe_3 \xrightarrow[\substack{CH_2Cl_2,\ -78^\circ \\ 2)\ 5\%\ K_2CO_3}]{1)\ TiCl_4} R^1COCH_2CR^2R^3CR^4R^5CO_2Me$$

R^1	R^2	R^3	R^4	R^5	% Yield
Ph	Ph	H	Me	Me	100
Me	Me	Me	$PhCH_2$	H	79

I.A.7.b.1-4 K. Narasaka, K. Soai, Y. Aikawa, and T. Mukaiyama, <u>Bull. Chem. Soc. Japan</u>, **49**, 779 (1976).

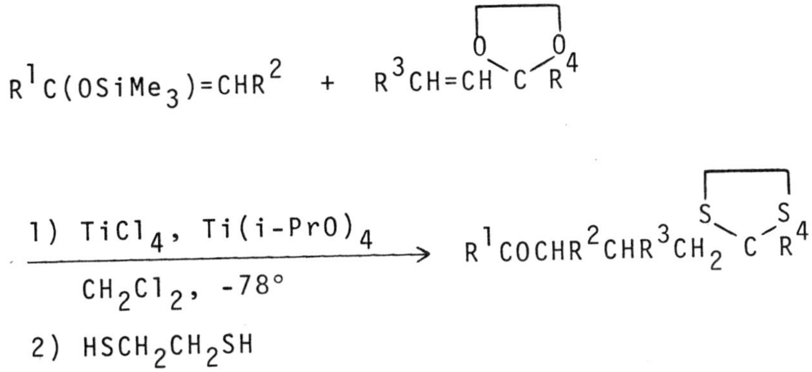

R^1	R^2	R^3	R^4	% Yield
Ph	H	H	Me	78
Me	Ph	H	Me	61
Me	Ph	$-(CH_2)_3-$		58

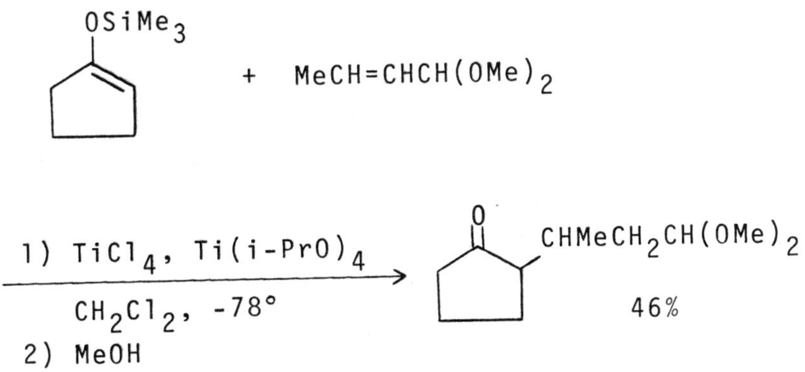

I.A.7.b.1-5 M. Miyashita, T. Yanami, and A. Yoshikoshi, J. Am. Chem. Soc., 98, 4679 (1976).

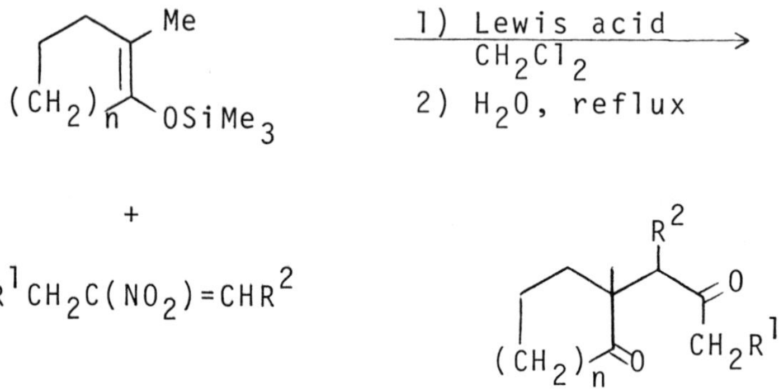

n	R^1	R^2	Lewis Acid	% Yield
2	H	H	$SnCl_4$	85
2	Me	H	$TiCl_4$	76
2	H	Me	$AlCl_3$	63
1	H	H	$SnCl_4$	41

I.A.7.b.1-6 H. Hagiwada, T. Kodama, H. Kosugi, and H. Uda, J. Chem. Soc., Chem. Commun., 1976, 413.

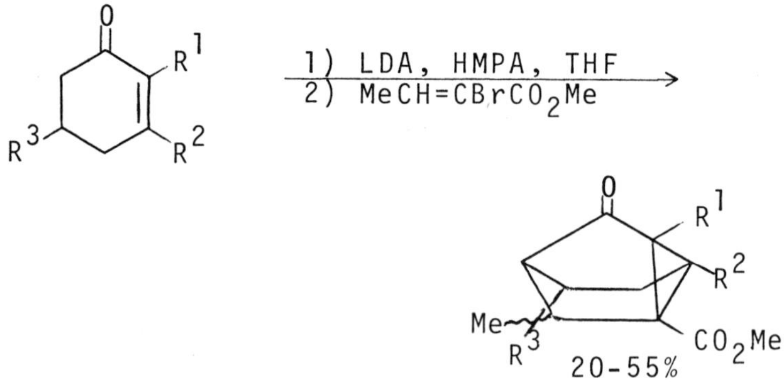

I.A.7.b.2-1 W. G. Kofron and J. Mathew, J. Org. Chem., 41, 114 (1976).

$$R^1R^2C{=}CR^3CO_2Me \xrightarrow[Et_2O]{NaCHPh_2} Ph_2CHCR^1R^2CHR^3CO_2Me$$

R^1	R^2	R^3	% Yield
H	H	H	40
H	H	Me	40
Me	H	H	87
Me	Me	H	90
Ph	H	Me	76

I.A.7.b.2-2 G. Stork, A.Y.W. Leong, and A. M. Touzin, J. Org. Chem., 41, 3491 (1976).

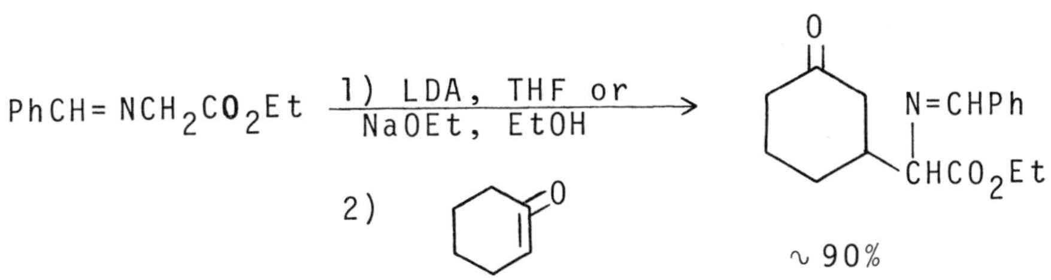

∿ 90%

I.A.7.b.2-3 I.G.C. Coutts and M. Hamblin, <u>J. Chem. Soc., Chem. Commun.</u>, 1976, 58.

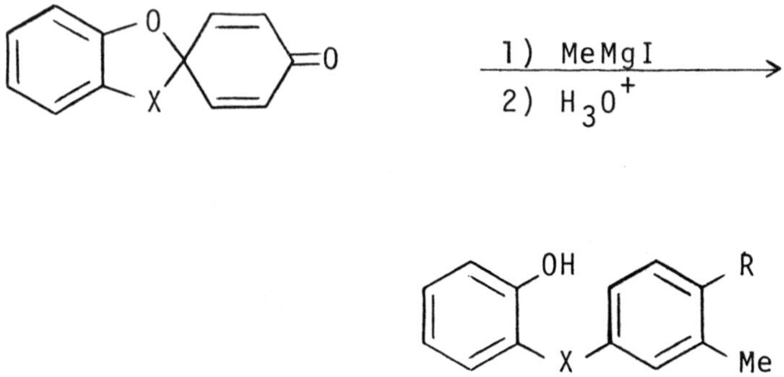

R = OH and Me

R = O and NTs

Unusual 1,3-addition of Grignard to the enone system.

I.A.7.b.2-4 E. Elkik and M. Imbeaux-Oudotte, <u>Bull. Soc. Chim. Fr.</u>, 1976, 439.

R = Ph, 82%

R = Et, 72%

I.A.7.b.2-5 Y. Ittah and I. Shahak, Synthesis, 1976, 320.

$$(MeS)_2C{=}C(CO_2Et)_2 \xrightarrow[Et_2O]{RMgX,\ CuBr\ (cat.)} \underset{A}{R_3CCH(CO_2Et)_2} \text{ or } \underset{B}{R_2C{=}C(CO_2Et)_2}$$

R	% Yield
Me	78-A
Et	72-A
Ph	60-A
i-Pr	60-B
$PhCH_2$	42-B

I.A.7.b.2-6 R. D. Clark and C. H. Heathcock, J. Org. Chem., 41, 636 (1976).

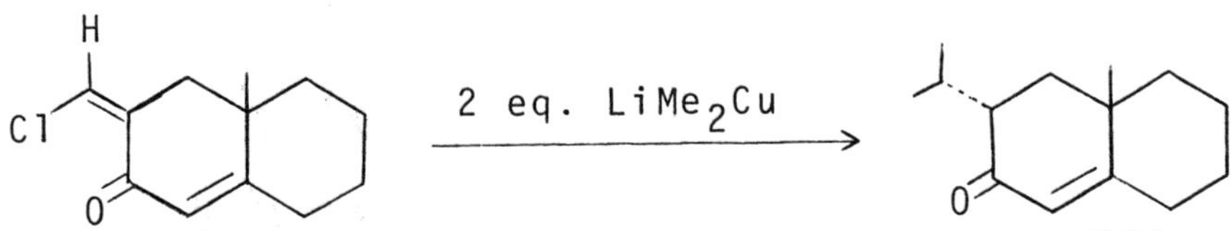

88%

Other reactions of β-chloro-α,β-unsaturated ketones are described.

I.A.7.b.2-7 A. Alexakis, A. Commercon, J. Villiéras and J. F. Normant, Tetrahedron Lett., 1976, 2313.

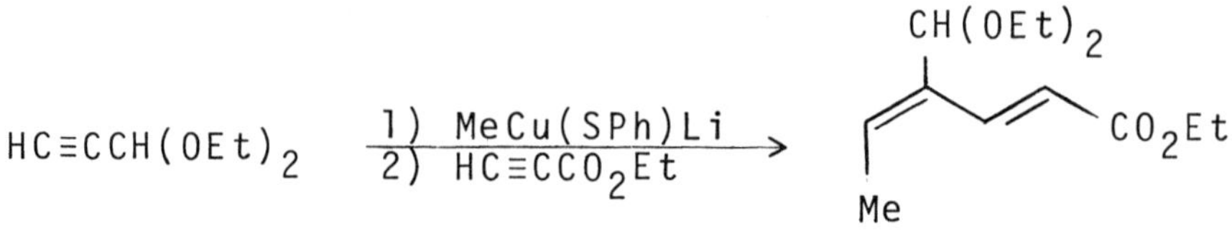

82%

I.A.7.b.2-8 A. Alexakis, J. Normant, and J. Villiéras, Tetrahedron Lett., 1976, 3461.

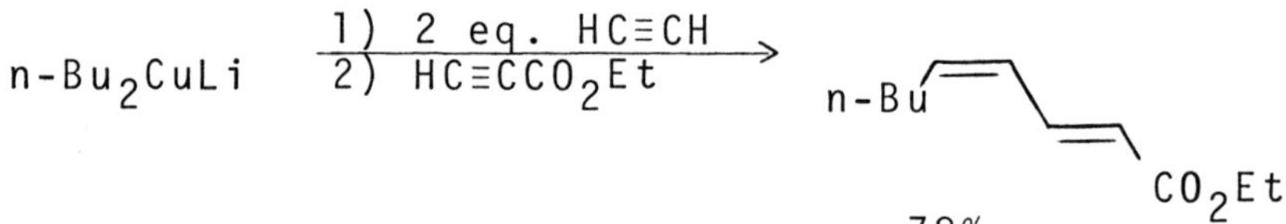

78%

I.A.7.b.2-9 N. Miyaura, M. Itoh, and A. Suzuki, Tetrahedron Lett., 1976, 255.

$$CH_2{=}CHCN \xrightarrow{[R_3BMe]Cu} RCH_2CH_2CN$$

R = n-Pr, 84%

R = i-Bu, 88%

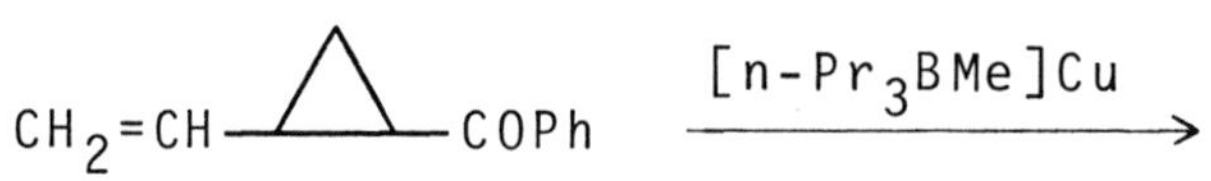

n-$PrCH_2CH{=}CH(CH_2)_2COPh$

84%

I.A.7.b.3-1 D. R. Hicks, R. C. Anderson, and B. Fraser-Reid, <u>Synth. Commun.</u>, 6, 417 (1976).

$$R^1R^2C{=}CHCOR^3 \xrightarrow[h\nu]{R^4CHO} R^4COCR^1R^2CH_2COR^3$$

R^1	R^2	R^3	R^4	% Yield
Me	Me	Me	Me	31
H	$\leftarrow CH_2 \rightarrow_3$		n-Pr	42
H	$CH(OEt)OCH(CH_2OTr)$		Ph	58

I.A.7.b.3-2 H. Stetter and H. Kuhlmann, <u>Chem. Ber.</u>, <u>109</u>, 2890 and 3426 (1976); see also: H. Stetter, M. Schreckenberg, and K. Wiemann, *ibid*, 541.

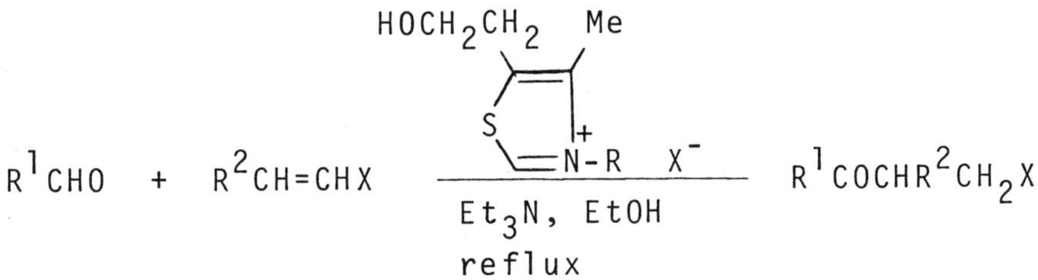

R^1	R^2	X	% Yield
Me	H	CN	30
Pr	Ph	COPh	70
2-Furyl	H	CO_2Et	31
2-Furyl	Ph	COMe	80

I.A.7.b.3-3 P. Knittel and J. Warkentin, Can. J. Chem., 54, 1341 (1976).

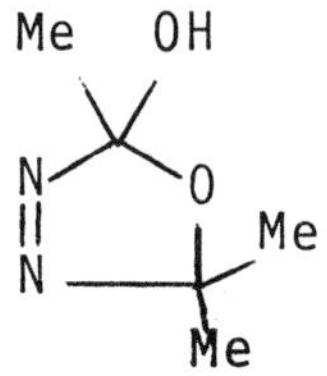

+ RCH=CHX $\xrightarrow[60°]{PhH}$ $AcOCMe_2CHRCH_2X$

R = H, Me or CO_2Me

R = CN, CHO, COME or CO_2Me

I.A.7.b.3-4 D.W.K. Yeung and J. Warkentin, Can. J. Chem., 54, 1345 (1976).

$Ph_2C(OH)N{=}Nt\text{-}Bu$ + RCH=CHX $\xrightarrow[50°]{PhH}$ $t\text{-}BuCHRCH_2X$

R	X	% Yield
H	CN	63
H	COMe	31
Me	CHO	32

I.A.7.b.3-5 K. Yamamura, J. Chem. Soc., Chem. Commun., 1976, 438.

$$PhCH{=}C(COPh)_2 \xrightarrow[Pd(OAc)_2]{PhH\text{-}HOAc} PhCHCH(COPh)_2$$

52%

I.A.7.b.3-6 D. A. White, J. Chem. Soc., Perkin Trans. I, 1976, 1926.

$$R^1CH{=}CR^2(X) \xrightarrow[2)\ MeY]{1)\ Et_4\overset{+}{N}\ NC\cdot CO_2^-} NCCHR^1CR^2(X)CO_2Me$$

R^1	R^2	X	Y	% Yield
H	H	CO_2Me	I	77
H	Me	CO_2Me	I	82
Me	H	CO_2Me	I	68
H	H	CN	SO_4Me	78

I.A.8-1 W. E. Parham and Y. A. Sayed, <u>Synthesis</u>, <u>1976</u>, 116.

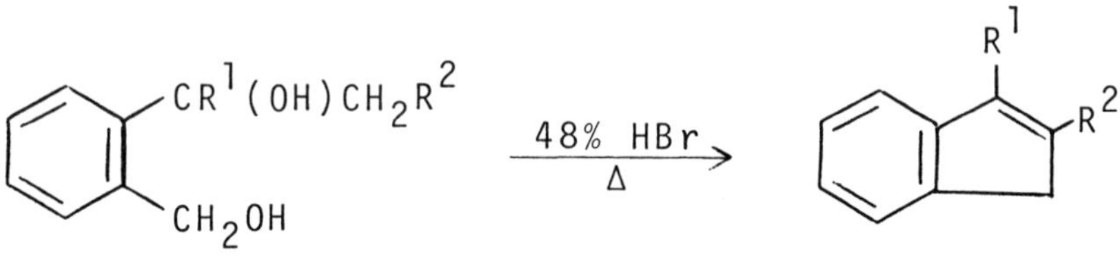

R^1	R^2	% Yield
Me	Me	84
Et	Me	79
$-(CH_2)_4-$		81
Ph	H	52

Not useful for the preparation of 2- or 3-monoalkylindenes.

I.A.8-2 I. Fleming, A. Pearce and R. L. Snowden, <u>J. Chem. Soc., Chem. Commun.</u>, <u>1976</u>, 182.

$$PhCH(OH)C(P(O)Ph_2)EtCH_2SiMe_3 \xrightarrow[PhH]{p\text{-}TsOH} Ph_2P(O)CHPhCEt{=}CH_2$$

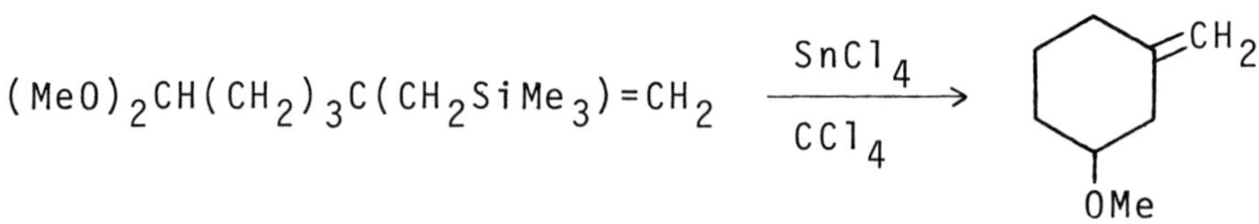

Use of Me_3Si- group to control products in carbonium ion reactions is demonstrated.

I.A.8-3 I. A. Savost'yanova, M. S. Burmistrova, and Y. I. Baukov, J. Gen. Chem. USSR, 45, 2308 (1975).

$$Me_3SiC{\equiv}COEt \quad + \quad RCHO \xrightarrow[Et_2O]{Et_2O \cdot BF_3} RCH{=}C(SiMe_3)CO_2Et$$

R = CCl_3, 76%
R = Ph, 72%

I.A.8-4 O. V. Lubinskaya, A. S. Shashkov, V. A. Chertkov, and W. A. Smit, Synthesis, 1976, 742.

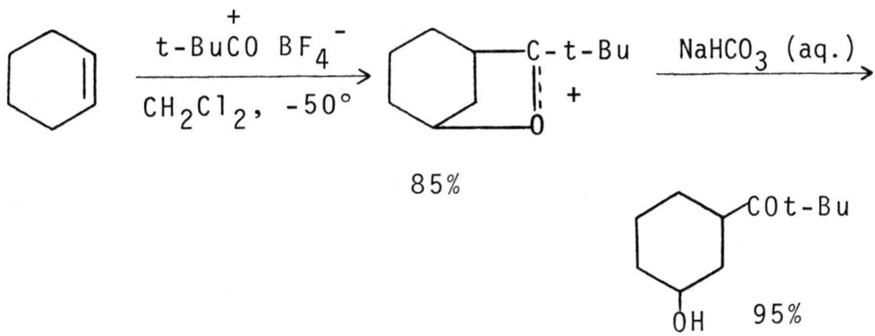

I.A.8-5 M. Grignon-Dubois, J. Dunogues, and R. Calas, Synthesis, 1976, 737.

R	% Yield
Me	60
Ph	50
$Me_2C{=}CH$	75

I.A.8-6 M. Hojo, R. Masuda, Y. Kokuryo, H. Shioda, and S. Matsuo, Chem. Lett., 1976, 499.

$$RYCH{=}CH_2 \xrightarrow{(CX_3CO)_2O} RYCH{=}CHCOCX_3$$

RY	X	% Yield
EtO	F	100
EtO	Cl	71
p-$MeC_6H_4SO_2NPh$	F	92
$MeSO_2NPh$	Cl	94

I.A.8-7 M. Hojo, R. Masuda, and Y. Kamitori, Tetrahedron Lett., 1976, 1009.

$$RSCY{=}CH_2 \xrightarrow[\text{py., } CHCl_3]{(CX_3CO)_2O} RSCY{=}CHCOCX_3$$

R	Y	X	% Yield
p-MeC_6H_4	p-MeC_6H_4S	F	93
p-MeC_6H_4	p-MeC_6H_4S	Cl	96
p-ClC_6H_4	H	F	97
Me	H	F	100
Ph	Ph	F	79

I.A.8-8 V. I. Gorbutenko, and N. V. Mel'nichenko, J. Org. Chem. USSR, 11, 2261 (1975).

$$ROCH{=}CH_2 \xrightarrow[\text{2) } Et_3N,\ Et_2O,\ -20°]{\text{1) } ClC(O)NCO,\ Et_2O} ROCH{=}CHC(O)NCO$$

36-43%

I.A.8-9 L. Capuano, T. Tammer, and R. Zander, Chem. Ber., 109, 3497 (1976).

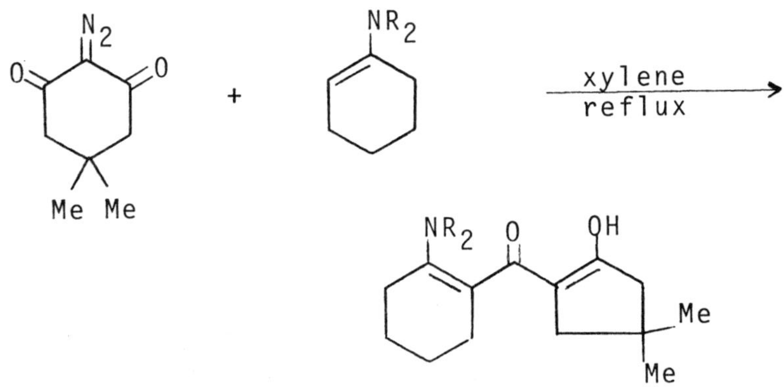

$R_2 = \text{-}(CH_2)_2O(CH_2)_2\text{-}$, 62%

I.A.8-10 B. Sket and M. Zupan, <u>Synth. Commun.</u>, 6, 309 (1976).

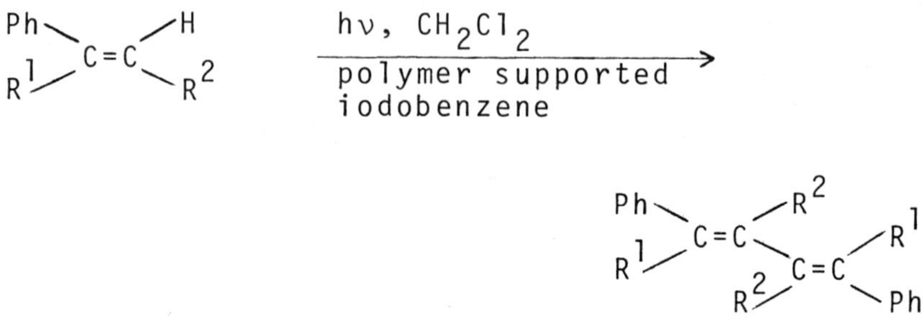

R^1	R^2	% Yield
Ph	H	80
Ph	Me	65
H	Me	50

I.A.8-11 A. Couture, K. Ho, M. Hoshino, P. de Mayo, R. Suau, and W. R. Ware, <u>J. Am. Chem. Soc.</u>, 98, 6218 (1976).

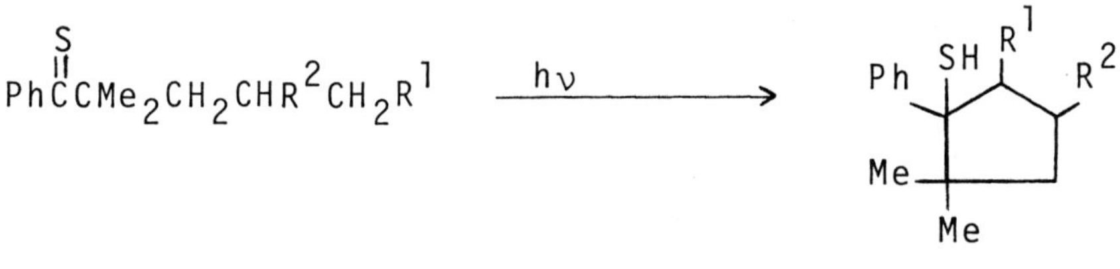

70-90%

I.A.8-12 T. Sato, G. Izumi, and T. Imamura, J. Chem. Soc., Perkin Trans. I, 1976, 788.

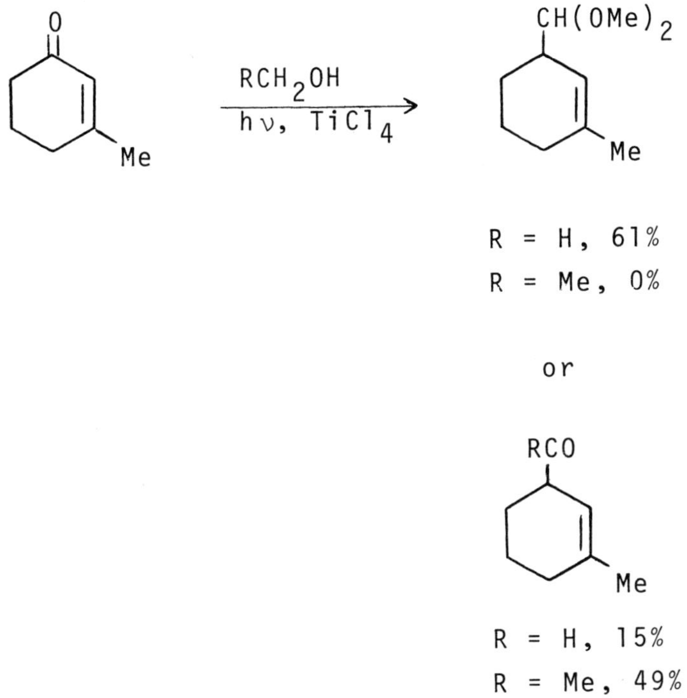

Dihydrofurans are isolated in cases with a cisoid α,β-unsaturated ketone and in some open chain systems.

I.A.8-13 T. Shono, I. Nishiguchi, and H. Omizu, Chem. Lett., 1976, 1233.

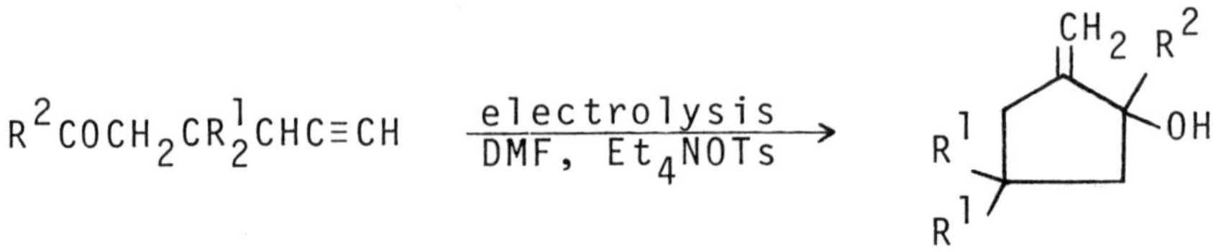

85-95%

I.A.8-14 J. J. Habeeb and D. G. Tuck, J. Chem. Soc., Chem. Commun., 1976, 698.

$$PhCOBr + EtBr \xrightarrow[\text{Cd anode, } Et_3NClO_4\text{, diglyme}]{\text{electrolysis}} PhCOEt$$

64% (based on Cd consumed.

I.A.8-15 E. Yoshii, T. Koizumi, and T. Kawazoe, Chem. Pharm. Bull., 24, 1957 (1976).

$$R^1CH_2CR^2C{=}C(OMe)OSiEt_3 \xrightarrow[\text{PhH or DME}]{PhHgCBrCl_2} R^1CH_2CR^2{=}C(Cl)CO_2Me$$

R^1	R^2	% Yield
H	H	80
Me	Me	90
Ph	H	52

I.A.8-16 P. Kuhl, M. Mühlstädt, and J. Graefe, Synthesis, 1976, 825.

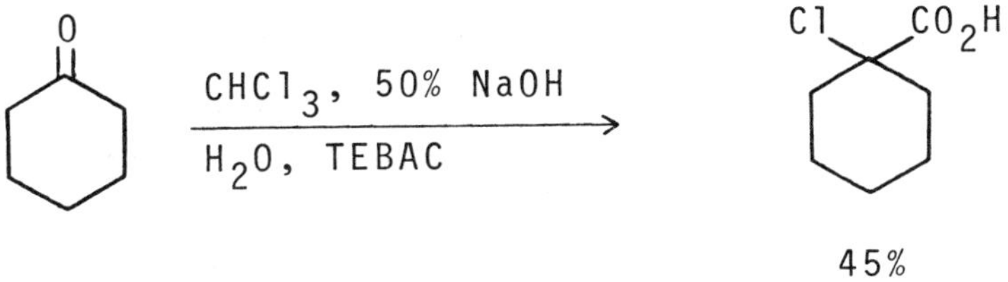

I.A.8-17 G. F. Luteri and W. T. Ford, J. Organomet. Chem., 105, 139 (1976).

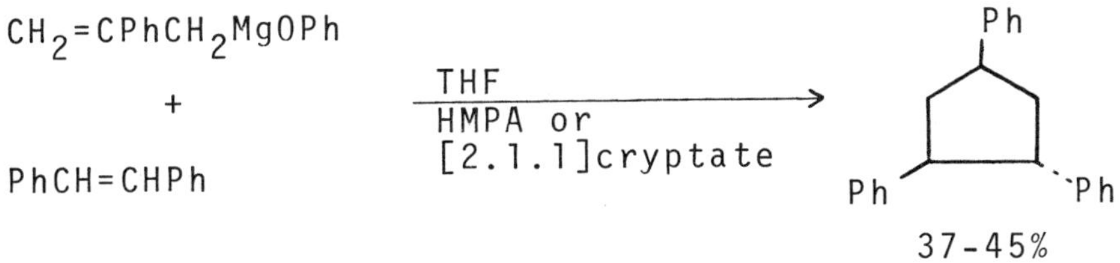

In the absence of HMPA or [2.1.1]cryptate the reaction does not proceed.

I.A.8-18 P. L. Fuchs, J. Org. Chem., 41, 2935 (1976); see also: G. Stork and A. A. Ponaras, *ibid*, 2937.

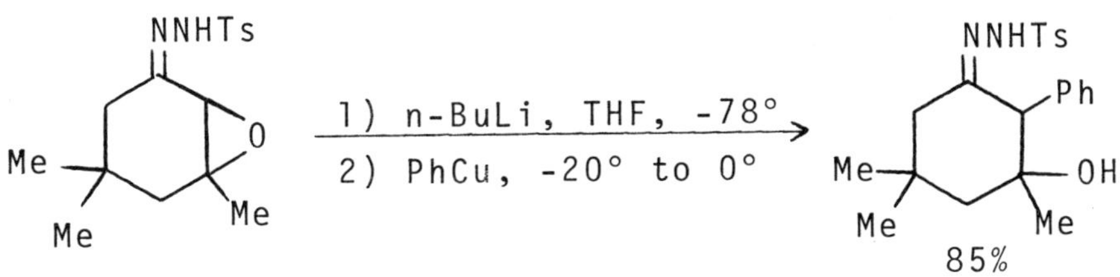

Products can be dehydrated and carbonyl exchanged to give α-aryl-α,β-unsaturated ketones.

I.A.8-19 F. Mathey and P. Savignac, Synthesis, 1976, 766.

$$(EtO)_2P(O)Me \xrightarrow[\substack{3)\ n\text{-BuLi},\ -60^\circ \\ 4)\ CuI,\ -35^\circ,\ 1\ hr \\ 5)\ R^2COCl,\ Et_2O,\ -40^\circ}]{\substack{1)\ n\text{-BuLi, THF},\ -65^\circ \\ 2)\ R^1X,\ -78^\circ\ to\ 25^\circ}} (EtO)_2P(O)CHR^1COR^2$$

R^1X	R^2	% Yield
MeI	Me	87
$CH_2=CHCH_2Br$	Me	75
MeI	i-Bu	70

A "one-pot" synthesis of diethyl 1-alkyl-2-oxoalkanephosphonates.

I.A.8-20 P.-L. Compagnon and B. Grosjean, Synthesis, 1976, 448.

$$PhC{\equiv}CCu \xrightarrow[Et_2O,\ MeCN]{BrCN} PhC{\equiv}CCN$$

60%

I.A.8-21 N. DeKimpe, R. Verhe , L. DeBuyck, H. Hasma, and N. Schamp, Tetrahedron, 32, 3063 (1976).

$$R^1R^2C(Cl)CH{=}Nt\text{-}Bu \xrightarrow[reflux]{KCN,\ MeOH} R^1R^2C{=}C(CN)NHt\text{-}Bu$$

R^1	R^2	% Yield
Me	Me	95
Me	Ph	17-36 (depends on conditions)
$(CH_2)_5$		94

I.B.1-1 H. J. Bestmann, W. Stransky, and O. Vostrowsky, Chem. Ber., 109, 1694 (1976).

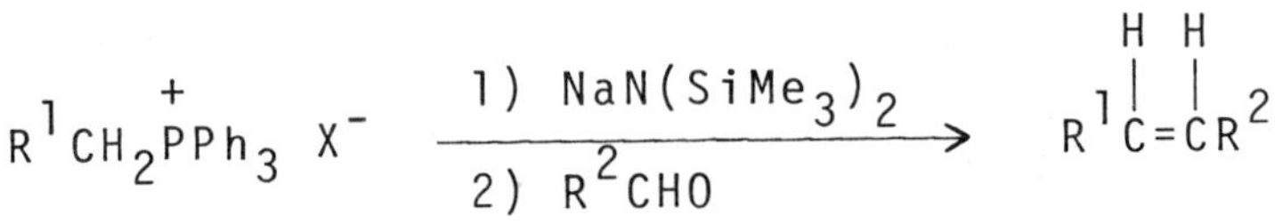

Use of the sodium *bis*(trimethylsilyl)amide base allows preparation of lithium-salt free ylides.

I.B.1-2 R. Broos and M. Ateunis , Synth. Commun., 6, 53 (1976).

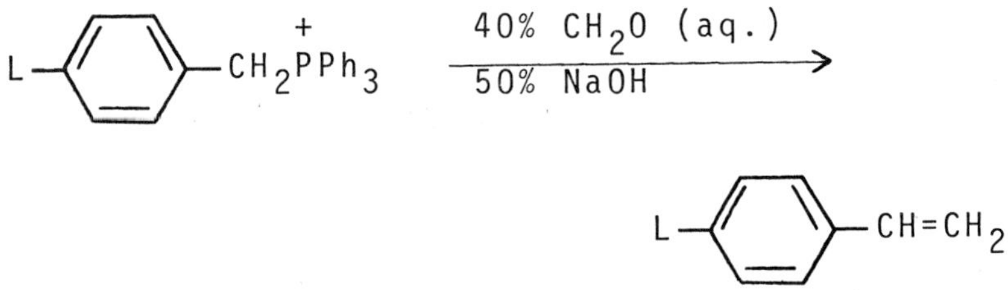

L	% Yield
CN	80
Br	82
Me	90
MeO	80

I.B.1-3 V. Subramanyam, E. H. Silver, and A. H. Soloway, J. Org. Chem., 41, 1272 (1976).

$$Ph_3P{=}CHR + HCO_2Et \longrightarrow RCH{=}CHOEt$$

R = CO_2Et, 95%
R = Ph, 90%
R = i-Bu, 22%

I.B.1-4 L. Kalvoda, Coll. Czech. Chem. Commun., 41, 2034 (1976).

$$RCOX \xrightarrow[Ph_3P{=}CHCO_2Et]{HCN,\ CHCl_3} RC(CN){=}CHCO_2Et$$

R	X	% Yield
Ph	Cl	81
Me	OAc	58
t-Bu	Cl	83 (only *cis* isomer)

I.B.1-5 P. Bravo, A. Ricca, C. Ticozzi, and O. Vajna de Pava, *Gazz. Chim. Ital.*, **106**, 743 (1976).

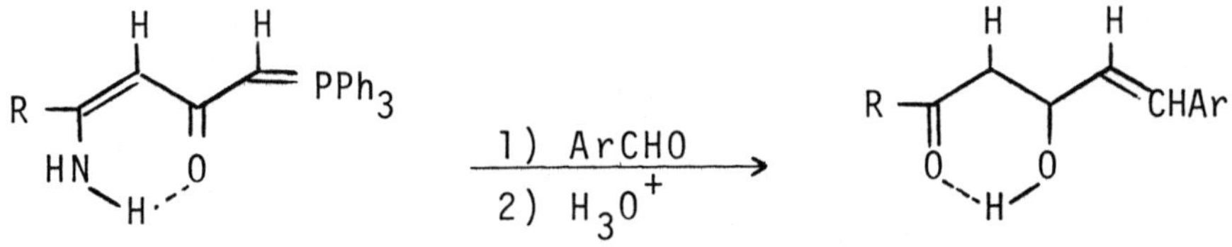

R	Ar	% Yield
Me	p-$NO_2C_6H_4$	60
i-Pr	p-$MeOC_6H_4$	78
Ph	p-$MeOC_6H_4$	75

I.B.1-6 H.-J. Bestmann and R. W. Saalfrank, *Chem. Ber.*, **109**, 403 (1976).

$$Ph_3P{=}CHC(OEt){=}CR^1COR^2 \xrightarrow{R^3CHO} R^3CH{=}CHC(OEt){=}CR^1COR^2$$

R^1	R^2	R^3	% Yield
H	Ph	Ph	57
H	Ph	Me	71
H	OEt	Ph	66
$-(CH_2)_4-$		Ph	51

I.B.1-7 N. Petragnani, R. Rodrigues, and J. V. Comasseto, J. Organomet. Chem., 114, 281 (1976).

$$Ph_3P{=}CR^1SePh + R^2CHO \xrightarrow{THF} R^2CH{=}CR^1SePh$$

R^1	R^2	% Yield
H	Ph	98
Me	Ph	99
Me	Et	65

I.B.1-8 G. A. Wheaton and D. J. Burton, Tetrahedron Lett., 1976, 895.

$$Ph_3P{=}CR^1R^2 \xrightarrow[Et_2O]{HCF_2Cl} F_2C{=}CR^1R^2$$

R^1,R^2 = H or alkyl

81-100%

The by-product $Ph_3\overset{+}{P}CHR^1R^2Cl^-$ can be recovered in >90% yield.

I.B.1-9 A. Hercouet and M. LeCorre, *Tetrahedron Lett.*, **1976**, 825.

$$Ph_3P{=}CHCOCH_2P(O)(OEt)_2 \xrightarrow{RCHO}$$

$$\xrightarrow{PhH} (EtO)_2P(O)CH_2COCH{=}CHR$$

$$\xrightarrow[DME]{NaH} Ph_3P{=}CHCOCH{=}CHR$$

I.B.1-10 M. Mikolajczyk, S. Grzejszczak, and A. Zatorski, *Tetrahedron Lett.*, **1976**, 2731.

$$(R^1O)_2P(O)CH(SR^2)_2 \xrightarrow[2)\ R^3COR^4]{1)\ n\text{-}BuLi,\ THF} (R^2S)_2C{=}CR^3R^4$$

R^3 and R^4 = H, alkyl and aryl

66-96%

I.B.1-11 P. Coutrot, C. Laurenco, J. Petrova, and P. Savignac, Synthesis, 1976, 107.

$$PhSCHClP(O)(OEt)_2 \xrightarrow[2)\ R^1COR^2]{1)\ LiCCl_3} R^1R^2C{=}C(Cl)SPh$$

R^1	R^2	% Yield
p-ClC_6H_4	H	65
PhCH=CH	H	46
$-(CH_2)_5-$		60

I.B.1-12 M. Mikolajczy, S. Grzejszczak, W. Midura, and A. Zatorski, Synthesis, 1976, 396.

$$(EtO)_2P(O)CH_2X \ + \ RCHO \xrightarrow[CH_2Cl_2]{50\%\ NaOH} XCH{=}CHR$$

X	R	% Yield
CN	Ph	80
CO_2H	Ph	95
MeSO	Ph	51
$(EtO)_2P(O)$	PhCH=CH	70

No phase-transfer catalyst necessary.

I.B.1-13 S. E. Dinizo, R. W. Freerksen, W. E. Pabst, and D. S. Watt, J. Org. Chem., 41, 2846 (1976).

$$R^1COR^2 \xrightarrow[\text{NaH, THF}]{(EtO)_2P(O)CH(OR^3)CN} R^1R^2C{=}C(OR^3)CN$$

R^1	R^2	R^3	% Yield
Ph	H	Me	81
Ph	H	t-Bu	94
Ph	Me	t-Bu	93
$\text{-(}CH_2\text{)}_5\text{-}$		Me	81

I.B.1-14 B. Lythgoe, T. A. Moran, M.E.N. Nambudiry, and S. Ruston, J. Chem. Soc., Perkin Trans. I, 1976, 2386; see also: A. H. Davidson and S. Warren, *ibid*, 639.

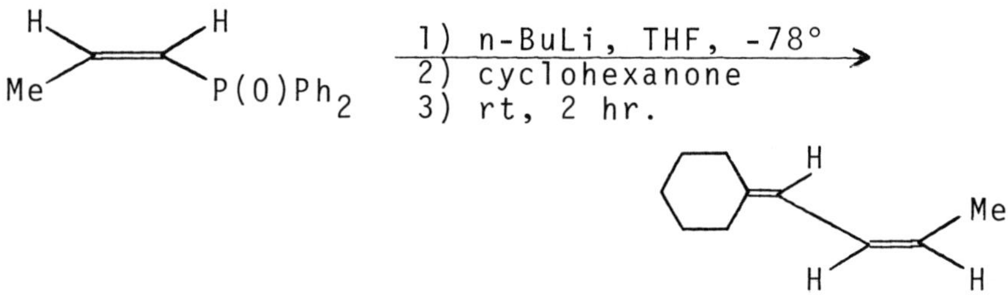

(∿3% Z-isomer) 48%

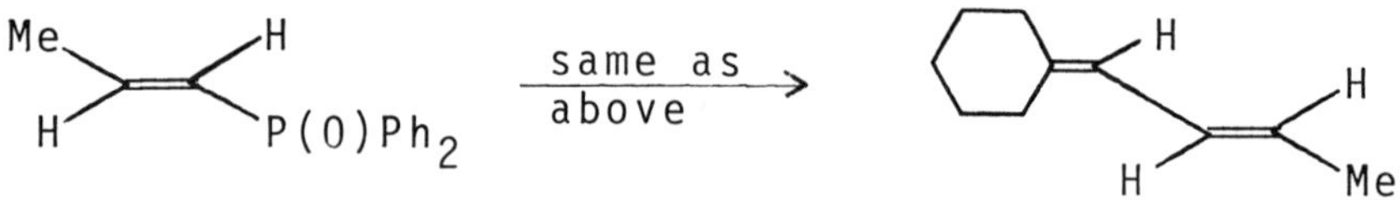

(<2% E-isomer) 72%

I.B.1-15 E. J. Corey, D. Enders and M. G. Bock, Tetrahedron Lett., 1976, 7.

$$Me_3SiCHR^1CH{=}NNMe_2 \xrightarrow[2)\ R^2COR^3,\ -78°]{1)\ LDA,\ -20°} R^2R^3C{=}CR^1CH{=}NNMe_2$$

85-95%

I.B.1-16 R. R. Schrock, J. Am. Chem. Soc., 98, 5399 (1976).

$$Ta[CH_2t\text{-}Bu]_3[CHt\text{-}Bu] \xrightarrow{R^1COR^2} R^1R^2C{=}CHt\text{-}Bu$$

R^1	R^2	% Yield
Me	Me	80
Ph	H	90
Me	OEt	60
H	NMe_2	77

Analogously Nb compounds give similar results.

I.B.2.a-1 C. W. Spangler and T. W. Hartford, Synthesis, 1976, 108; see also: *ibid*, J. Chem. Soc., Perkin Trans. I, 1976, 1792.

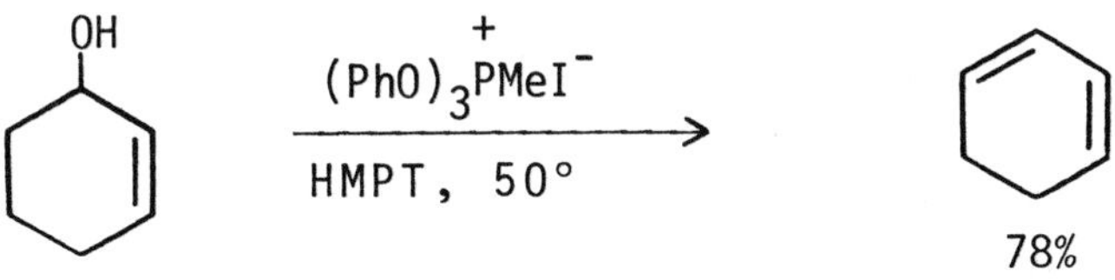

$$CH_2{=}CMeCH(OH)CH_2CH{=}CH_2 \xrightarrow[\text{HMPT, 50°}]{(PhO)_3\overset{+}{P}MeI^-} CH_2{=}CMeCH{=}CHCH{=}CH_2$$

53%

I.B.2.a-2 C. Descoins and D. Samain, Tetrahedron Lett., 1976, 745.

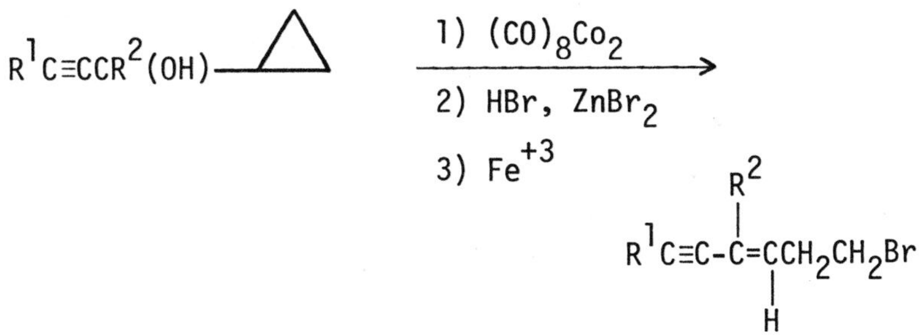

R^1,R^2 = alkyl or H

53-65%

Use of the cobalt complex gives a high stereoselectivity in the formation of the double bond (usually >97% E-isomer).

I.B.2.a-3 K. Utimoto, M. Obayashi, and H. Nozaki, J. Org. Chem., 41, 2940 (1976).

$$R^1COCH(SiMe_3)Am \xrightarrow[\text{2) t-BuOK, reflux or glacial HOAc}]{\text{1) } R^2Li\text{, THF}} R^1R^2C{=}CHAm$$

Stereochemistry of the product is determined by the method of elimination (step 2).

I.B.2.a-4 J. Remion, W. Dumont, and A. Krief, Tetrahedron Lett., 1976, 1385; see also: A. M. Leonard-Coppens and A. Krief, *ibid*, 3227, and J. Remion and A. Krief, *ibid*, 3743.

$$R^1CH(SePh)C(OH)R^2R^3 \xrightarrow[C_5H_{11}\text{, reflux}]{\text{p-TsOH}} R^1CH{=}CR^2R^3$$

R^1	R^2	R^3	% Yield
H	H	n-$C_{10}H_{21}$	71
Me	H	n-$C_{10}H_{21}$	65
H	$\leftarrow(CH_2)_2CH(t\text{-}Bu)(CH_2\rightarrow_2$		65

I.B.2.a-5 P. Brownbridge, I. Fleming, A. Pearce, and S. Warren, J. Chem. Soc., Chem. Commun., 1976, 751.

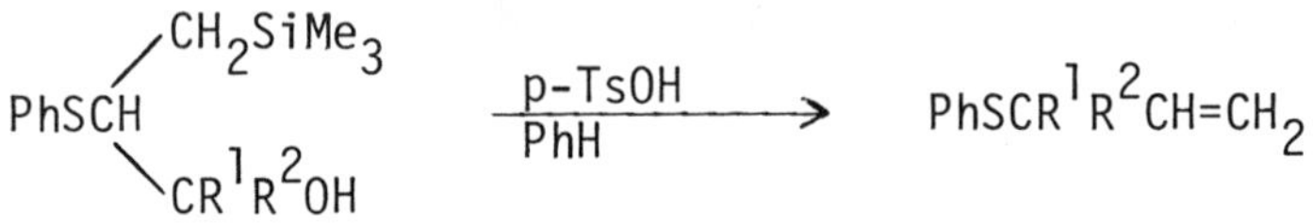

R^1 = Me, R^2 = H, 93%

R^1 = R^2 = Me, 96%

I.B.2.a-6 P. Entmayr and G. Köbrich, Chem. Ber., 109, 2175 (1976).

$$R^1COR^2 \xrightarrow[\text{2) } Me_3SiCl]{\text{1) } LiCHX_2} \underset{\text{A}}{R^1R^2C(OSiMe_3)CHX_2} \xrightarrow{R^3Li} \underset{\text{B}}{R^1R^2C{=}CX_2}$$

X	R^1	R^2	R^3Li	% Yield A	% Yield B
Cl	Ph	Me	n-BuLi	64	51
Cl	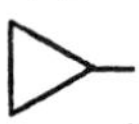	Me	n-BuLi	48	48
Cl	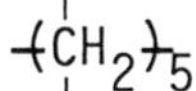		n-BuLi	75	46
Br	Ph	Me	LDA	64	59
Br	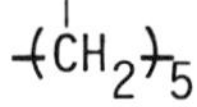		LDA	46	61

$$R^1R^2C(OSiMe_3)CHBr_2 \xrightarrow{n\text{-BuLi}} R^1R^2C{=}CHBr$$

52-93%

I.B.2.a-7 K. Nyberg, Synthesis, 1976, 545.

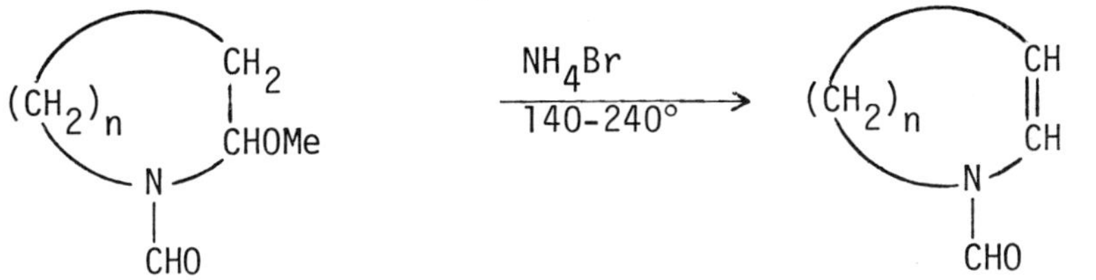

n = 2-4, 76-89%

I.B.2.a-8 G. H. Posner and G. M. Gurria, J. Org. Chem., 41, 578 (1976).

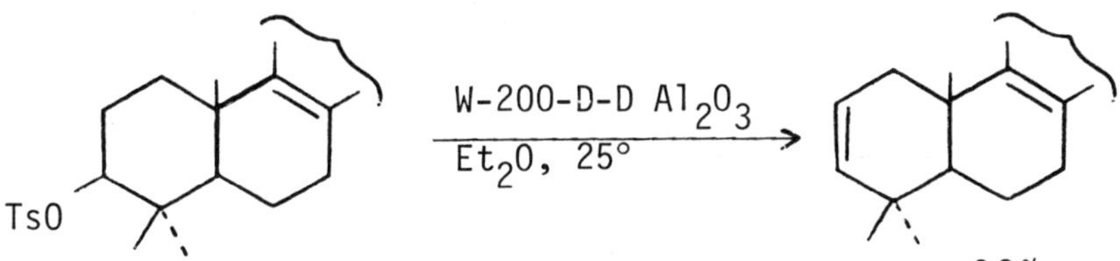

90%

Use of a vacuum-dehydrated alumina enhanced its activity in elimination reactions.

I.B.2.a-9 J. H. Babler and A. E. Moormann, J. Org. Chem., 41, 1477 (1976).

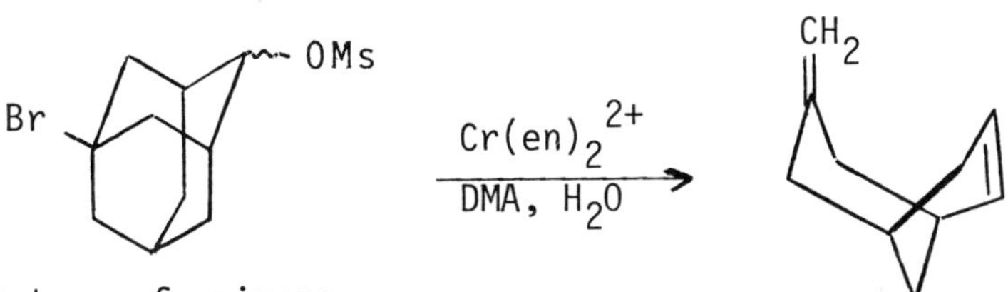

mixture of epimers

40%

I.B.2.b-1 R. N. Warrener, G. J. Collin, G. I. Hutchison, and M. N. Paddon-Row, J. Chem. Soc., Chem. Commun., 1976, 373.

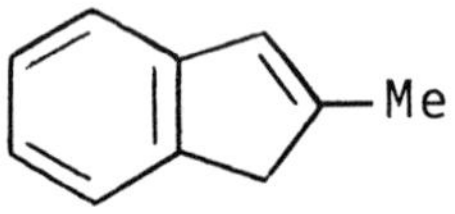

1) NBS, PhH or CCl_4

2) LiBr, Li_2CO_3, DMF, ref.

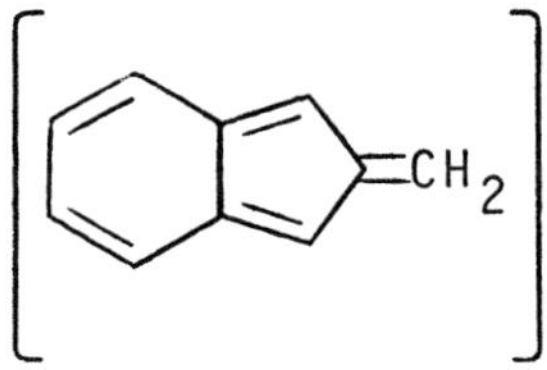

Trapped as N-methyl maleimide adduct.

I.B.2.b-2 P. E. Sonnet and J. E. Oliver, J. Org. Chem., 41, 3284 (1976).

$$R^1CHXCH(Cl)R^2 \xrightarrow[DMF]{NaI} R^1CH{=}CHR^2$$

X = Br or Cl

Highly stereoselective *cis*-eliminations were observed.

I.B.2.b-3 T. Kempe, T. Norin, and R. Caputo, *Acta Chem. Scand.*, **B30**, 366 (1976).

$$RCHBrCHBrR \xrightarrow[\text{DMF, } 140°]{Na_2S_2O_4} R^1CH{=}CHR^2$$

$R^1 = R^2 = Ph$, 90%

$R^1, R^2 = (CH_2)_4$, 60%

I.B.2.b-4 A. Kasal, *Coll. Czech. Chem. Commun.*, **41**, 2040 (1976).

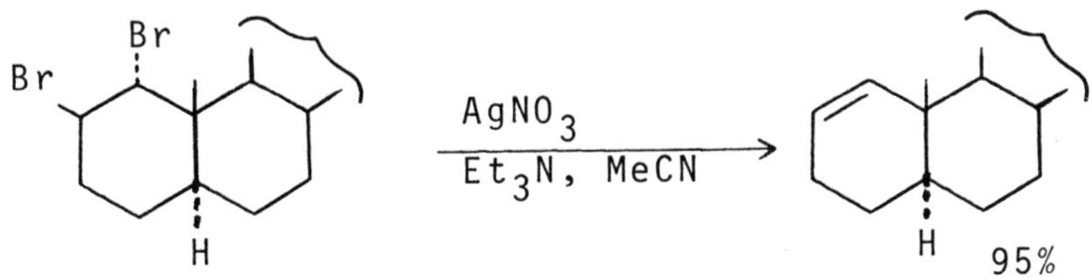

95%

A series of steroidal vicinal dibromides were studied.

I.B.2.b-5 G. A. Olah and G. K. SuryaPrakash, *Synthesis*, **1976**, 607.

$$R^1CHXCHXR^2 \xrightarrow[\text{THF}]{\text{Ti(II)}} R^1CH{=}CHR^2$$

X	R^1	R^2	% Yield
Br	Ph	Ph	88
Br	n-C_6H_{13}	H	84
Cl	$(CH_2)_4$		72
Br	$(CH_2)_6$		72

I.B.2.b-6 P. Dowd and L. K. Marwaha, *J. Org. Chem.*, **41**, 4035 (1976).

$$BrCH_2C(CO_2H)BrCH_2CHBrCO_2H \xrightarrow[\text{THF}]{\text{Sn-Cu couple}} CH_2{=}C(CO_2H)CH_2CHBrCO_2H$$

81%

$$(BrCH_2)_2C{=}C(CH_2Br)_2 \xrightarrow[\text{THF}]{\text{Sn-Cu couple}} CH_2{=}C(CH_2Br)C(CH_2Br){=}CH_2$$

Nonactivated dibromides ($BrCH_2CH_2Br$) react very slowly.

I.B.2.c-1 P. B. Dervan and M. A. Shippey, *J. Am. Chem. Soc.*, **98**, 1265 (1976).

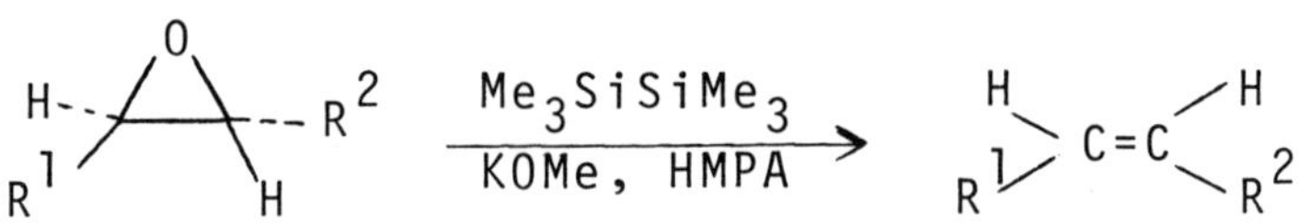

75-99%

(>99% stereospecific)

I.B.2.c-2 M. T. Reetz and M. Plachky, Synthesis, 1976, 199.

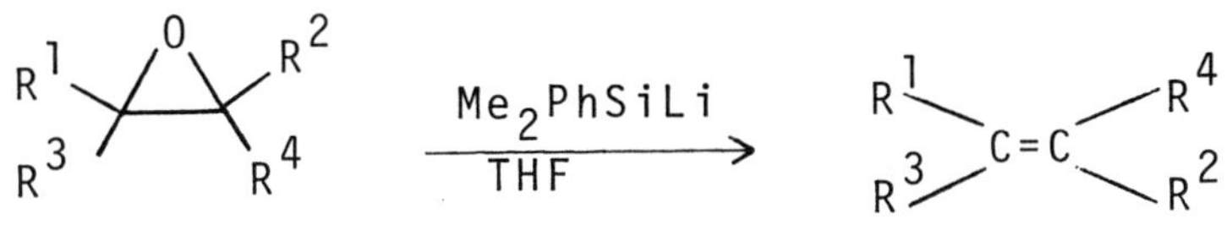

R^1	R^2	R^3	R^4	% Yield
Ph	H	H	Ph	75
Ph	Ph	H	H	83
Pr	H	H	H	60

I.B.2.c-3 V. Calo, L. Lopez, A. Mincuzzi, and G. Pesce, Synthesis, 1976, 200.

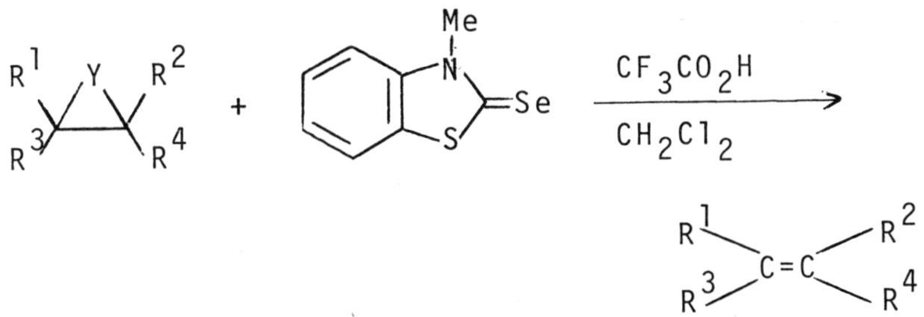

R^1	R^2	R^3	R^4	Y	% Yield
Cl	H	H	H	O	100
Ph	H	H	Ph	O	97
Ph	H	H	H	S	87
$-(CH_2)_4-$		H	H	O or S	100

I.B.2.c-4 T. Cohen, A. J. Mura, Jr., D. W. Shull, E. R. Fogel, R. J. Ruffner, and J. R. Falck, J. Org. Chem., 41, 3218 (1976).

$$PhSCHR^1CH_2CR^2(SPh_2) \xrightarrow[i\text{-}Pr_2NEt,\ PhH]{Cu_2Ph(CF_3SO_3)_2} CH_2{=}CHCR^3{=}CHR^4$$

R^1 = Me, R^2 = R^3 = H, R^4 = PhS, 84%

R^1 = R^4 = H, R^2 = Me, R^3 = PhS, 76%

I.B.2.c-5 E. Vedejs and D. A. Engler, Tetrahedron Lett., 1976, 3487.

$$R^1CH_2CR^2(SMe)CO_2R^3 \xrightarrow[2)\ DBU,\ DMF,\ 50°]{1)\ CF_3SO_3CH_2CO_2Et} R^1CH{=}CR^2CO_2R^3$$

70-94%

$$R^1CH_2CHR^2(SPh) \xrightarrow[2)\ DBU,\ DMF,\ 50°]{1)\ CF_3SO_3CH_2CO_2Et} R^1CH{=}CHR^2$$

31-89%

I.B.2.c-6 B. M. Trost and K. Hiroi, J. Am. Chem. Soc., 98, 4313 (1976); see also: B. M. Trost, T. N. Salzmann, and K. Hiroi, *ibid*, 4887.

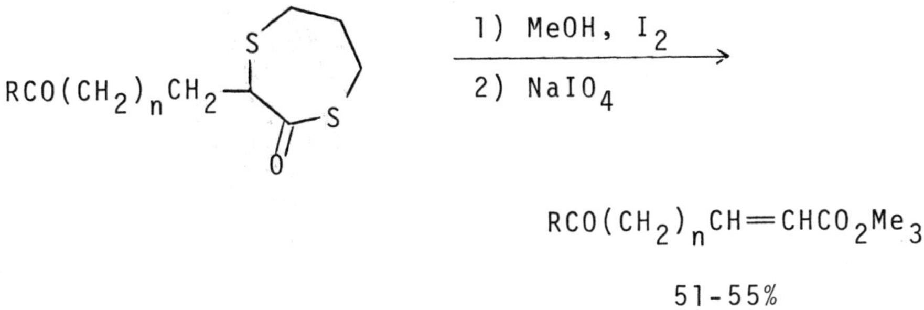

I.B.2.c-7 G.M. Ksander and J. E. McMurry, Tetrahedron Lett., 1976, 4691.

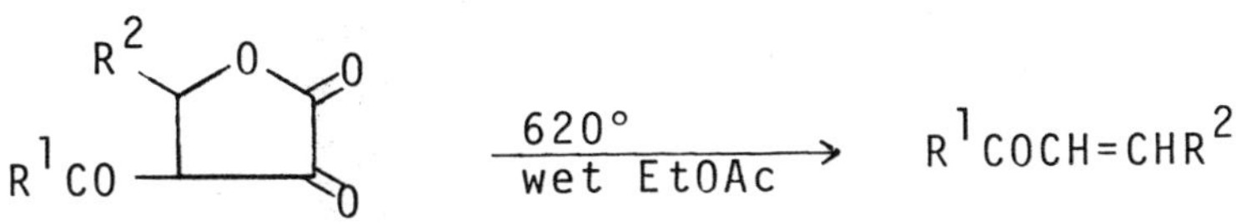

R^1 = Ph, R^2 = H, 82%

R^1 = n-C_6H_{13}, R^2 = Me, 84%

R^1 = cyclopropyl, R^2 = H, 65%

I.B.2.c-8 T. Kato, T. Chiba, and T. Chiba, Chem. Pharm. Bull., 24, 3034 (1976).

R^1	R^2	% Yield
Ph	OMe	75
H	OMe	50
H	Me	51

I.B.2.c-9 V. Jäger and H. Grund, Angew. Chem. Int. Ed. Engl., 15, 50 (1976).

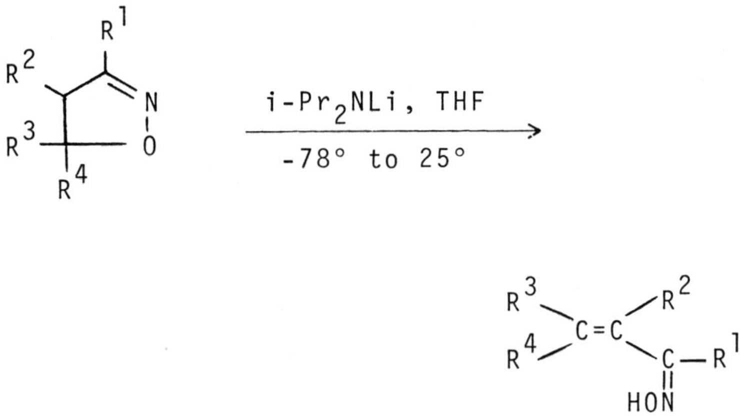

∿ 60%

I.B.3-1 J. E. McMurry and M. P. Fleming, J. Org. Chem., 41, 896 (1976); see also: A. L. Baumstark, E.J.H. Bechara, and M. J. Semigran, Tetrahedron Lett., 1976, 3265.

$$R^1R^2CO \xrightarrow{Ti^\circ} R^1R^2C{=}CR^1R^2$$

$R^1,R_2 = \text{-(}CH_2\text{)}_5\text{-}$, 85%

$R^1 = R_2 = $ i-Pr, 40%

$$R^1R^2C(OH)\text{-}C(OH)R^1R^2 \xrightarrow{Ti^\circ} R^1R^2C{=}CR^1R^2$$

$R^1,R^2 = \text{-(}CH_5\text{)}_5\text{-}$, 85%

$R^1 = H$, $R^2 = $ Bu, 80%

The active Ti° is prepared from $TiCl_3$ and K metal.

I.B.3-2 G. A. Olah and G. K. Surya Prakash, Synthesis, 1976, 607.

$$PhCRCl_2 \xrightarrow[THF]{Ti(II)} PhCR{=}CRPh$$

R = Ph, 96%

R = H, 79%

I.B.3-3 T. Minami, M. Matsumoto, and T. Agawa, J. Chem. Soc., Chem. Commun., 1976, 1053.

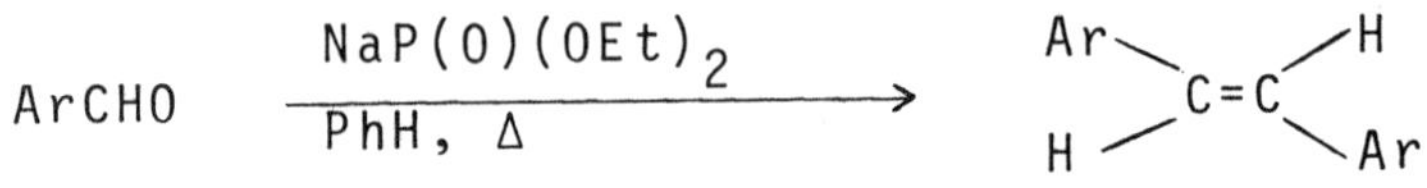

22-85%

I.B.3-4 A. J. Fry and L.-L. Chung, Tetrahedron Lett., 1976, 645.

$$(R^1R^2CBr)_2P(O)OMe \xrightarrow[\text{DMSO, TEAB, }20^\circ]{\text{electrolysis}} R^1R^2C{=}CR^1R^2$$

$R^1 = Ph$, $R^2 = H$, 76%

$R^1 = R^2 = Me$, 40%

I.B.3-5 H. J. Bestmann, L. Kisielowski, and W. Distler, Angew. Chem. Int. Ed. Engl., 15, 298 (1976).

$$RCH{=}PPh_3 \quad + \quad (PhO)_3P{-}O_3 \xrightarrow{\text{PhMe or } CH_2Cl_2} RCH{=}CHR$$

R	% Yield
$n\text{-}C_{11}H_{23}$	61
Ph	89
MeCO	87
MeO_2C	83

I.B.3-6 T. G. Back, D.H.R. Barton, M.R. Britten-Kelly, and F. S. Guziec, Jr., J. Chem. Soc., Perkin Trans. I, 1976, 2079.

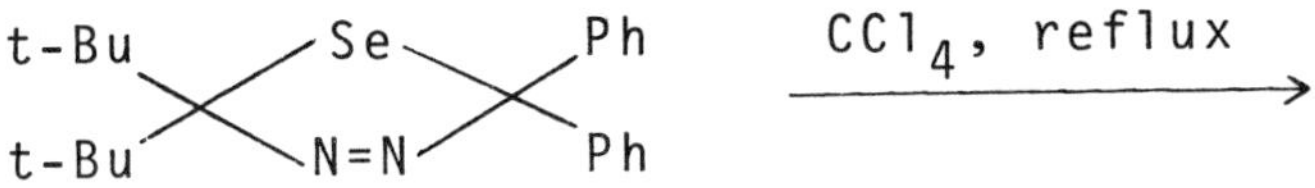

$(t\text{-}Bu)_2C{=}C(Ph)_2$

67%

Attempts to synthesize tetra-t-butylethylene by this procedure were unsuccessful.

I.B.3-7 A. L. Fridman and F. A. Gabitov, J. Org. Chem. USSR, 11, 1113 (1975).

$PhCOC(I)(Ph)C(NO_2)_3 \xrightarrow[\text{reflux}]{PhCl} PhCOCPh{=}CPhCOPh$

95%

I.B.3-8 A. G. Hortmann and A. Bhattacharjya, J. Am. Chem. Soc., 98, 7081, (1976).

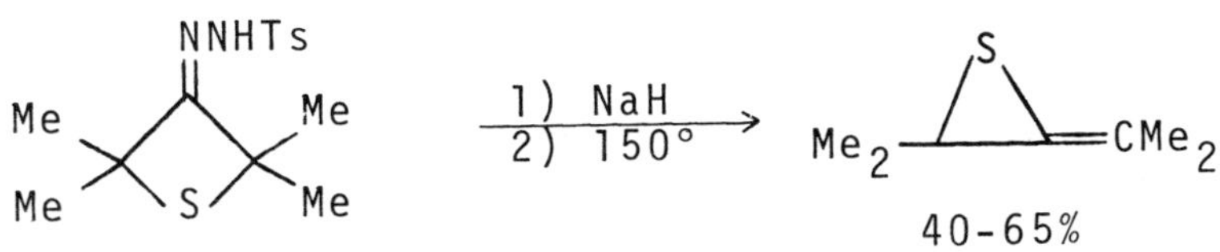

40-65%

I.B.3-9 D. A. Konen, P. E. Pfeffer, and L. S. Silbert, Tetrahedron, 32, 2507 (1976).

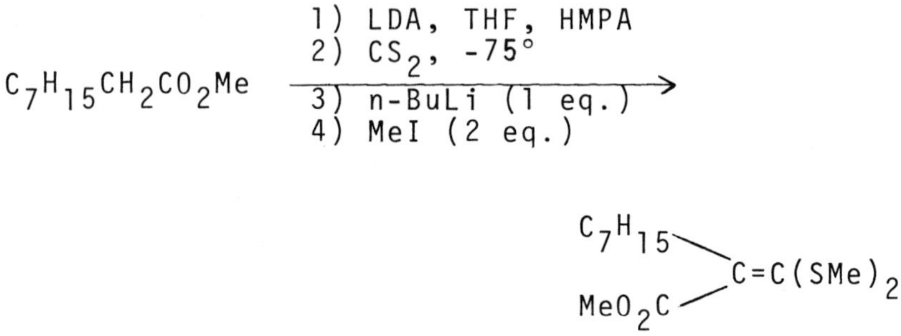

78%

I.B.3-10 A. C. Brouwer and H.J.T. Bos, Tetrahedron Lett., 1976, 209.

30-70%

R^1,R^2 = aryl

I.B.3-11 M. V. Bhatt, C. G. Rao, and S. Rengaraju, J. Chem. Soc., Chem. Commun., 1976, 103.

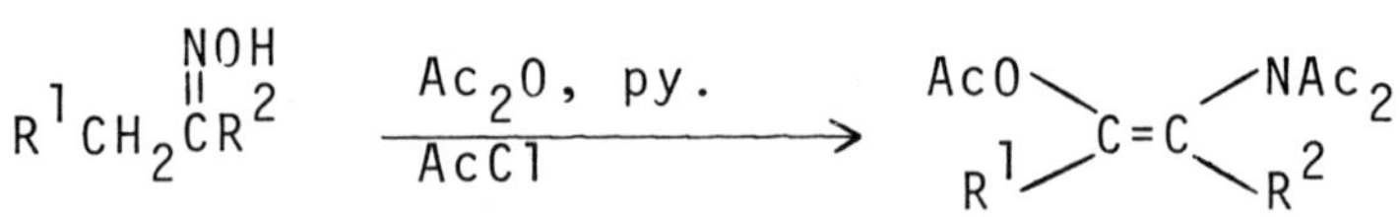

17-58%

I.B.3-12 J. J. Eisch and J. E. Galle, J. Am. Chem. Soc., 98, 4646 (1976).

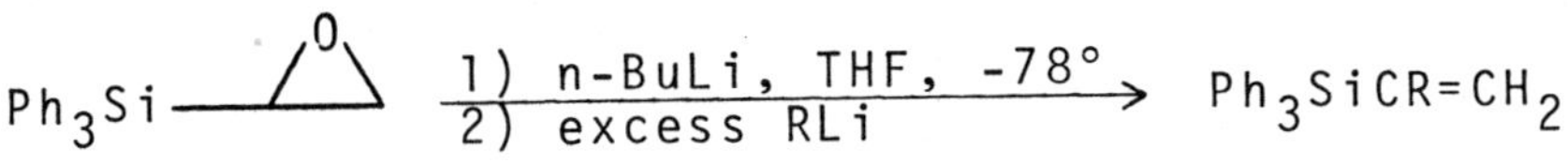

R = n-Bu, 70%
R = Ph, 66%

I.B.3-13 S. Iwasaki, Helv. Chim. Acta, 59, 2753 (1976).

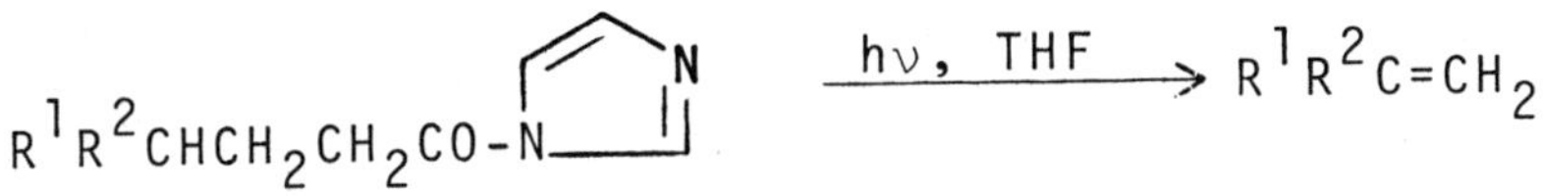

$R^1 = C_{14}H_{29}$, $R^2 = H$, 40%
$R^1 = C_{14}H_{29}$, $R^2 = Me$, 62%

Applicable to side-chain degradation of bile acids and lanosterol.

I.B.3-14 G. Ferdinand and K. Schank, Synthesis, 1976, 404.

$$RCH_2O_2SCH(OMe)COPh \xrightarrow[2)\ CH_2O\ (g)]{1)\ NaH\ or\ LiH,\ MeCN} RCHO_2SC(OMe)=CH_2$$

R = H, 70%
R = Ph, 80%

I.B.4-1 H. Yamazaki, J. Chem. Soc., Chem. Commun., 1976, 841.

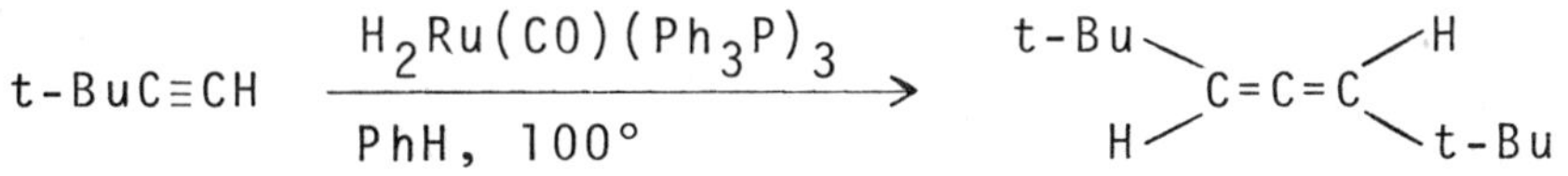

I.B.4-2 F. Delbecq and J. Gore, Angew.Chem. Int. Ed. Engl., 15, 496 (1976); see also: *ibid*, Bull. Soc. Chim. Fr., 1976, 533 and 541.

$$R^1R^2C(OAc)C{\equiv}CH \xrightarrow[MgBr_2,\ Et_2O]{R^3MgX} R^1R^2C{=}C{=}CHR^3$$

R^1	R^2	R^3	% Yield
Me	Me	n-C_8H_{17}	64
$\text{-}(CH_2)_5\text{-}$		n-Pr	72
$\text{-}(CH_2)_5\text{-}$		i-Pr	44
$\text{-}(CH_2)_5\text{-}$		t-Bu	80

I.B.4-3 D. J. Pasto, G. F. Hennion, R. H. Shults, A. Waterhouse, and S.-K. Chou, <u>J. Org. Chem.</u>, **41**, 3496 (1976).

$$R^1R^2C(Cl)C{\equiv}CR^3 \xrightarrow[0.5 \times 10^{-5}\ M\ FeCl_3]{R_4MgX}$$

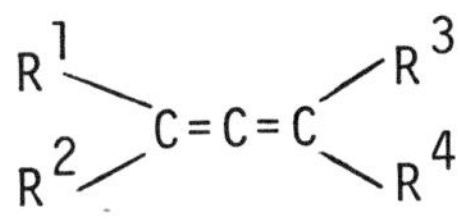

40-90%

I.B.4-4 G. Tadema, P. Vermeer, J. Meijer, and L. Bransdma, <u>Recl. Trav. Chim. Pays-Bas</u>, **95**, 66 (1976).

$$(EtO)_2CHC{\equiv}CH \xrightarrow[CuBr\ (cat.),\ Et_2O]{RMgBr} EtOCH{=}C{=}CHR$$

R	% Yield
n-Bu	78
cyclopentyl	80
t-Bu	79

I.B.4-5 J.-L. Moreau and M. Gaudemar, J. Organomet. Chem., 108, 159 (1976).

$$MeOCR^1R^2C{\equiv}CR^3 \xrightarrow[Et_2O,\ CuBr\ (10\ mol\%)]{R^4MgBr}$$

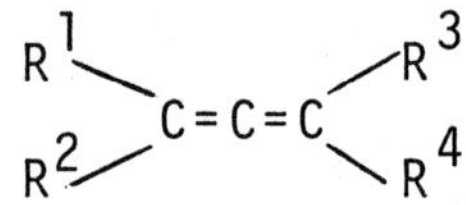

R^1	R^2	R^3	R^4	% Yield
H	H	H	C_7H_{15}	70
H	H	Me	Ph	82
Me	H	Et	$i\text{-}C_5H_{11}$	45
Me	Me	Me	C_8H_{17}	52

I.B.4-6 P. Crabbé, E. Barreiro, J.-M. Dollat, and J.-L. Luche, J. Chem. Soc., Chem. Commun., 1976, 183.

$$R^1_2CuLi \quad + \quad R^2C{\equiv}CC(OAc)R^3R^4 \longrightarrow$$

$$R^1R^2C{=}C{=}CR^3R^4\ (A) \quad \text{or} \quad R^2CH{=}C{=}CR^3R^4\ (B)$$

Conditions are described for the selective formation of product (A) or (B).

I.B.4-7 J. L. Ripoll, J. Chem. Soc., Chem. Commun., 1976, 235.

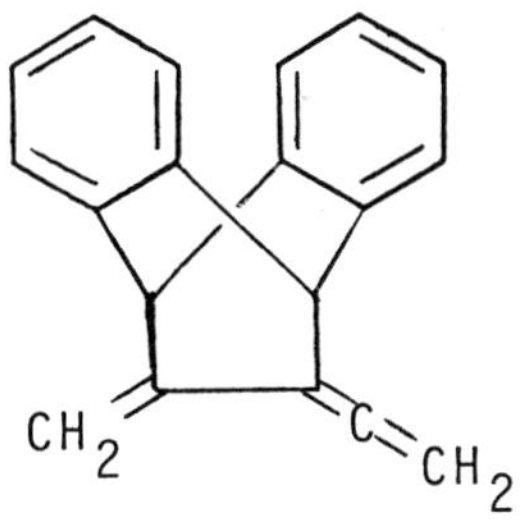

$\xrightarrow[10^{-3}\ \text{Torr}]{700°}$

$CH_2=C=C=C=CH_2$ (A)

+

MeC≡C=C≡CH (B)

85% (A:B = 7:3)

I.B.4-8 T. L. Gilchrist and D.P.J. Pearson, J. Chem. Soc., Perkin Trans. I, 1976, 989.

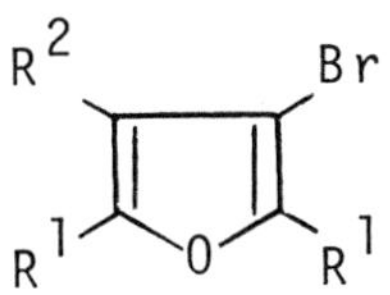

$\xrightarrow[2)\ H^+]{1)\ n\text{-BuLi}}$

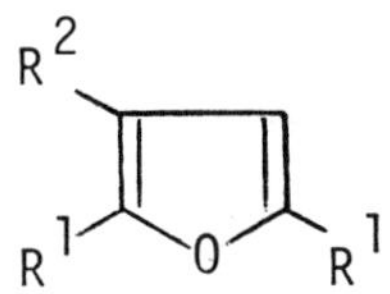

+

$R^1COCR^2=C=CHR^1$

+

$R^1COCHR^2C≡CR^1$

Product distribution is dependent on solvent and reaction conditions.

I.C-1 C. B. Reese and A. Shaw, J. Chem. Soc., Perkin Trans. I, 1976, 890.

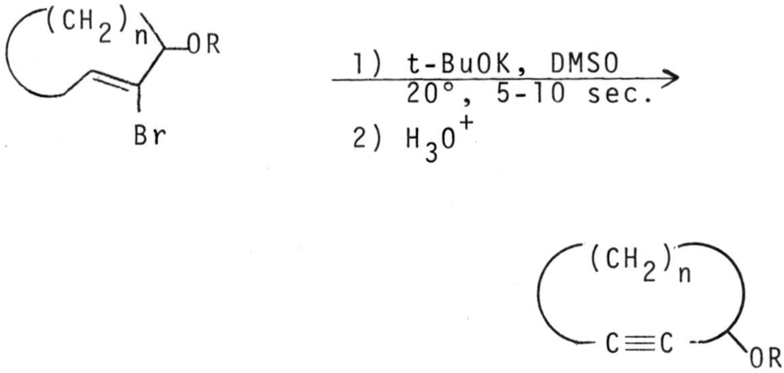

n = 5, 74%

n = 6, 70%

I.C-2 D. P. Bauer and R. S. Macomber, J. Org. Chem., 41, 2640 (1976).

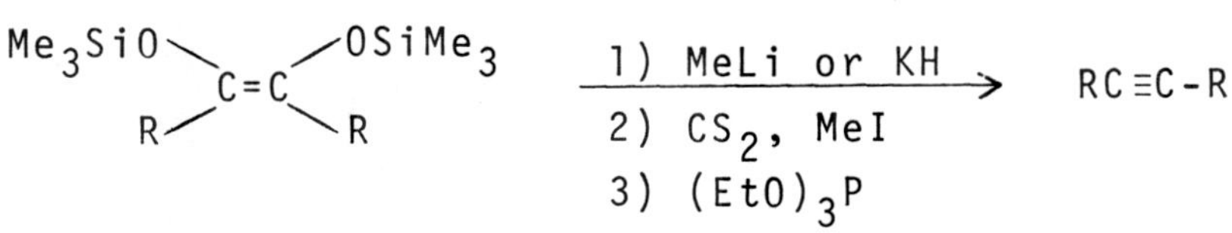

R = Ph, 35%

R = n-Pr, 28%

I.C-3 A. Gorgues and A. LeCoq, Tetrahedron Lett., 1976, 4723.

$$R^1CHBrCHBrR^2 \xrightarrow[50\%\ NaOH]{Bu_4NHSO_4,\ n\text{-}C_5H_{12}} R^1C{\equiv}CR^2$$

R^1 = H, R^2 = $CH(OEt)_2$, 80%

R^1 = R^2 = Ph, 75%

$$(EtO)_2CHCH{=}CCl_2 \xrightarrow[50\%\ NaOH]{BuNHSO_4,\ n\text{-}C_5H_{12}} (EtO)_2CHC{\equiv}CCl$$

70%

I.C-4 A. Gorgues and A. Le Coq, Bull. Soc. Chim. Fr., 1976, 125.

$$R^1CX{=}CHR^2 \xrightarrow[PhH\ or\ PhH,\ CH_2Cl_2]{Triton\ B} R^1C{\equiv}CR^2$$

R^1	R^2	X	% Yield
p-MeC_6H_4	H	Cl	70
Br	$CCl_2{=}CCl$	Br	70
Br	$CH(OEt)_2$	Br	85
COMe	PhC≡C	I	64
COMe	$CBr_2{=}CCl$	Br	51

I.C-5 T. Suzuki, T. Sonoda, S. Kobayashi, and H. Taniguchi, *J. Chem. Soc., Chem. Commun.*, **1976**, 180.

$$Ar_2C{=}CHBr \xrightarrow{h\nu} ArC{\equiv}CAr$$

50-86%

Ar must be oxygen substituted ($Ar = MeOC_6H_4$)

I.C-6 P. J. Kocienski, J. M. Ansell, and B. E. Norcross, *J. Org. Chem.*, **41**, 3650 (1976).

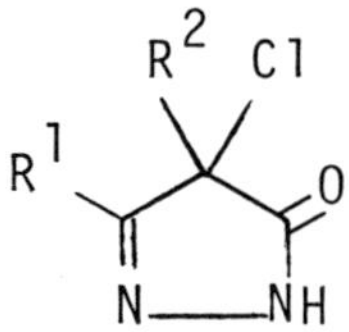

$$\xrightarrow[K_3Fe(CN)_6]{NaOH\ (aq.)} R^1C{\equiv}CR^2$$

$R^1 = R^2 = Ph$, 80%

$R^1 = Ph$, $R^2 = Me$, 60%

$R^1 = Me$, $R^2 = Ph$, 12%

I.C-7 C. Wentrup and W. Reichen, *Helv. Chim. Acta*, **59**, 2615 (1976).

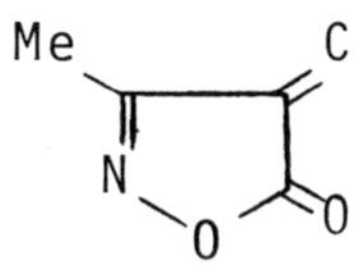

$$\xrightarrow[0.01\text{-}0.03\ Torr]{520\text{-}650^\circ} ArC{\equiv}CH$$

45-95%

I.C-8 J. Levisalles and H. Rudler, J. Organomet. Chem., 122, C17 (1976).

$$\text{n-BuCOCl} \xrightarrow[\text{HMPA}]{Na_2W(CO)_5} \text{n-}C_5H_{12}\ (26\%) + \text{n-BuC}\equiv\text{CH}\ (15\%)$$

I.C-9 K. Okuhara, J. Org. Chem., 41, 1487 (1976).

$$\text{ArLi} \xrightarrow[\text{2) n-BuLi, 60°; 3) } H_2O]{\text{1) } CF_2{=}CCl_2} \text{ArC}\equiv\text{CH}\quad 52\text{-}78\%$$

I.D.1-1 E. V. Dehmlow, Tetrahedron Lett., 1976, 91.

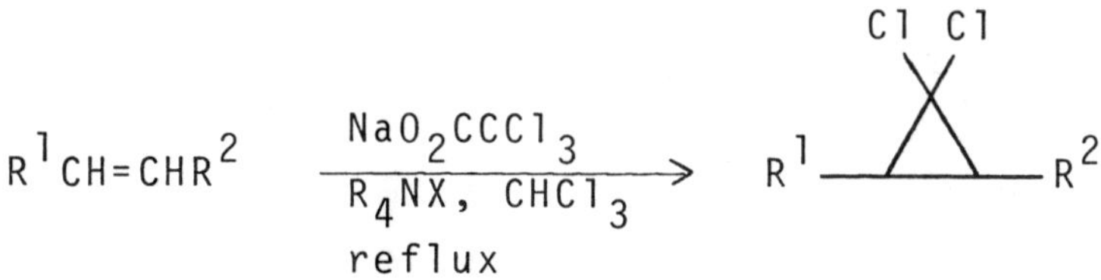

R^1	R^2	% Yield
Ph	Ph	7
$-(CH_2)_4-$		57
$-(CH_2)_6-$		73

I.D.1-2 I. Tabushi, Y. Kuroda, and Z. Yoshida, Tetrahedron, 32, 997 (1976).

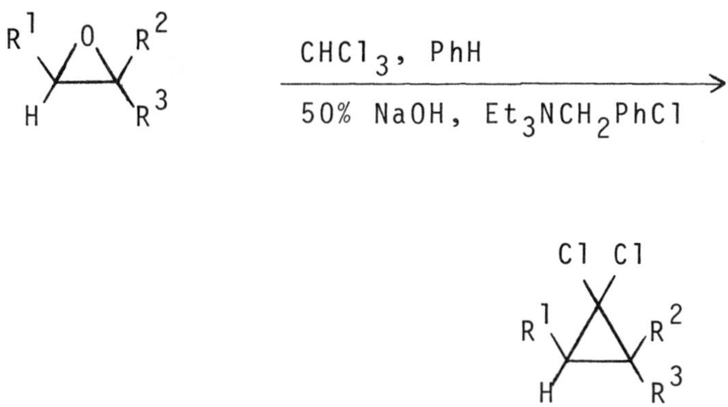

15-39%

I.D.1-3 R. J. Kricks and A. A. Volpe, Synthesis, 1976, 313.

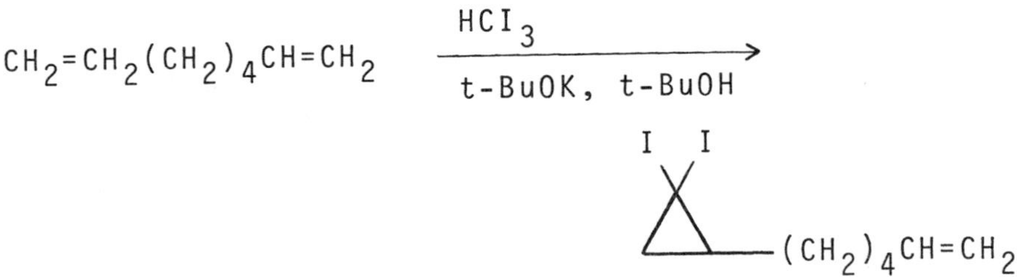

29%

I.D.1-4 M. S. Baird, J. Chem. Soc., Perkin Trans. I, 1976, 54.

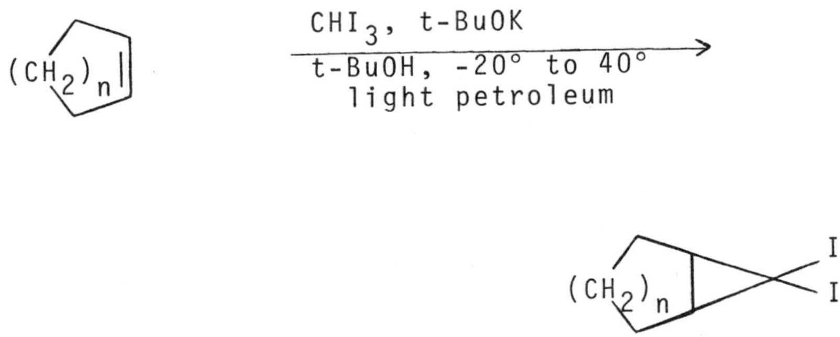

40-65%

(n = 5-8)

Many of the products are distillable contrary to earlier reports.

I.D.1-5 M. Kamel, W. Kimpenhaus, and J. Buddrus, Chem. Ber., 109, 2351 (1976).

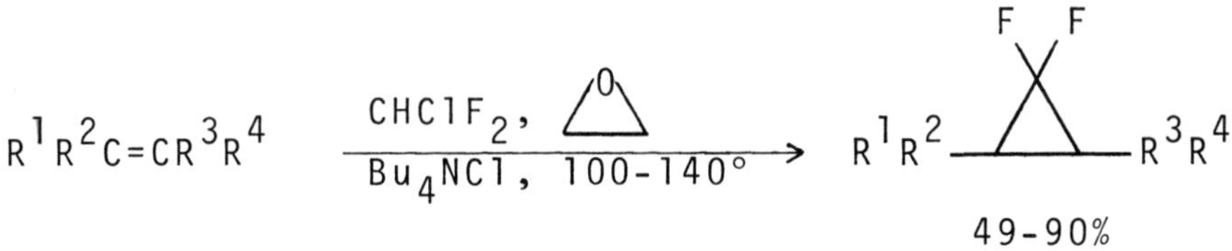

I.D.1-6 R. A. Olofson, K. D. Lotts, and G. N. Barber, Tetrahedron Lett., 1976, 3381 and 3379; see also: G. N. Barber and R. A. Olofson, *ibid*, 3783.

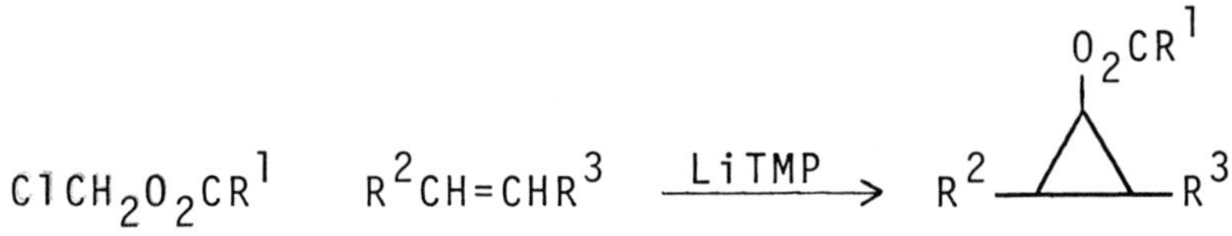

R^1	R^2	R^3	% Yield
t-Bu	H	OEt	30
MeO	$-(CH_2)_4-$		10
Me_2N	$-(CH_2)_4-$		25

I.D.1-7 A. Jonczyk and M. Makosza, Synthesis, 1976, 387; see also: I. Artaud, J. Seyden-Penne, and P. Viout, Comptes Rendus, Ser. C, 285, 503 (1976).

$$R^1C(X)=CH_2 + R^2CHX^2Cl \xrightarrow[PhCH_2\overset{+}{N}Et_3Cl^-]{50\%\ NaOH}$$

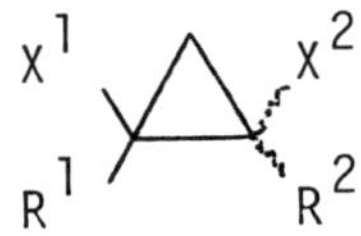

R^1	R^2	X^1	X^2	% Yield
H	H	CN	CO_2t-Bu	45
H	Me	CO_2t-Bu	CN	67
Me	Me	CN	CN	75

The isomeric ratio was determined in all cases.

I.D.1-8 N. Kawabata, M. Naka, and S. Yamashita, J. Am. Chem. Soc., 98, 2676 (1976).

$$R^1CH=CHR^2 \xrightarrow[Cu,\ I_2\ (cat.)]{R^3CHX_2,\ PhH}$$

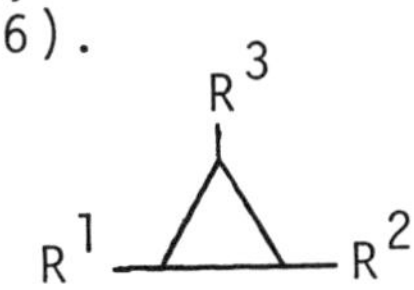

R^1	R^2	R^3	X	% Yield
Ph	H	H	I	90
C_6H_{13}	H	H	I	86
$-(CH_2)_4-$		CO_2Me	Br	31
$-(CH_2)_4-$		Cl	I	48
$-(CH_2)_4-$		H	I	85-87

I.D.1-9 T. Sasaki, S. Eguchi, M. Ohno, and F. Nakata, J. Org. Chem., 41, 2408 (1976).

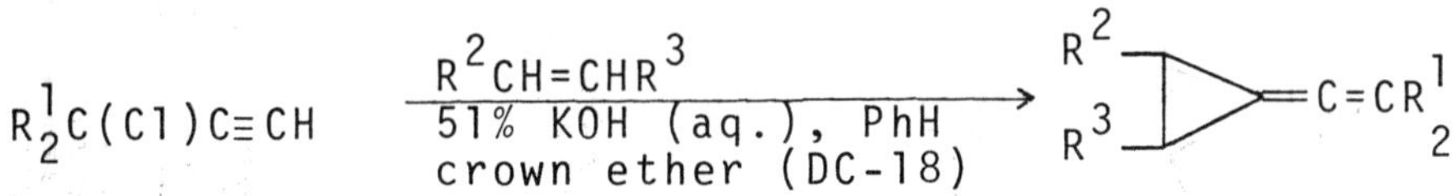

37-65%

I.D.1-10 A. Doutheau and J. Gore, Bull. Soc. Chim. Fr., 1976, 1189.

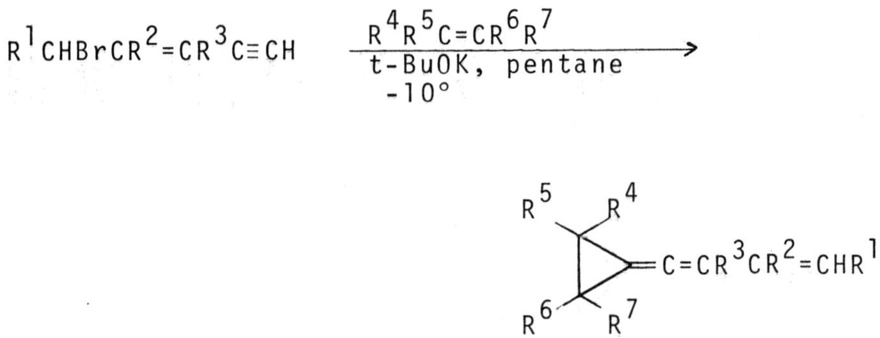

R^1	R^2	R^3	R^4	R^5	R^6	R^7	% Yield
Me	H	H	Ph	H	H	H	41
Me	H	H	Me	Me	Me	Me	43
H	H	Me	Ph	H	H	H	42
Me	Me	Me	Me	Me	Me	Me	21

Does not work well for the chloroenynes.

I.D.1-11 R. F. Cunico and Y.-K. Han, *J. Organomet. Chem.*, **105**, C29 (1976).

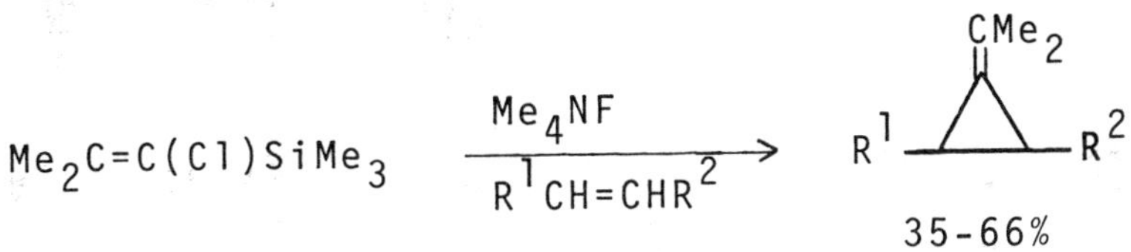

I.D.1-12 D. Seyferth and D. Dagani, *J. Organomet. Chem.*, **104**, 145 (1976).

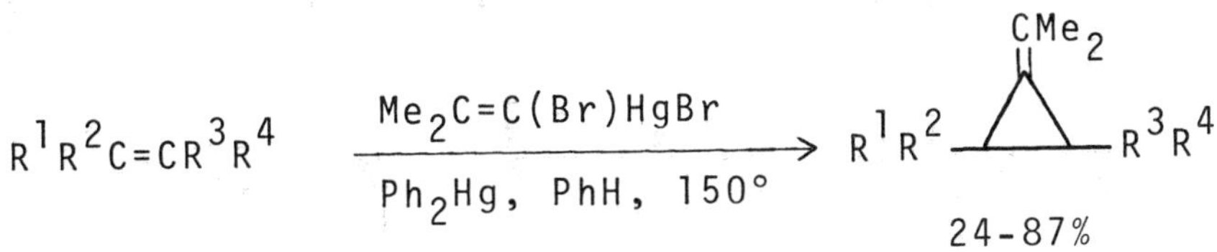

I.D.1-13 D. Seyferth, C. K. Haas, and D. Dagani, *J. Organomet. Chem.*, **104**, 9 (1976).

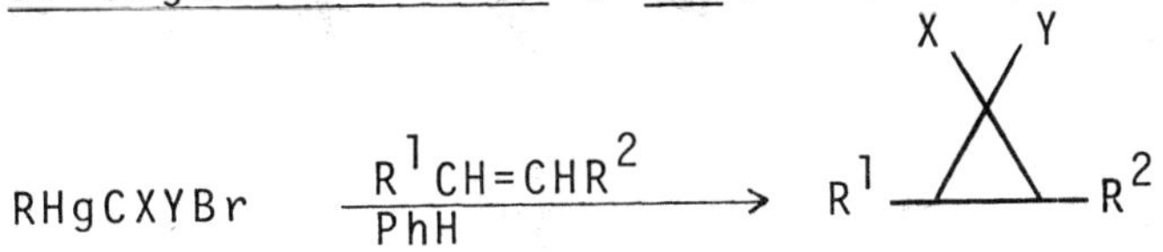

(R = cyclohexyl and $PhCH_2CH_2$, X,Y = Cl, Cl; Cl, Br; Br, Br)

Use of alkyl group R allows the reaction to occur slowly at room temperature.

I.D.1-14 A. J. Hubert, A. F. Noels, A. J. Anciaux, and P. Teyssie, Synthesis, 1976, 600.

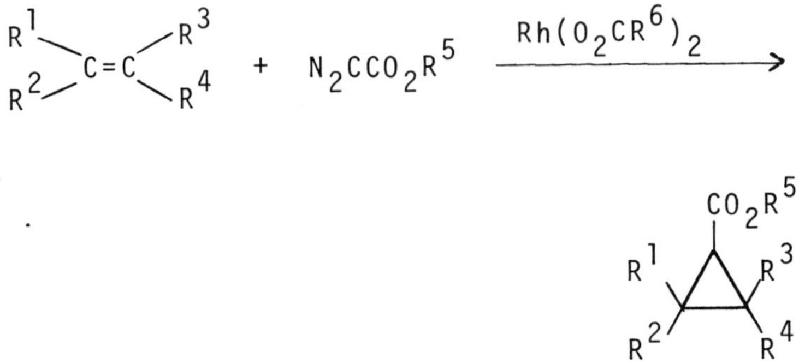

A comparison of various rhodium(II) carboxylates with copper(II) triflate and palladium(II) acetate was made.

I.D.1-15 S. H. Goh, K. C. Chan, and H. L. Chong, Aust. J. Chem., 29, 1699 (1976).

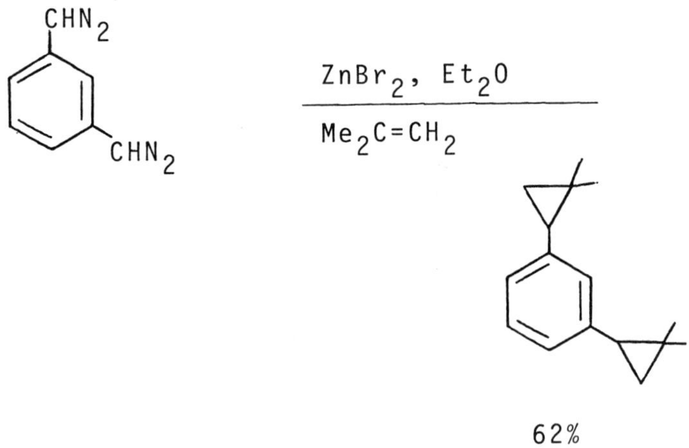

I.D.2-1 S. Cohen and A. Vogev, J. Am. Chem. Soc., 98, 2013 (1976).

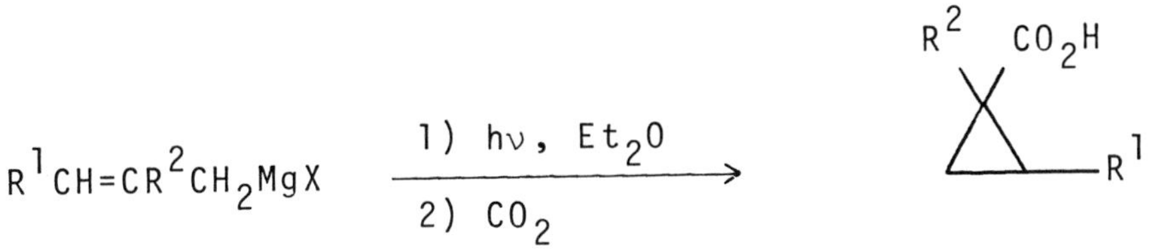

$R^1 = R^2 = H$, 45%
$R^1 = H$, $R^2 = Me$, 70%
$R^1 = Ph$, $R^2 = H$, 75%

I.D.2-2 A. Couture, M. Hoshino, and P. deMayo, J. Chem. Soc., Chem. Commun., 1976, 131.

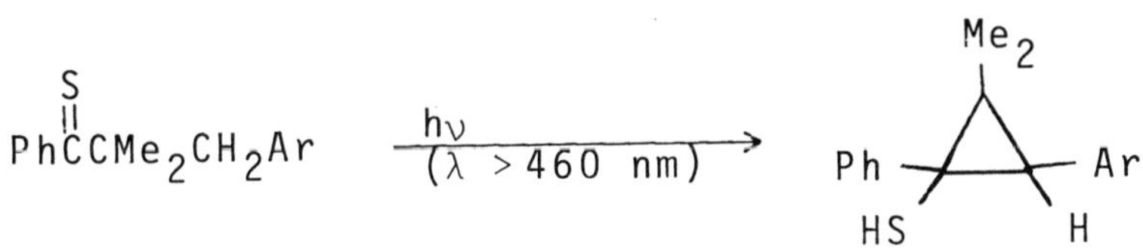

high yields

I.D.2-3 P. Baeckstrom, J. Chem. Soc., Chem. Commun., 1976, 476; see also: M. J. Bullivant and G. Pattenden, J. Chem. Soc., Perkin Trans. I, 1976, 256.

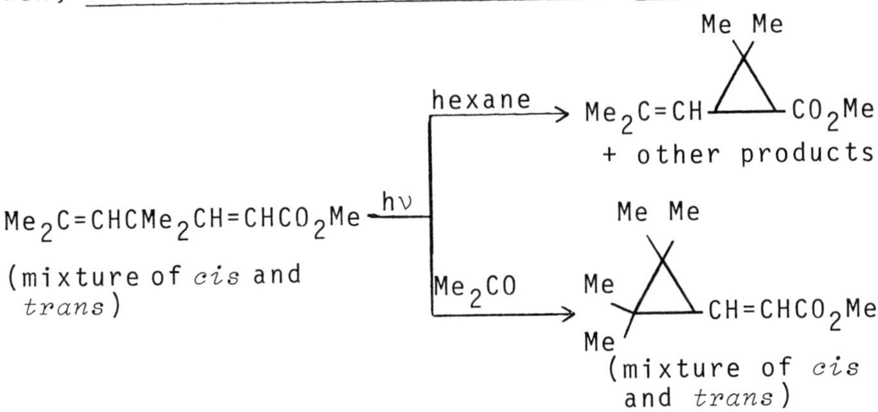

I.D.2-4 G. Jones, II and L. P. McDonnell, J. Chem. Soc., Chem. Commun., 1976, 18.

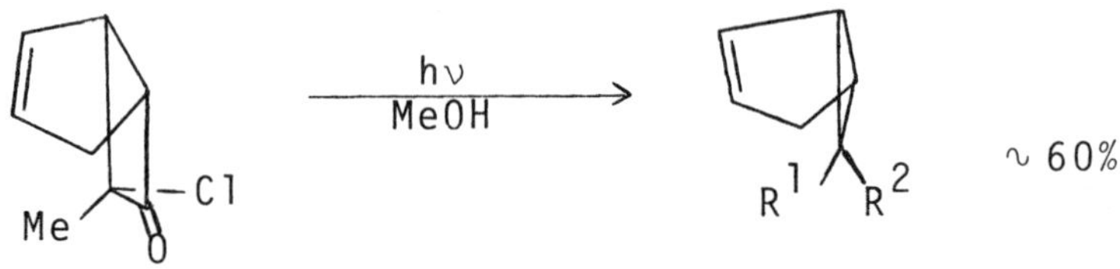

(R^1 = Me, R^2 = CO_2Me/R^1 = CO_2Me/R^2 = Me = 6/2)

Other stereoisomer at α-carbonyl position gives mainly openchain product upon photolysis.

I.D.2-5 M. Braun and D. Seebach, Chem. Ber., 109, 669 (1976).

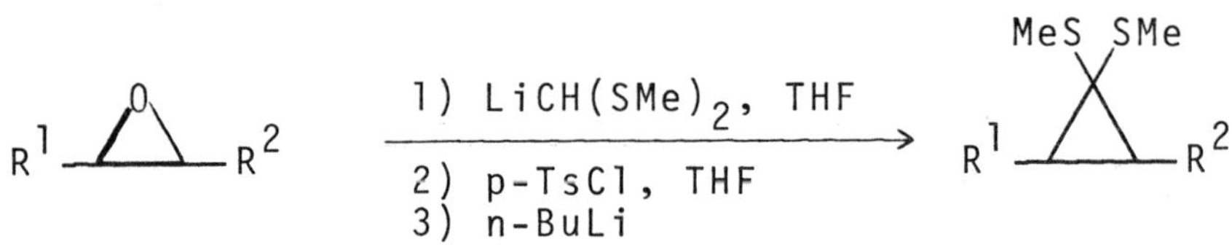

R^1	R^2	% Yield
Me	Me	66
n-$C_{10}H_{21}$	H	77
CH_2=CH	H	60
$-(CH_2)_4-$		81

An "improved one-pot-modification" is described.

I.D.2-6 J.-L. Burgot, J. Masson, P. Metzner, and J. Vialle, Tetrahedron Lett., 1976, 4297; see also: J.-L. Burgot, J. Masson, and J. Vialle, *ibid*, 4775.

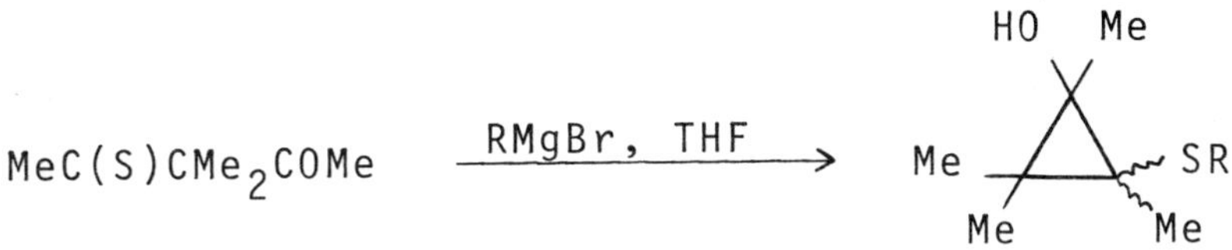

R = Et, 60%

I.D.2-7 A. Hercouet and M. LeCorre, Tetrahedron Lett., 1976, 825; see also: A. Turcant and M. LeCorre, *ibid*, 1277.

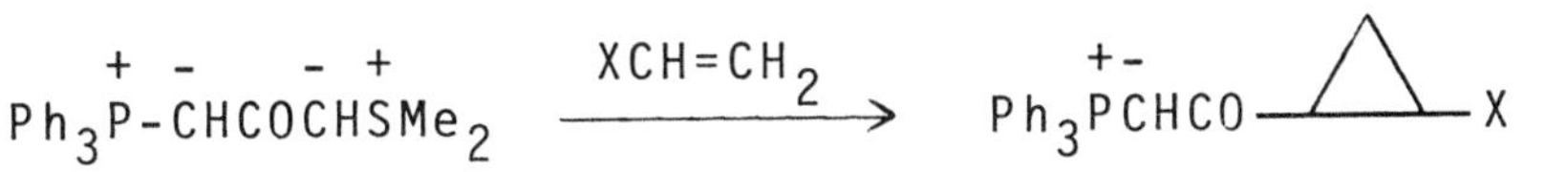

X = CO_2Me, 60%
X = COMe, 53%

I.D.2-8 H. Braun and G. Huber, Tetrahedron Lett., 1976, 2121.

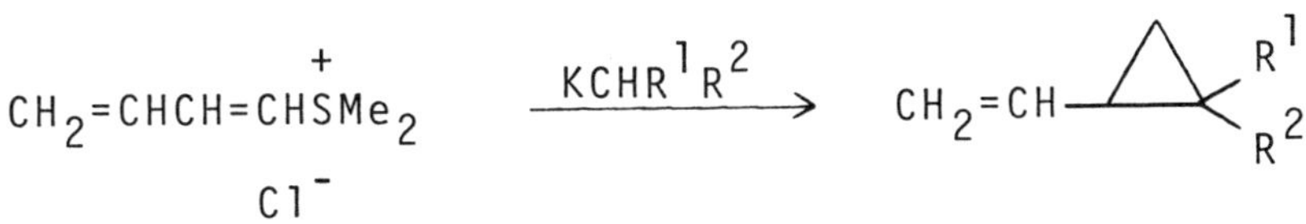

$R^1 = R^2 = CO_2Et$, 45%
$R^1 = R^2 = COMe$, 60%
$R^1 = CO_2Me$, $R^2 = COMe$, 60%

I.D.2-9 C. W. Jefford and A. F. Boschung, Helv. Chim. Acta, 59, 962 (1976).

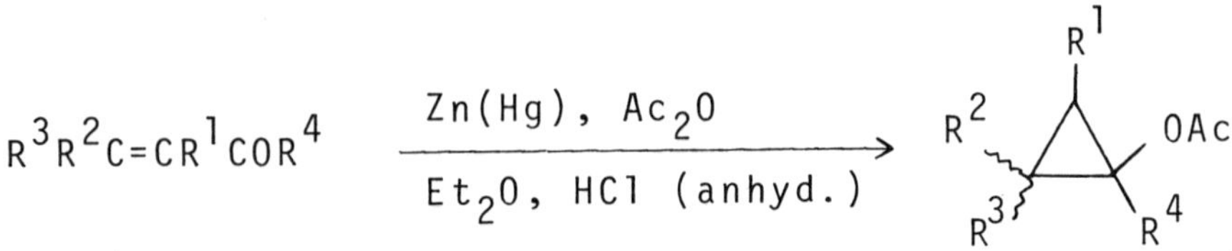

R^1	R^2	R^3	R^4	% Yield
H	Me	Me	Me	65
H	H	H	Me	32
H	H	$-(CH_2)_4-$		40
Me	H	$-(CH_2)_3-$		40

I.D.2-10 W. Tam and M. F. Rettig, J. Organomet. Chem., 108, C1 (1976).

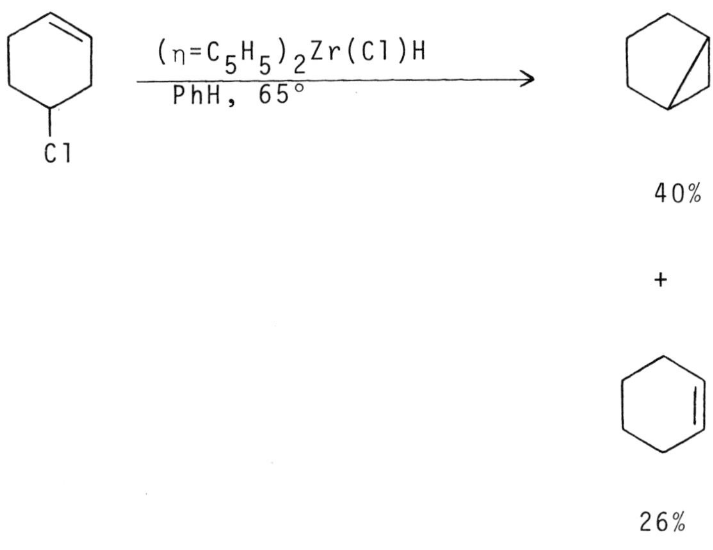

I.D.2-11 H.-G. Heine and D. Wendisch, Justus Liebigs Ann. Chem., 1976, 463.

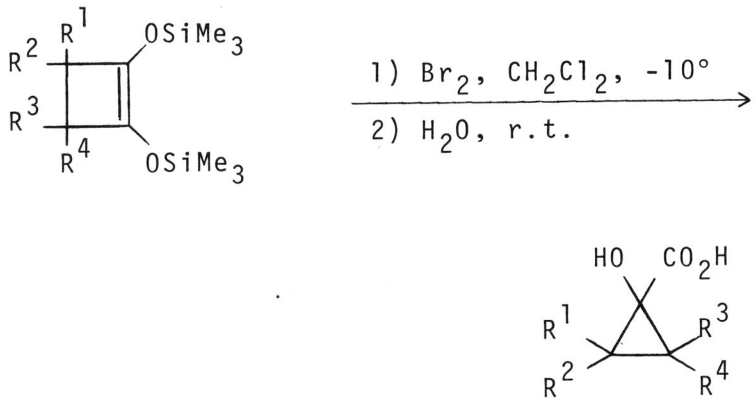

~ 90%

I.D.2-12 U. Langer and H. Musso, Justus Liebigs Ann. Chem., 1976, 1180.

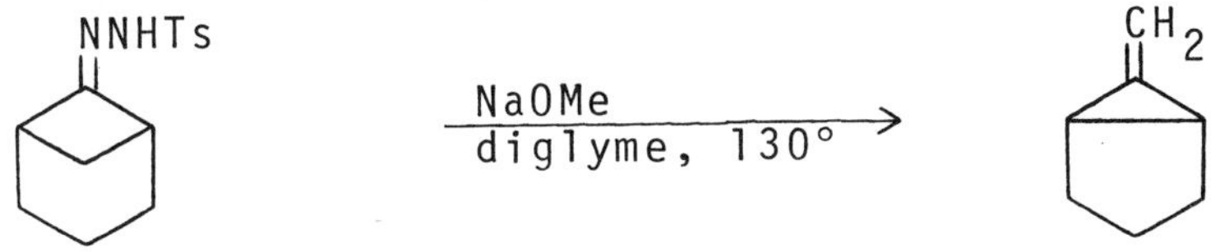

39%

I.E.1-1 M. Kakushima, J. Espinosa, and Z. Valenta, Can. J. Chem., 49, 3304 (1976).

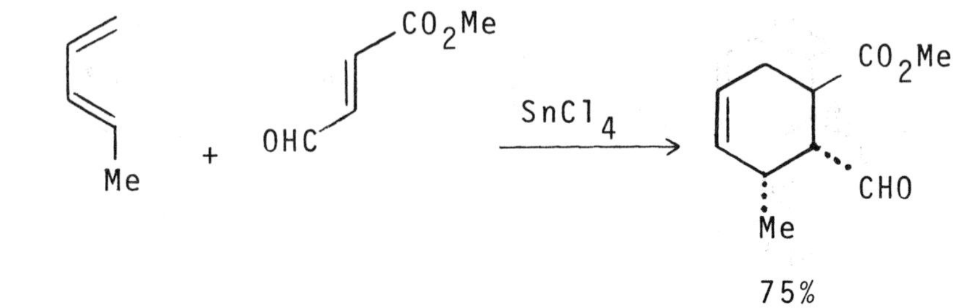

Lewis acid catalyzed Diels-Alder reaction of simple dienophiles exhibits clean orientation reversal.

I.E.1-2 M. E. Jung and C. A. McCombs, Tetrahedron Lett., 1976, 2935.

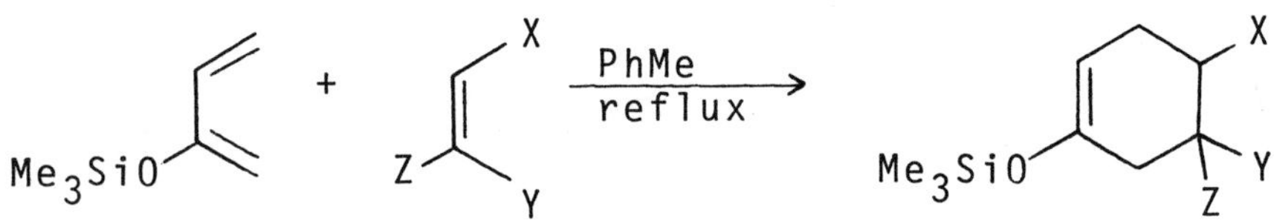

X	Y	Z	% Yield
COMe	H	H	60
CO_2Et	H	CO_2Et	77
CO_2Et	CO_2Et	H	39

I.E.1-3 B. M. Trost and A. J. Bridges, J. Am. Chem. Soc., 98, 5017 (1976).

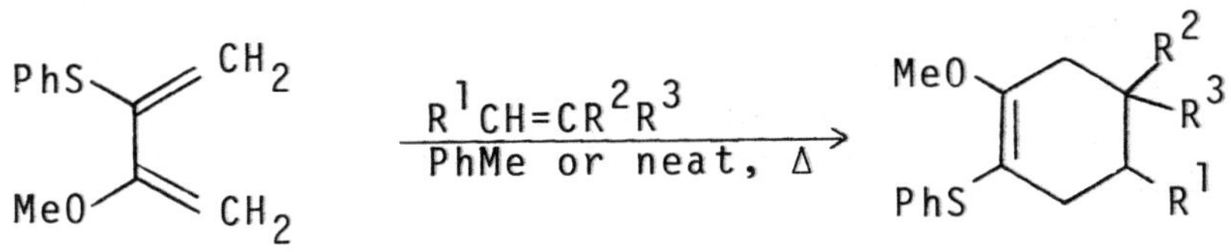

R^1	R^2	R^3	% Yield
H	CN	H	63
H	COMe	H	75
H	CHO	Me	72
-C(O)OCO-		H	48

I.E.1-4 I. Fleming and A. Percival, J. Chem. Soc., Chem. Commun., 1976, 681.

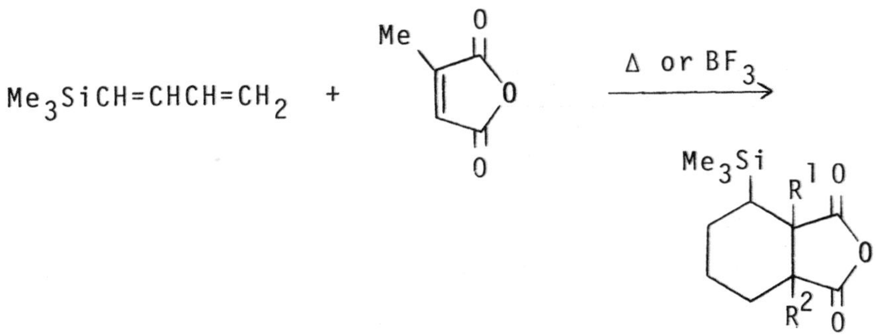

R^1 = Me, R^2 = H, 28%

R^1 = H, R^2 = Me, 24%

Little regioselectivity shown with other dienophiles.

I.E.1-5 L. C. Dunn, Y.-M. Chang, and K. N. Houk, J. Am. Chem. Soc., 98, 7095 (1976).

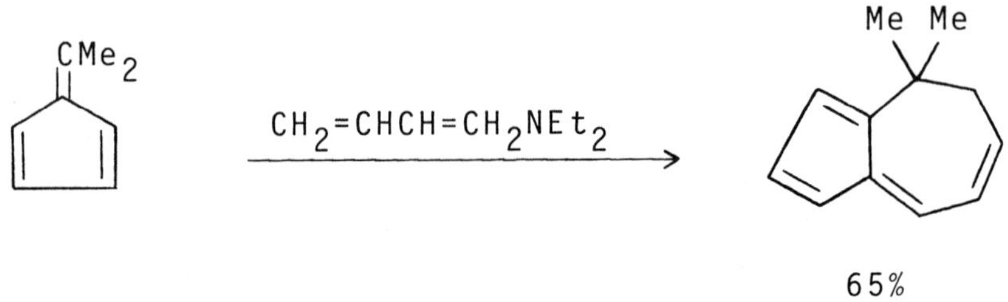

I.E.1-6 J. S. Baum and H. G. Viehe, J. Org. Chem., 41, 183 (1976).

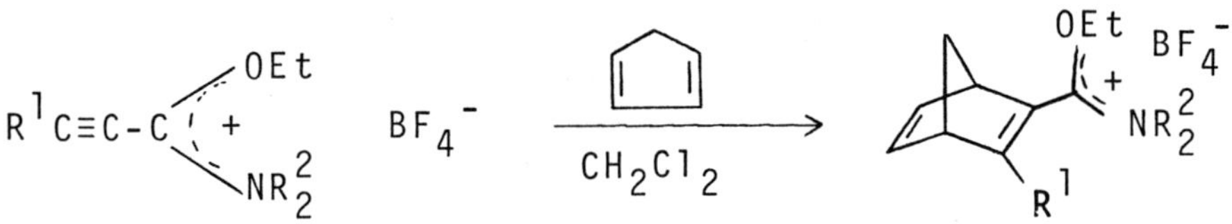

R^1	R^2	% Yield
Ph	Me	85
t-Bu	Me	32
H	Me	87
H	H	71

The acetylenic iminium compounds are more reactive in Diels-Alder cycloadditions than the corresponding esters, acid chlorides or nitriles.

I.E.1-7 R. G. Salomon, J. R. Burns, and W. J. Dominic, J. Org. Chem., 41, 2918 (1976).

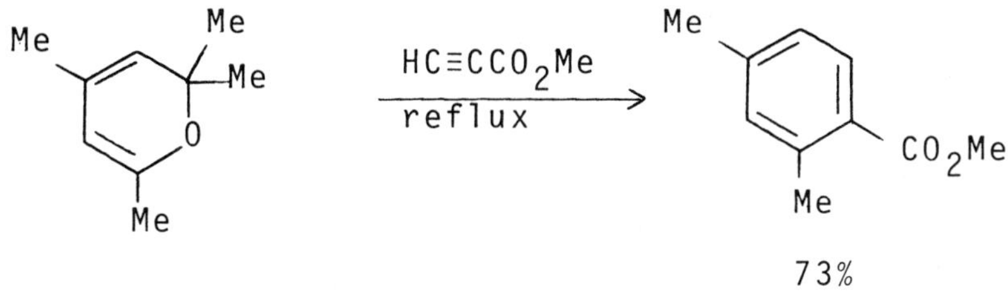

Other 2H-pyrans were studied and found to undergo similar regiospecific reactions.

I.E.1-8 J. H. Lukas, F. Baardman, and A. P. Kouwenhoven, Angew. Chem. Int. Ed. Engl., 15, 369 (1976).

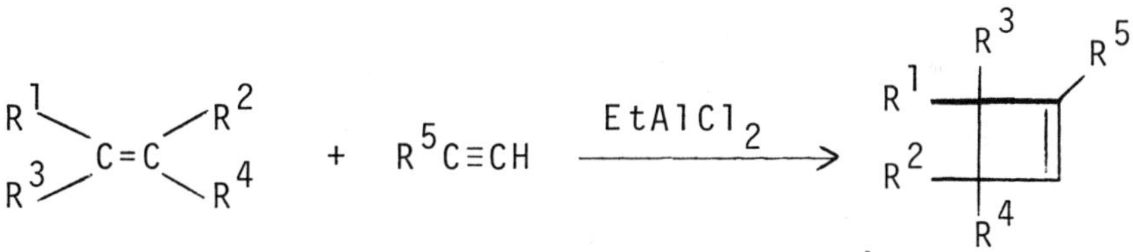

R^1	R^2	R^3	R^4	R^5	% Yield
Me	Me	Me	Me	Me	27
Me	Me	Me	Me	C_5H_{11}	50
H	Me	H	Me	Et	30

I.E.1-9 P.H.J. Ooms, M. A. Bertisen, H. W. Scheeren, and R.J.F. Nivard, J. Chem. Soc., Perkin Trans I, 1976, 1538.

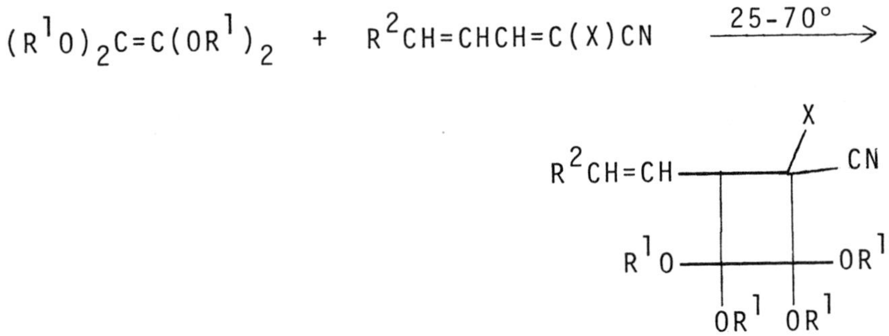

R^1	R^2	X	% Yield
Me	i-Pr	CN	70
Me	Ph	CN	88
Et	p-$NO_2C_6H_4$	CN	80
Me	Ph	CO_2Et	82
Et	Ph	SO_2Ph	80

I.E.1-10 R. Huisgen, R. Schug, and G. Steiner, Bull. Soc. Chim. Fr., 1976, 1813.

Review: Cycloadditions 2 + 2 Par Des Intermediaires Dipolaires-1,4.

I.E.1-11 E. M. Mil'vitskaya, A. V. Tarakanova, and A. F. Plate, *Russ. Chem. Rev.*, **45**, 469 (1976).

Review: Thermal Rearrangements of Vinylcyclopropanes.

I.E.1-12 E. Piers, C. K. Lau, and I. Nagakura, *Tetrahedron Lett.*, **1976**, 3233.

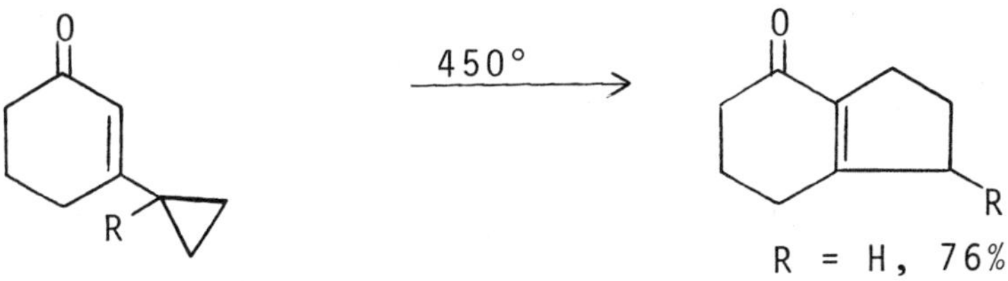

R = H, 76%

R = Me, 79%

I.E.1-13 E. Piers and I. Nagakura, *Tetrahedron Lett.*, **1976**, 3237; see also: J. P. Marino and L. J. Browne, *ibid*, 3245; P. A. Wender and M. P. Filosa, *J. Org. Chem.*, **41**, 3490 (1976).

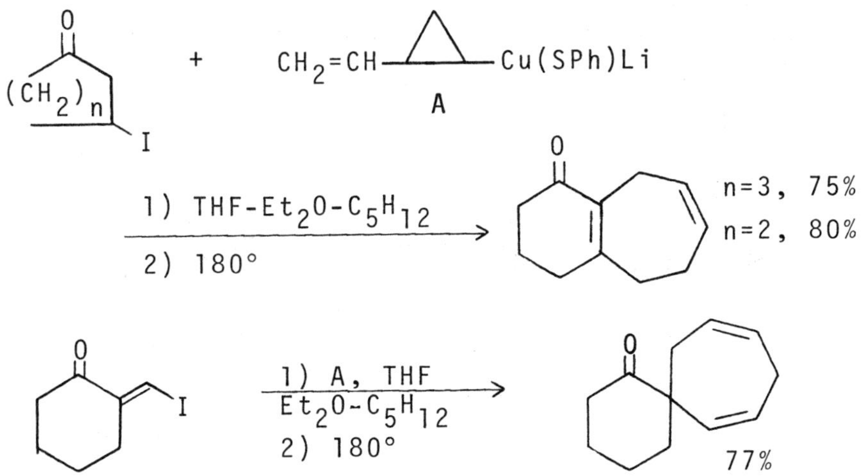

I.E.2-1 J. Haslouin and F. Rouessac, Tetrahedron Lett., 1976, 4651.

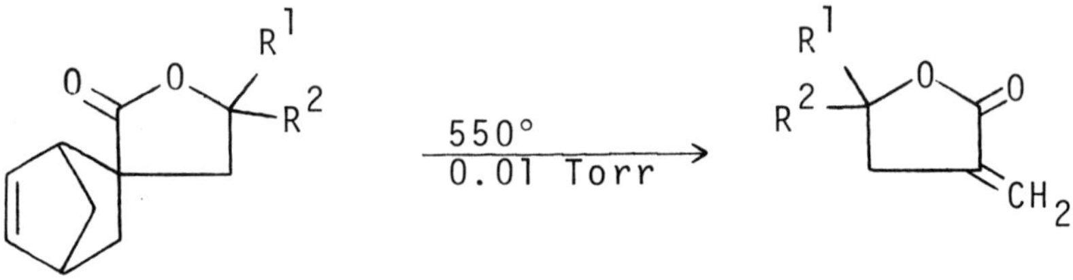

quantitative yield

I.E.2-2 J. Haslouin and F. Rouessac, Bull. Soc. Chim. Fr., 1976, 1122; see also: O. Gringore, J. Haslouin, and F. Rouessac, *ibid*, 1523.

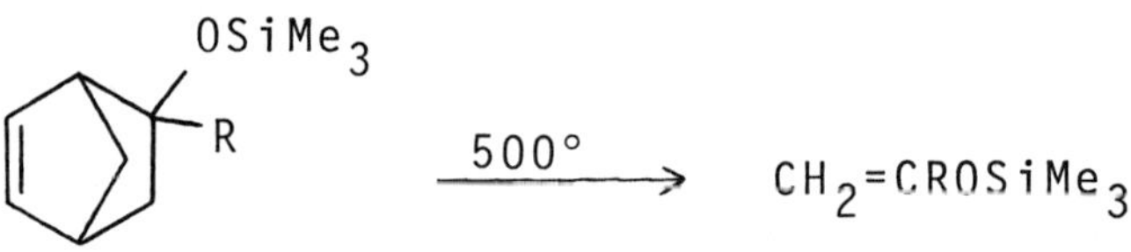

quantitative yield

I.E.2-3 F. Jung, J. Chem. Soc., Chem. Commun., 1976, 525.

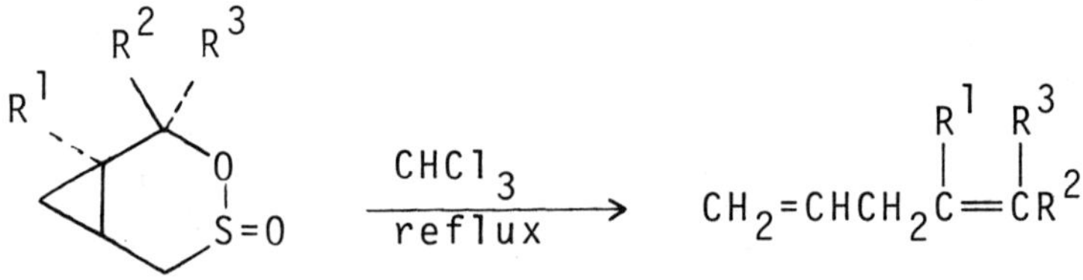

100% yield

(>99.5% isomeric purity)

I.E.2-4 H. Krapf, P. Riedl, and J. Sauer, Chem. Ber., 109, 576 (1976); see also: M. Hirth, H. Krapf, P. Riedl, J. Sauer, and E. Oeser, *ibid*, 562.

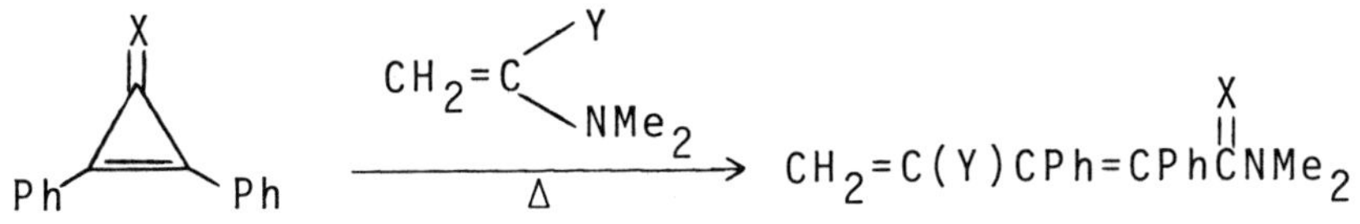

X = O and S

Y = NMe_2, OEt, and SMe

I.E.2-5 H. Wolfers, U. Kraatz, and F. Korte, Chem. Ber., 109, 1061 (1976).

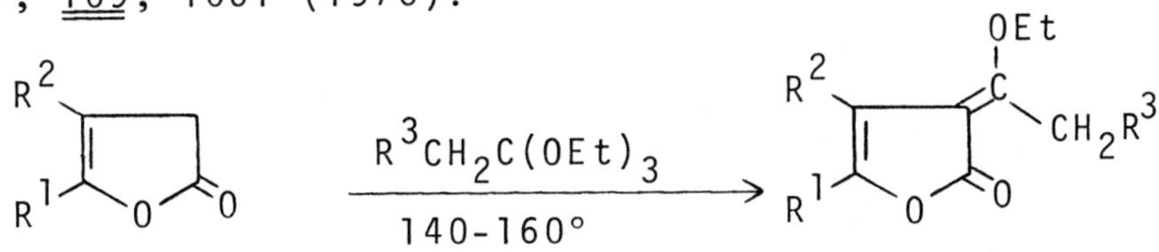

30-90%

I.E.2-6 O. Achmatowicz, Jr. And M. Pietraszkiewicz, J. Chem. Soc., Chem. Commun., 1976, 484.

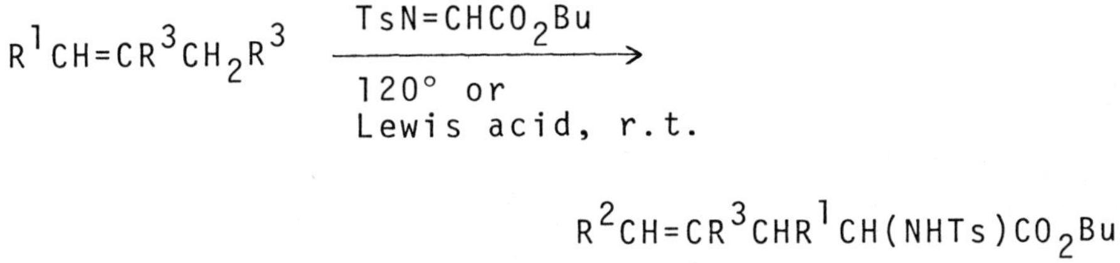

R^1	R^2	R^3	% Yield
H	H	H	70
H	Et	H	90
H	H	Ph	92
$-(CH_2)_3-$		H	81

I.E.2-7 D. S. Karanewsky and Y. Kishi, J. Org. Chem., 41, 3026 (1976).

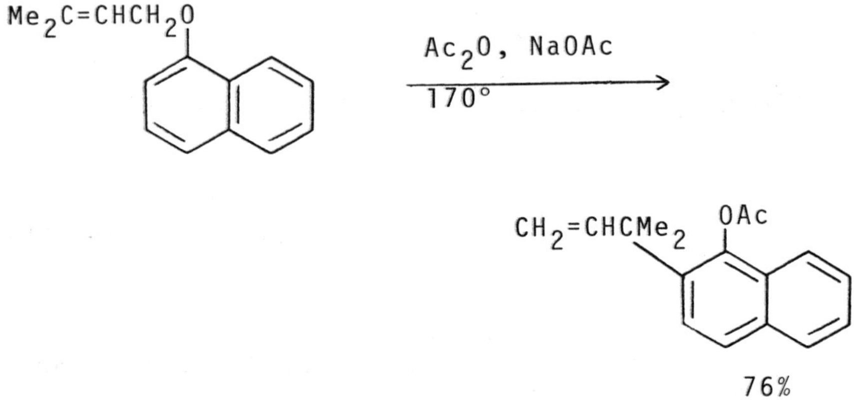

Conditions for formation of normal Claisen rearrangement products have been developed for cases where standard conditions give the abnormal product.

I.E.2-8 R. E. Ireland, R. H. Mueller, and A. K. Willard, J. Am. Chem. Soc., 98, 2868 (1976).

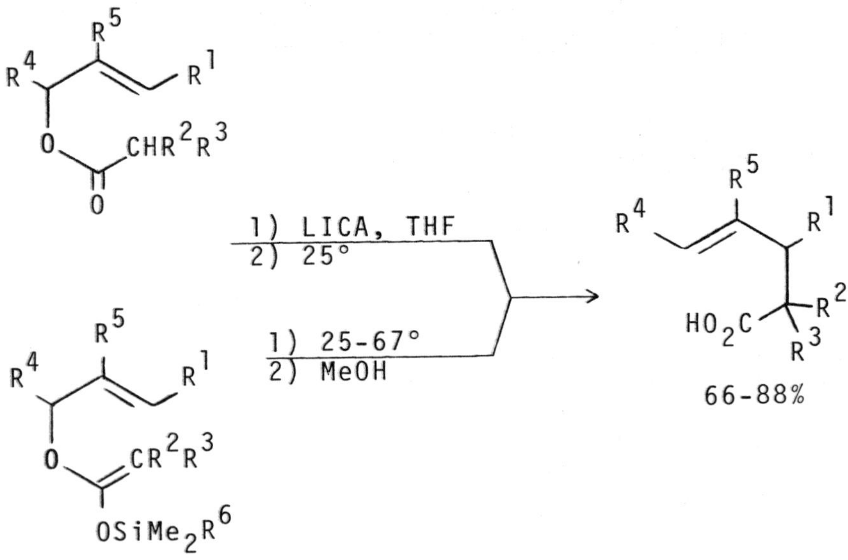

I.E.2-9 K. A. Parker and R. W. Kosley, Jr., Tetrahedron Lett., 1976, 341.

$$HC{\equiv}CC(OH)Me_2 \xrightarrow[\text{115°; 2) } H_3O^+]{\text{1) } (EtO)_2C(NMe_2)Me} Me_2CHCO(CH_2)_2CO_2Et$$

73%

$$Me_3SiC{\equiv}CC(OH)R^1R^2 \xrightarrow[\text{xylene, reflux}]{(EtO)_2C(NMe_2)Me} R^1R^2C{=}C{=}C(SiMe_3)CH_2CONMe_2$$

59-86%

I.E.2-10 R. W. Thies and E. P. Seitz, J. Chem. Soc., Chem. Commun., 1976, 846.

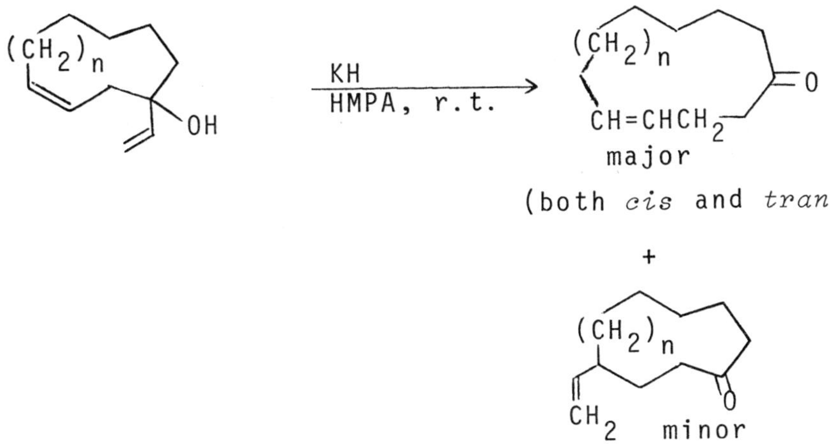

Use of the K salt allows the [1,3]shift of oxy-Cope system to occur at room temperature.

I.E.2-11 J. F. King and D.R.K. Harding, J. Am. Chem. Soc., **98**, 3312 (1976).

$$CH_2{=}CRCH_2SO_2CH{=}CH_2 \xrightarrow[\text{py, EtOH}]{165\text{-}175°} CH_2{=}CR(CH_2)_3SO_3^-$$

R = H, 70%
R = Me, 96%
R = Ph, 81%

I.E.2-12 F. Vögtle and E. Goldschmitt, Chem. Ber., **109**, 1 (1976).

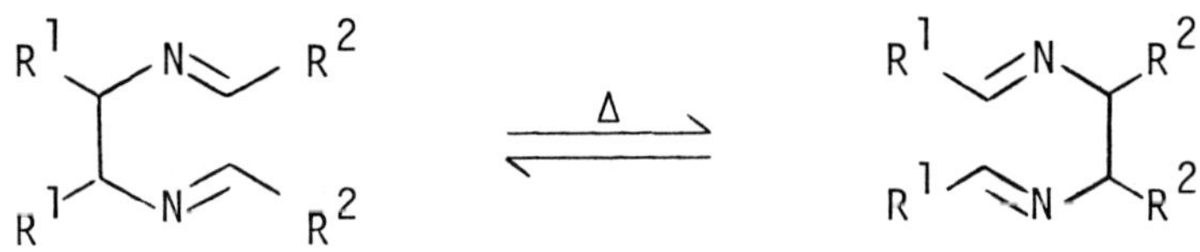

A full paper describing the general diaza-Cope rearrangement and the use of an o-hydroxy group to determine the equilibrium of the valence isomerization.

I.E.2-13 C. W. Spangler, Chem. Rev., 76, 187 (1976).

Review: Thermal [1,j]Sigmatropic Rearrangements

I.E.2-14 L. E. Overman and L. A. Clizbe, J. Am. Chem. Soc., 98, 2352 (1976).

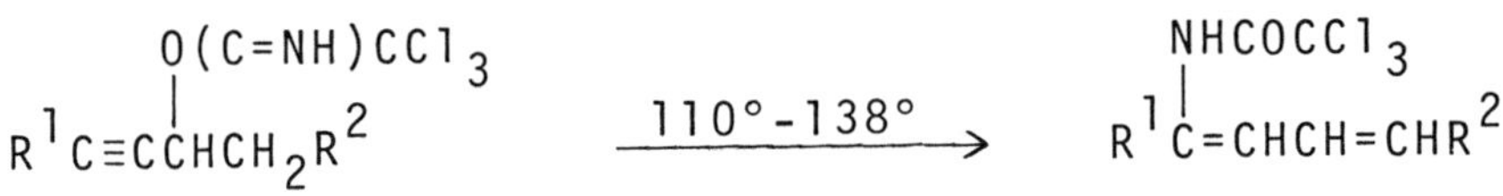

$R^1 = R^2 = H$, 38%

$R^1 = Ph$, $R^2 = Me$, 45%

$R^1 = R^2 = t\text{-}Bu$, 85%

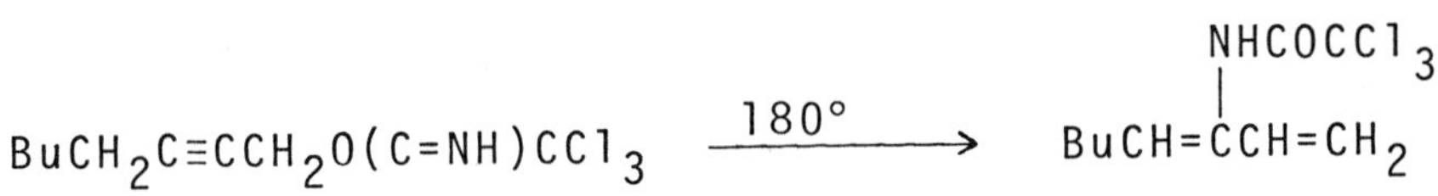

78%

The trichloroacetamido-1,3-dienes are useful in Diels-Alder reactions.

I.F.1-1 G. P. Schiemenz and U. Schmidt, *Justus Liebigs Ann. Chem.*, **1976**, 1514.

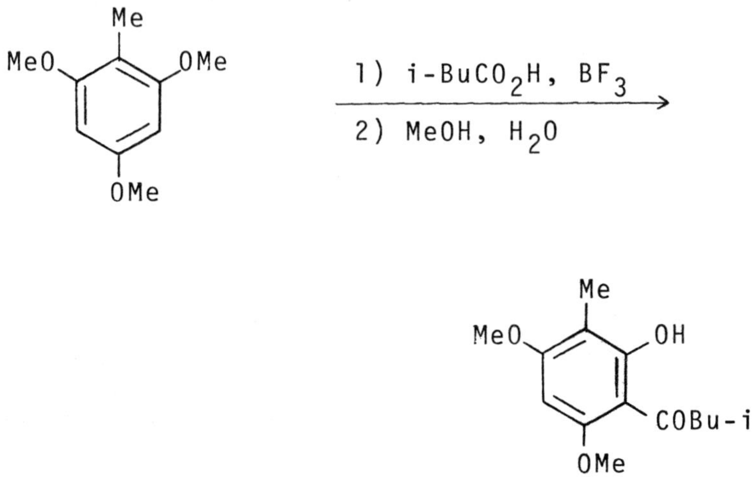

73%

I.F.1-2 E. P. Papadopoulos, *J. Org. Chem.*, **41**, 962 (1976).

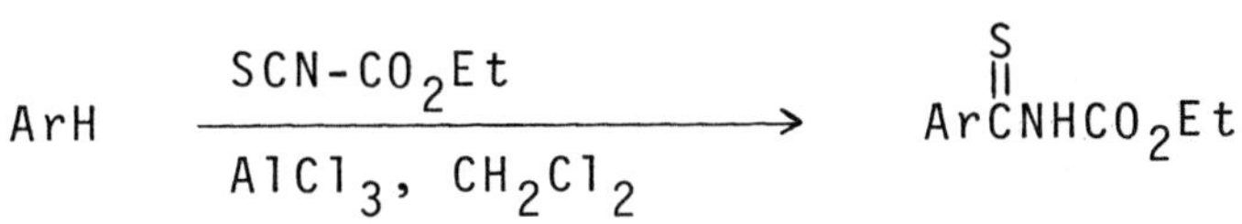

29-90%

By changing the workup procedure the aromatic thioamides can be isolated directly in approximately the same yield.

I.F.1-3 A. E. Feiring, J. Org. Chem., 41, 148 (1976).

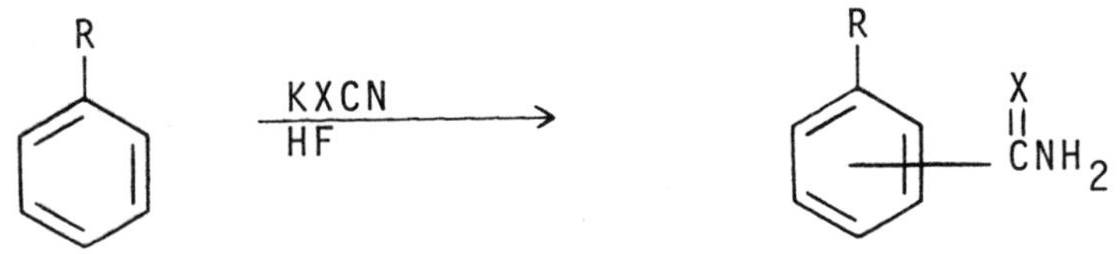

X	R	% Yield
O	H	63
O	Me	50 (o/p = 26/74)
S	Me	36 (only p-)

I.F.1-4 J. Zavada, M. Pankova, and Z. Arnold, Coll. Czech. Chem. Commun., 41, 1777 (1976).

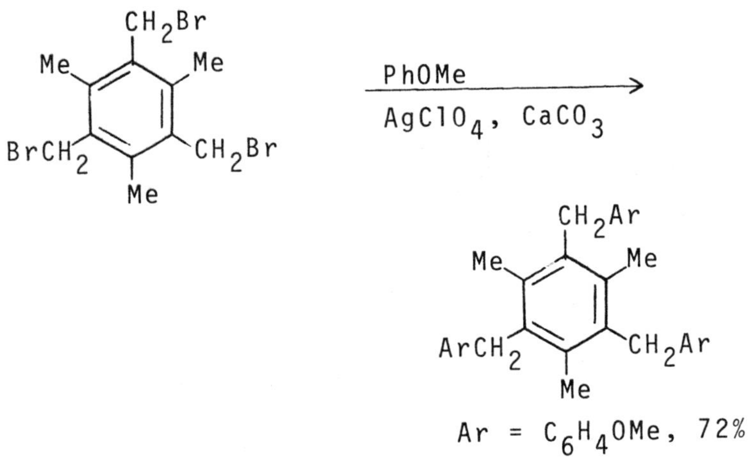

Use of $AgClO_4$-$CaCO_3$ allows the synthesis of systems that are unstable to normal Friedel-Crafts catalysts.

I.F.1-5 D. Ben-Ishai, I. Sataty, and Z. Bernstein, Tetrahedron, 32, 1571 (1976).

$$\text{ArH} + \text{RCONHCH(OH)CO}_2\text{H} \xrightarrow{\text{H}_2\text{SO}_4} \text{ArCH(NHCOR)CO}_2\text{H}$$

Ar	R	% Yield
Ph	Ph	91
$C_{10}H_7$	OMe	41
p-MeO_6H_4	Ph	72
C_4H_3S	Ph	84

I.F.1-6 J.-P. Gesson and J.-C. Jacquesy, J. Chem. Soc., Chem. Commun., 1976, 652.

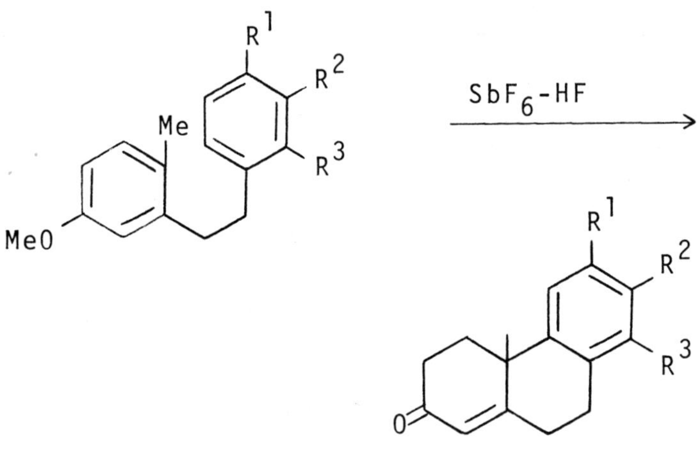

45-78%

I.F.1-7 J. Hocker, H. Giesecke, and R. Merten, Angew. Chem. Int. Ed. Engl., 15, 169 (1976).

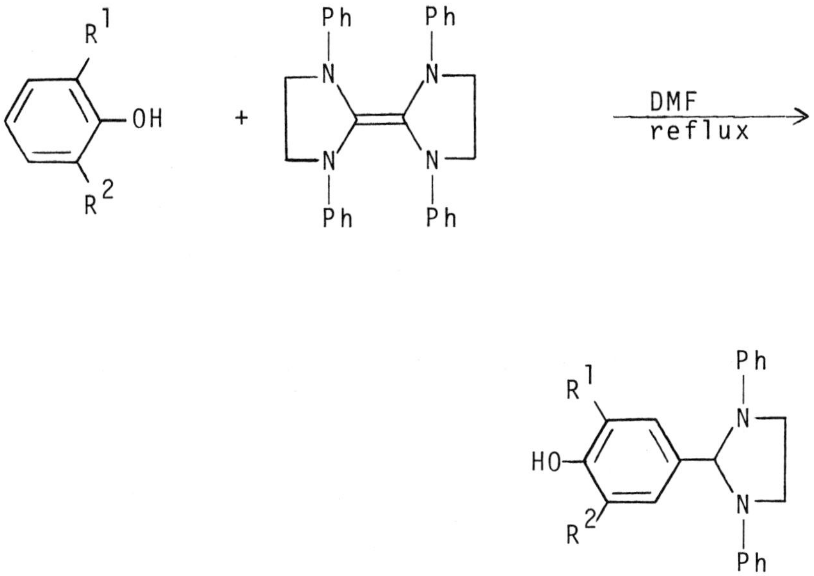

R^1	R^2	% Yield
H	H	55
Me	Me	58
t-Bu	t-Bu	84
Me	Cl	74

I.F.2-1 J. B. Melpolder and R. F. Heck, J. Org. Chem., 41, 265 (1976); see also: A. J. Chalk and S. A. Magennis, *ibid*, 273 and 1206.

$$PhI \quad + \quad CH_2{=}CHCHROH \xrightarrow[Et_3N,\ MeCN,\ 100°]{Pd(OAc)_2\ (cat.)} PhCH_2CH_2COR$$

Study of the palladium-catalyzed arylation of allylic alcohols.

I.F.2-2 A. Sekiya and N. Ishikawa, J. Organomet. Chem., 118, 349 (1976).

$$ArX \quad + \quad RMgBr \xrightarrow[Et_2O\ or\ THF]{(Ph_3P)_2Pd(Ph)I\ (cat.)} ArR$$

ArX	R	% Yield
PhI	PhC≡C	84
PhI	Et	32
p-ClC_6H_4Br	Ph	73

I.F.2-3 R. A. Kretchmer and R. Glowinski, J. Org. Chem., 41, 2661 (1976).

$$\mathrm{ArHgX} \xrightarrow[\text{py, }113^\circ]{\text{Cu, }PdCl_2\text{ (cat.)}} \underset{47\text{-}95\%}{\mathrm{Ar\text{-}Ar}}$$

I.F.2-4 F. E. Ziegler, K. W. Fowler, and S. Kanfer, J. Am. Chem. Soc., 98, 8282 (1976).

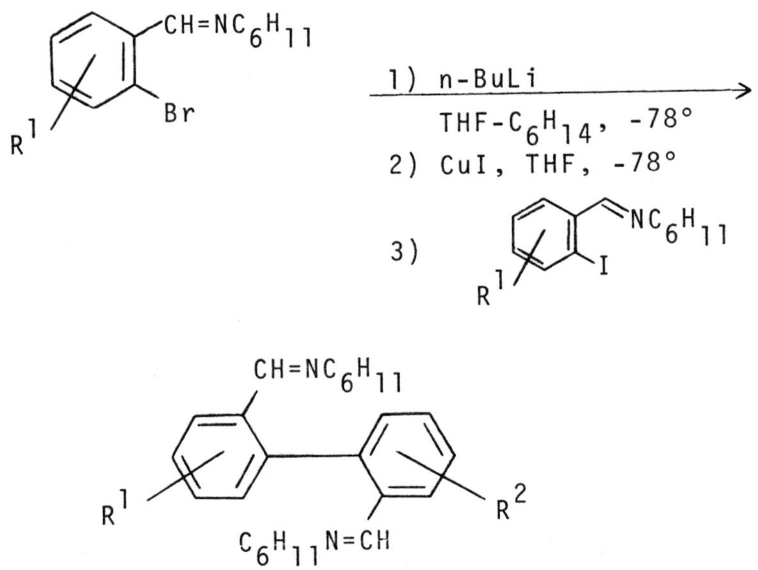

44-63%

I.F.2-5 E. Negishi and S. Baba, J. Chem. Soc., Chem. Commun., 1976, 596.

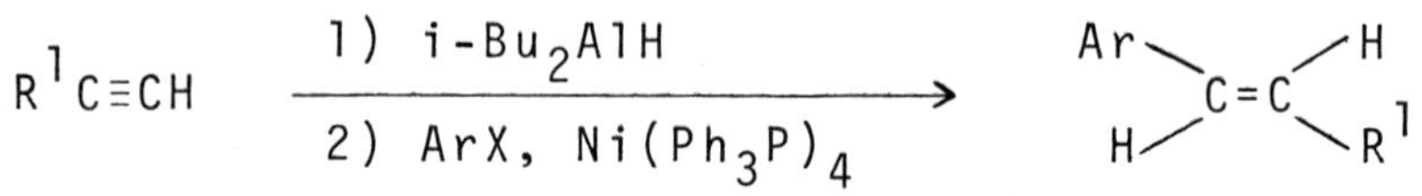

R^1	ArX	% Yield
n-Bu	PhI	89
n-Bu	PhBr	85
cyclohexyl	$p\text{-}MeC_6H_4Br$	75
n-Bu	$p\text{-}NCC_6H_4Br$	64

I.F.3-1 E. Havinga and J. Cornelisse, Pure and Appl. Chem., 47, 1 (1976).

Plenary Lecture: Aromatic Photosubstitution Reactions.

I.F.3-2 F. Minisci, Fortschr. Chem. Forsch., 62, 1 (1976).

Review: Recent Advances of Homolytic Aromatic Substitutions.

I.F.3-3 O. N. Chupakhin and I. Ya. Postovskii, Russ. Chem. Rev., 45, 454 (1976).

Review: Nucleophilic Substitution of Hydrogen in Aromatic Systems.

I.F.3-4 T. Sakakibara and Y. Odaira, J. Org. Chem., 41, 2049 (1976).

$$ArH \xrightarrow[Ac_2O \text{ or } CCl_4,\ HOAc]{Na\ Pd[CH_2(CO_2)_2]_2} ArCO_2H$$

R = H, 98%

R = OMe, 55%

I.F.3-5 M. E. Kurz and R.T.Y. Chen, J. Chem. Soc., Chem. Commun., 1976, 968.

$$ArH \xrightarrow[Mn(OAc)_3,\ HOAc]{MeNO_2} ArCH_2NO_2$$

Ar	% Yield
C_6H_5	78
MeC_6H_4	77
ClC_6H_4	20
$O_2NC_6H_4$	0

I.F.3-6 S. Uemura, S. Tanaka, and M. Okano, J. Chem. Soc., Perkin Trans. I, 1976, 1966.

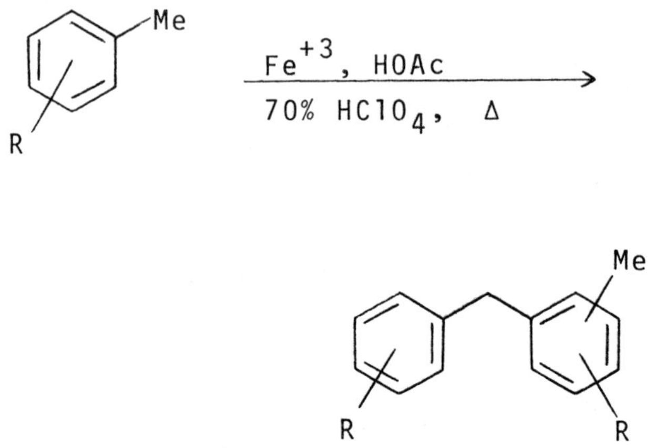

good yields

I.F.3-7 K. Tsujimoto, K. Miyake, and M. Ohashi, J. Chem. Soc., Chem. Commun., 1976, 386.

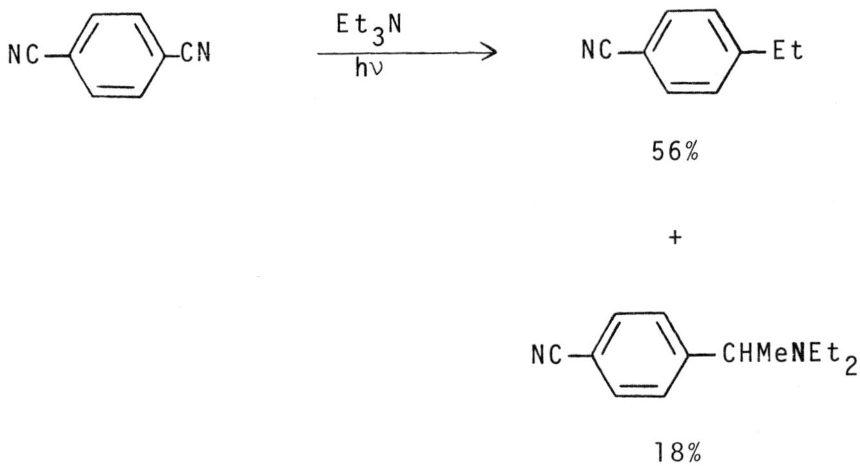

I.G.1-1 J. A. Sinclair and H. C. Brown, J. Org. Chem., 41, 1078 (1976).

$$Sia_2BC{\equiv}CR^1 \xrightarrow[2)\ I_2,\ THF,\ -78°]{1)\ LiC{\equiv}CR^2,\ THF,\ -78°} R^1C{\equiv}CC{\equiv}CR^2$$

R^1 = n-C_6H_{13}, R^2 = cyclohexyl, 73%

R^1 = cyclohexyl, R^2 = t-Bu, 70%

A "one-pot" procedure is described.

I.G.1-2 A. Pelter, R. Hughes, K. Smith, and M. Tabata, Tetrahedron Lett., 1976, 4385.

$$(cyclohexyl)_2BSMe \xrightarrow[2)\ LiC{\equiv}CR^2\ \ 3)\ I_2]{1)\ LiC{\equiv}CR^1} R^1C{\equiv}CC{\equiv}CR^2$$

R^1 = n-C_6H_{13}, R^2 = n-Bu, 61%

R^1 = t-Bu, R^2 = Ph, 51%

I.G.1-3 K. Utimoto, Y. Yabuki, K. Okada, and H. Nozaki, Tetrahedron Lett., 1976, 3969.

$$R^1_3B \xrightarrow[\text{3) } R^2X \quad \text{4) } I_2]{\text{1) LiC}\equiv\text{CH} \quad \text{2) n-BuLi}} R^1C\equiv CR^2$$

$R^1 = C_6H_{13}$, $R^2 = \text{n-Bu}$, 72%

$R^1 = \text{Bu}$, $R^2 = PhCH_2$, 87%

I.G.1-4 A. Pelter, T. W. Bentley, C. R. Harrison, C. Subrahmanyam, and R. J. Laub, J. Chem. Soc., Perkin Trans. 1, 1976, 2419; see also: A. Pelter, K. J. Gould, and C. R. Harrison, *ibid*, 2428.

$$R^1_3\overset{-}{B}C\equiv CR^2\ Li^+ \xrightarrow{R^3X} R^1_2BCR^1{=}CR^2R^3 \longrightarrow$$

mixture of *cis* and *trans*

$$\xrightarrow{\text{ox.}} R^1COCHR^2R^3$$

$$\xrightarrow{H^+} R^1CH{=}CR^2R^3$$

Full paper describes conditions favoring formation of dialkylvinylboranes and subsequent reactions; stereoselectivity in alkene formation is obtained using alkynyldialkyl(thexyl)borates.

I.G.1-5 K.-W. Chiu, E.-I. Negishi, M. S. Plante, and A. Silveira, Jr., J. Organomet. Chem., 112, C3 (1976).

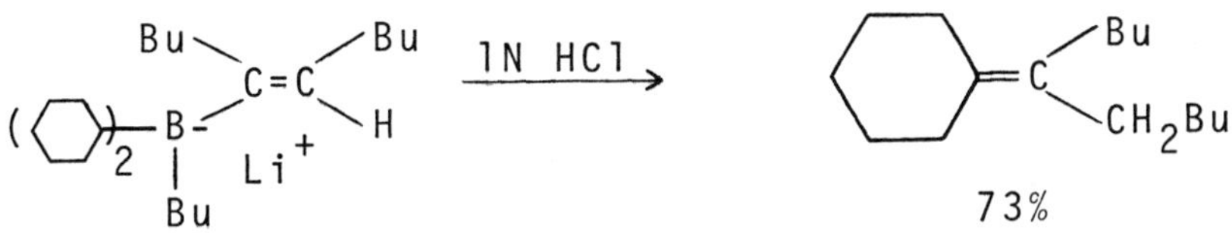

Other systems studied and the migrating group is the more highly substituted group.

I.G.1-6 Y. Yamamoto, H. Yatagai, A. Sonoda, and S.-I. Murahashi, J. Chem. Soc., Chem. Commun., 1976, 452.

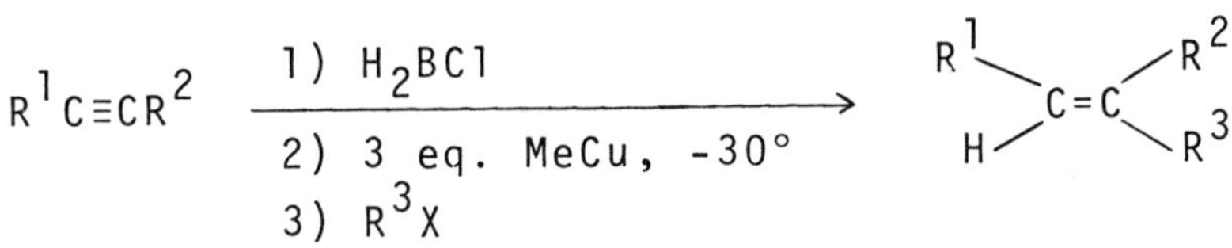

R^1	R^2	R^3X	% Yield
n-Bu	H	CH_2=$CHCH_2Br$	95
n-Bu	n-Bu	CH_2=$CHCH_2Br$	70
n-Bu	H	n-BuI	47

For the non-activated alkyl halides (R^3 = n-Bu) a phosphite was added.

I.G.1-7 J. Hooz and R. Mortimer, Tetrahedron Lett., 1976, 805.

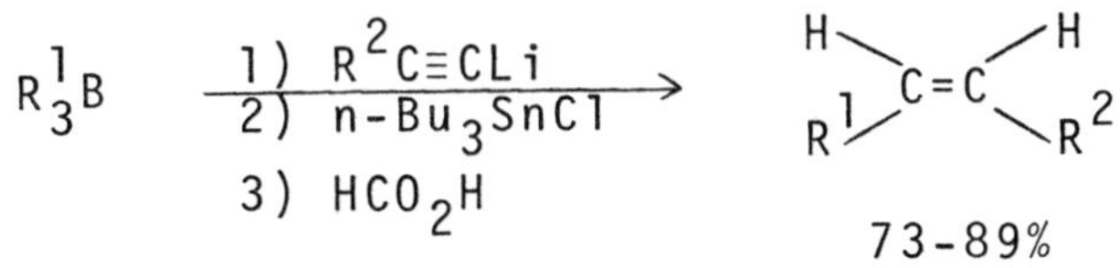

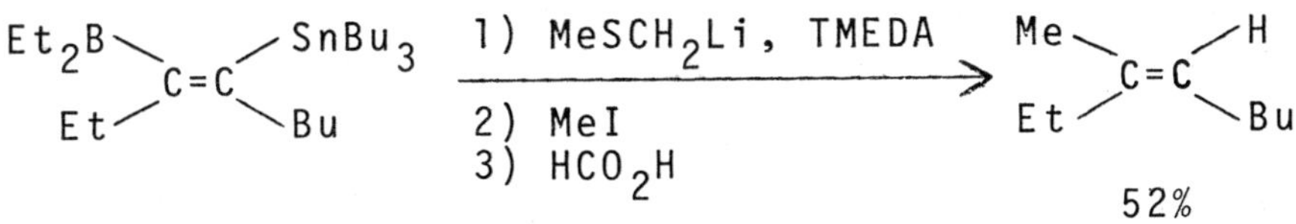

I.G.1-8 P. Jacob, III and H. C. Brown, J. Am. Chem. Soc., 98, 7832 (1976).

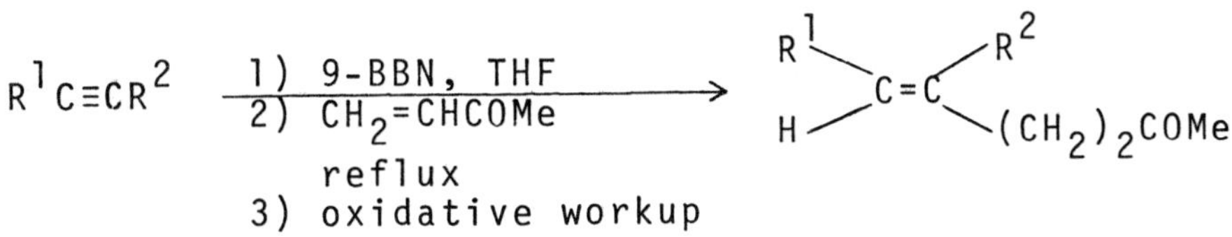

R^1	R^2	% Yield
n-Bu	H	87
t-Bu	H	85
Ph	H	93
t-Bu	Me	69
Et	Et	35

I.G.1-9 R. Köster, H.-J. Zimmerman, and W. Fenzl, Justus Liebigs Ann. Chem., 1976, 1116.

$$CH_2{=}CMeCHO \xrightarrow{R^1_2BR^2} \underset{A}{R^2CH_2CMe{=}CHOBR^1_2} \xrightarrow{PhCOCOPh} \underset{B}{R^2CHMeCHO}$$

R^1	R^2	% Yield A	% Yield B
Et	Et	89	-
Et	$MeCH{=}CHCH_2$	65	-
Et	$PhCH_2$	86	-
i-Pr	i-Pr	79	98

I.G.1-10 R. Murphy and R. H. Prager, Tetrahedron Lett., 1976, 463.

$$CH_2{=}CR^1(CH_2)_nCR^2{=}CH_2 \xrightarrow[\text{2) KOH, MeOH, aq. } AgNO_3]{\text{1) } BH_3\cdot THF}$$

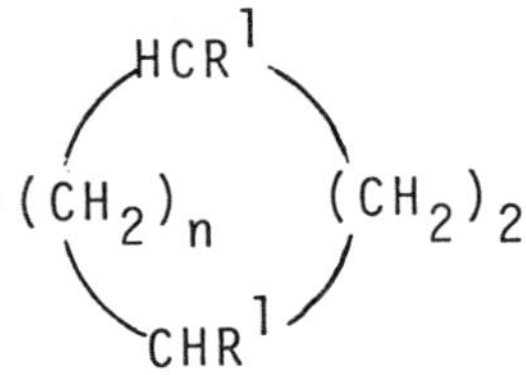

n	R^1	R^2	% Yield
0	Me	Me	79
1	Me	H	85
2	H	H	66 (17% methylcyclopentane)
3	H	H	67
4	H	H	42

I.G.1-11 M. W. Rathke, E. Chao, and G. Wu, J. Organomet. Chem., 122, 145 (1976).

$$Cl_2CHB(i\text{-}PrO)_2 \xrightarrow[\text{2) } H_2O_2\text{, pH 8.8}]{\text{1) RM}} RCHO$$

RM = n-BuLi, 60%
RM = t-BuLi, 35%
RM = PhMgBr, 70%

$$Cl_2CHB(i\text{-}PrO)_2 \xrightarrow[\text{2) } H_2O_2\text{, pH 8.8}]{\text{1) 2 eq. n-BuLi}} (n\text{-}Bu)_2CHOH$$

40%

I.G.1-12 R. J. Hughes, A. Pelter, K. Smith, E. Negishi, and T. Yoshida, Tetrahedron Lett., 1976, 87.

$$R^1R^2_2B \xrightarrow[\text{2) } HgCl_2 \text{ (3 eq.)}]{\text{1) } LiCH(SPh)_2} R^2_2CHOH$$
$$\text{3) } H_2O_2\text{, NaOH}$$

R^1	R^2	% Yield
n-C_6H_{13}	n-C_6H_{13}	85
cyclohexyl	cyclohexyl	88
thexyl	cyclopentyl	80

I.G.1-13 A. B. Levy and S. J. Schwartz, Tetrahedron Lett., 1976, 2201.

$$R_3B \xrightarrow[\substack{2)\ 2N\ HCl \\ 3)\ H_2O_2,\ NaOH}]{1)\ LiC(OMe)=CH_2} R_2C(OH)Me$$

R	% Yield
n-C_6H_{13}	100
sec-Bu	92
cyclopentyl	87

$$(RCH_2)_3B \xrightarrow[2)\ H_2O_2,\ NaOH]{1)\ LiC(OMe)=CH_2} RCH_2COMe$$

$R = C_5H_{11}$, 82%

I.G.1-14 N. Miyaura, M. Itoh, and A. Suzuki, Synthesis, 1976, 618.

$$n\text{-}Bu_3B \xrightarrow[\substack{2)\ CuCN,\ THF \\ 3)\ PhCH_2Br}]{1)\ MeLi} \underset{68\%}{PhC_5H_{11}} + \underset{16\%}{PhCH_2CH_2Ph}$$

I.G.1-15 Y. Takahashi, M. Tokuda, M. Itoh, and A. Suzuki, Synthesis, 1976, 616.

$$R_3B \ + \ MeNO_2 \xrightarrow{\text{electrolysis}} RCH_2NO_2$$

R	% Yield
n-Pr	84
n-C_8H_{17}	92
sec-Bu	76
cyclopentyl	150

I.G.1-16 K. Utimoto, N. Sakai, M. Obayashi, and H. Nozaki, Tetrahedron, 32, 769 (1976).

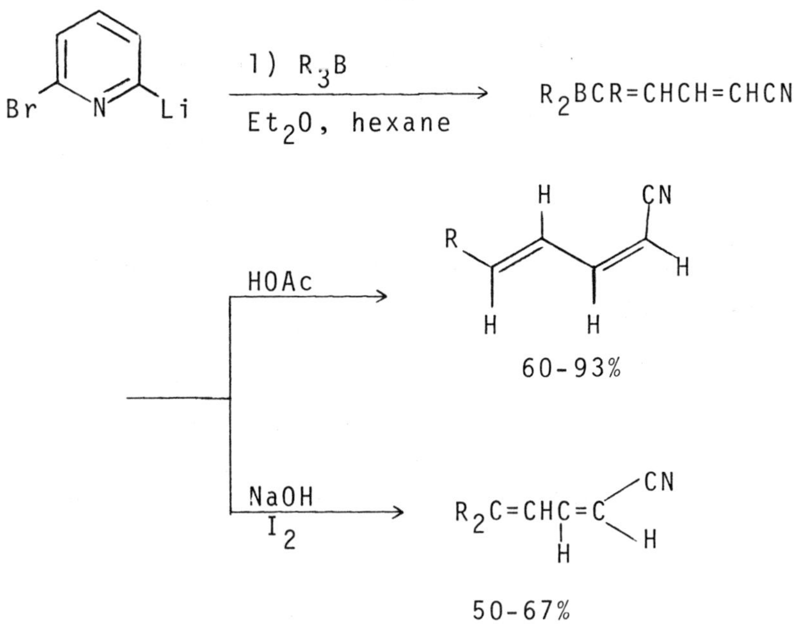

I.G.2-1 C. A. Bertelo and J. Schwartz, J. Am. Chem. Soc., 98, 262 (1976).

$$R^1R^2C{=}CR^3CH{=}CH_2 \xrightarrow[\text{2) CO (50 psi), PhH} \quad \text{3) } H_3O^+]{\text{1) } (Cp)_2Zr(H)Cl\text{, PhH}} R^1R^2C{=}CR^3(CH_2)_2CHO$$

R^1	R^2	R^3	% Yield (based on RZr)
Me	H	H	81
H	H	Me	98
Me	H	Me	91
Me	Me	H	98

I.G.2-2 T. Yamamoto, T. Kohara, and A. Yamamoto, Chem. Lett., 1976, 1217.

$$R_2Ni\ (bpy) \xrightarrow[-78^\circ \text{ to rt}]{\text{CO, THF}} RCOR$$

R = Me, Et, Pr (80-90%)

$$PhMgBr + PhBr \xrightarrow[NiEt_2\ (bpy)]{CO,\ Et_2O} PhCOPh + Ph_3COH$$

(bpy) = 2,2'-bipyridine

I.G.2-3 W. Best, B. Fell, and G. Schmitt, Chem. Ber., 109, 2914 (1976).

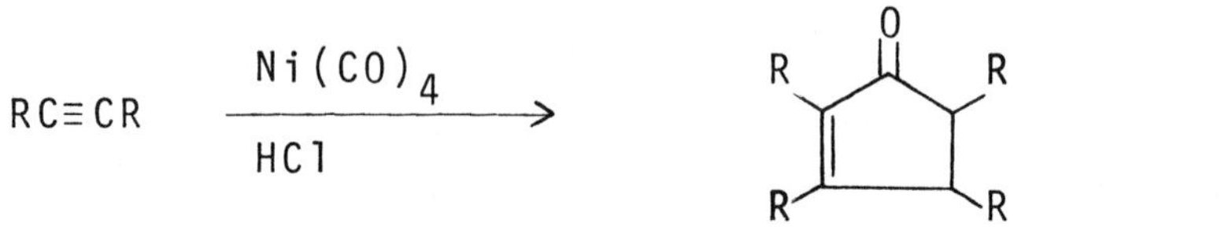

R = H, alkyl, or Ph

15-70%

I.G.2-4 J. X. McDermott, M. E. Wilson, and G. M. Whitesides, J. Am. Chem. Soc., 98, 6529 (1976).

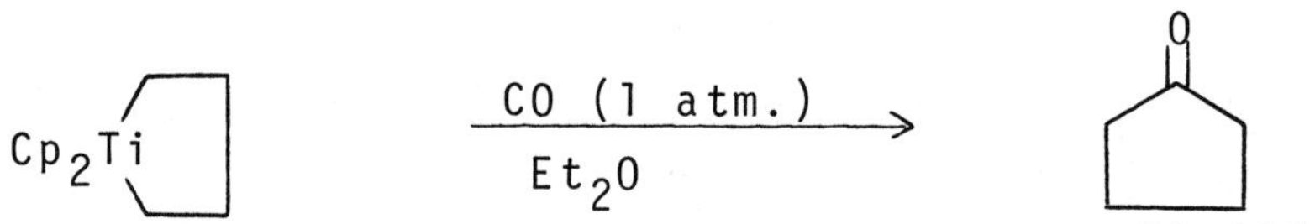

80-90%

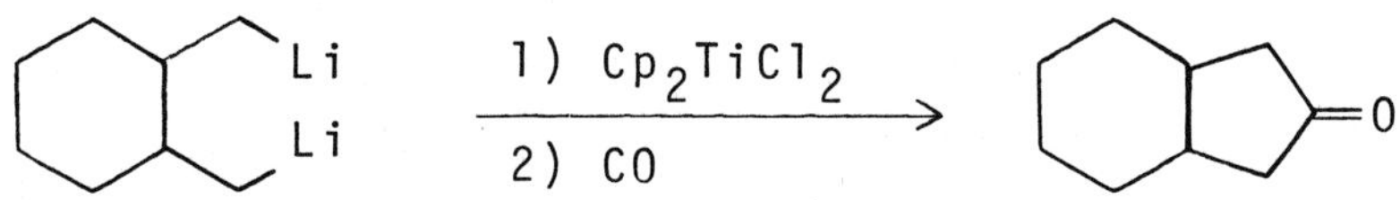

20%

I.G.2-5 R. Aumann and J. Knecht, Chem. Ber., 109, 174 (1976).

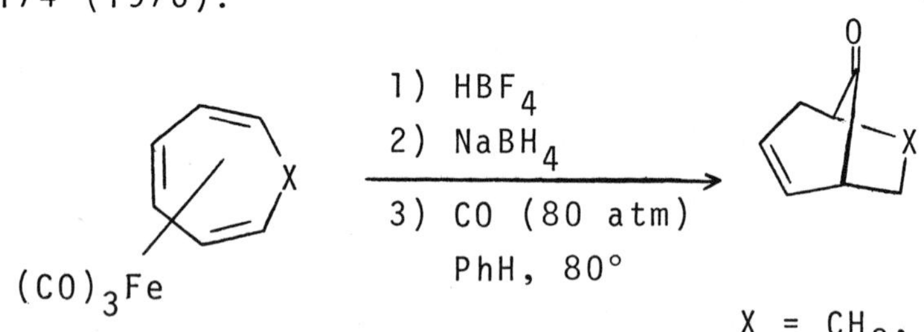

X = CH_2, 57%

X = CH=CH, 62%

I.G.2-6 R. F. Heldeweg and H. Hogeveen, J. Am. Chem. Soc., 98, 6040 (1976).

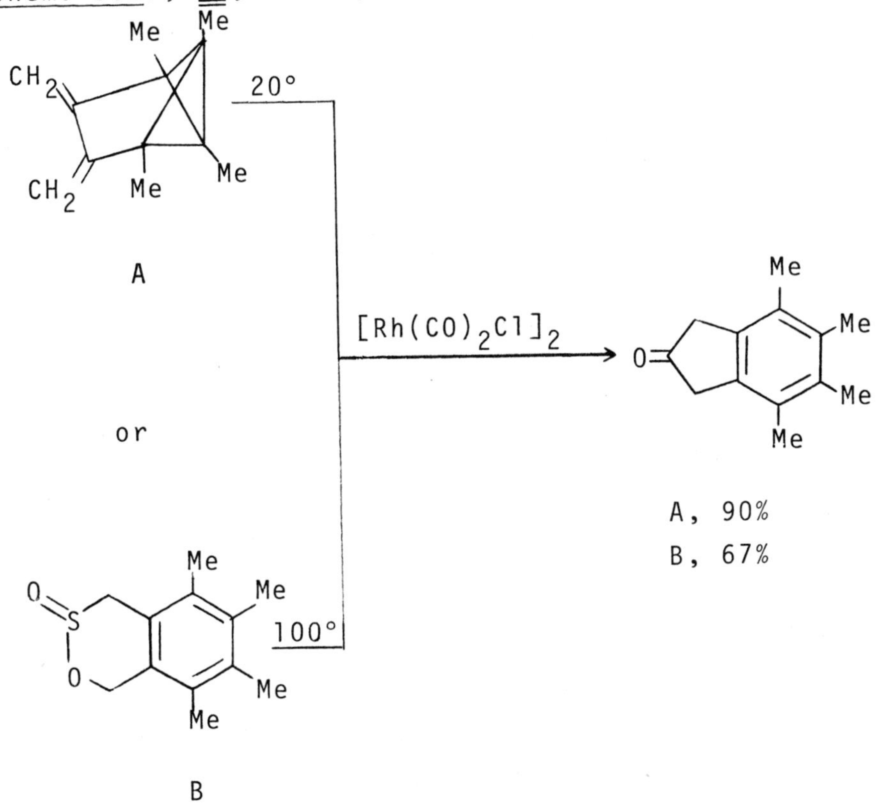

A, 90%
B, 67%

I.G.2-7 L. Cassar, M. Foa, and A. Gardano, J. Organomet. Chem., 121, C55 (1976).

$$RX \xrightarrow[\text{30\% NaOH, xylene, 95\%}]{\text{CO (5 atm), } Pd(Ph_3P)_4,\ Bu_4NI} RCO_2Na$$

R = $PhCH_2$, X = Cl, 83%
R = p-BrC_6H_4, X = Cl, 90%

The use of phase-transfer catalysis allows selectivity in polyhalogenated aromatic compounds.

I.G.3-1 S. Danishefsky, T. Kitahara, M. Tsai, and J. Dynak, J. Org. Chem., 41, 1669 (1976).

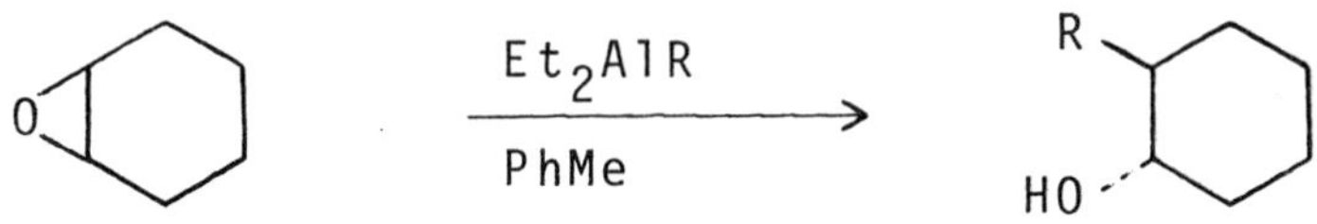

R = CH_2CO_2t-Bu, 68%
R = C≡COEt, 95%

The RLi derivatives give no or very poor yields of product.

I.G.3-2 Y. Kitagawa, S. Hashimoto, S. Iemura, H. Yamamoto, and H. Nozaki, J. Am. Chem. Soc., 98, 5030 (1976).

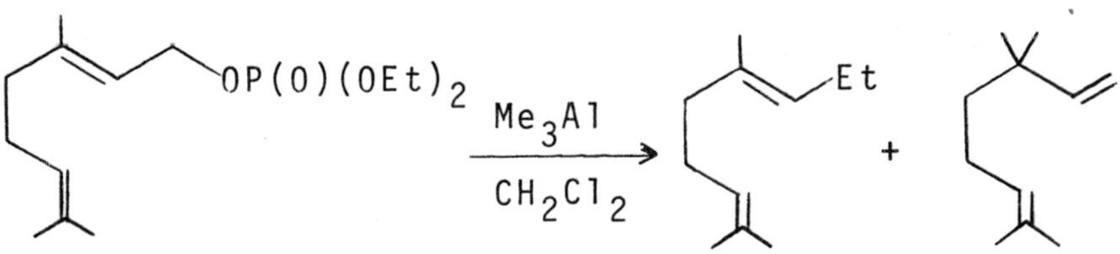

9 : 1

> 90%

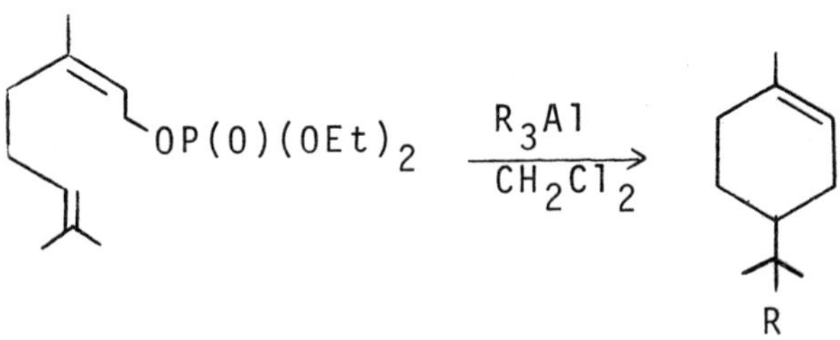

R = Me, 68%
R = Et, 70%
R = i-Bu, 68%

I.G.3-3 S. Hashimoto, Y. Kitagawa, S. Iemura, H. Yamamoto, and H. Nozaki, Tetrahedron Lett., 1976, 2615.

$$Me_2C{=}CH(CH_2)_2CMe{=}CHCH_2OR^1 \xrightarrow[\text{hexane}]{R^2_3Al}$$

$$\underset{A}{Me_2C{=}CH(CH_2)_2CMe{=}CHCH_2R^2} + \underset{B}{Me_2C{=}CH(CH_2)_2CR^2MeCH{=}CH_2}$$

R^1	R^2	% Yield	A:B
Ac	Me	74	92:8
Ac	i-Bu	66	97:3
CO_2Et	Et	69	85:13
CO_2Et	i-Bu	73	97:3
OTHP	Me	80	96:4
OTHP	i-Bu	<3	-

I.G.3-4 P. W. Chum and S. E. Wilson, Tetrahedron Lett., 1976, 1257.

$$CH_2{=}CH(CH_2)_2CH{=}CH_2 \xrightarrow[\text{mineral oil, 115°}]{i\text{-}Bu_2AlH\ (10\ mol\ \%)}$$

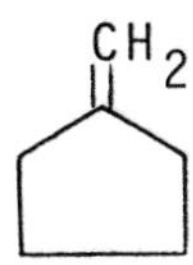

81%

$$[CH_2{=}CH(CH_2)_2]_2C{=}CH \xrightarrow[\text{mineral oil, 110°}]{i\text{-}Bu_2AlH\ (cat.)}$$

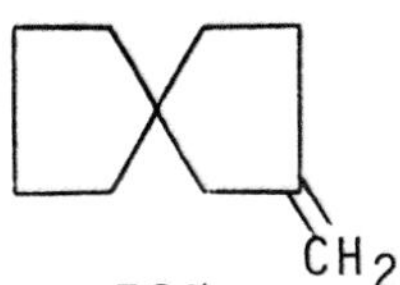

78%

I.G.3-5 E. Negishi, S. Baba, and A. O. King, J. Chem. Soc., Chem. Commun., 1976, 17.

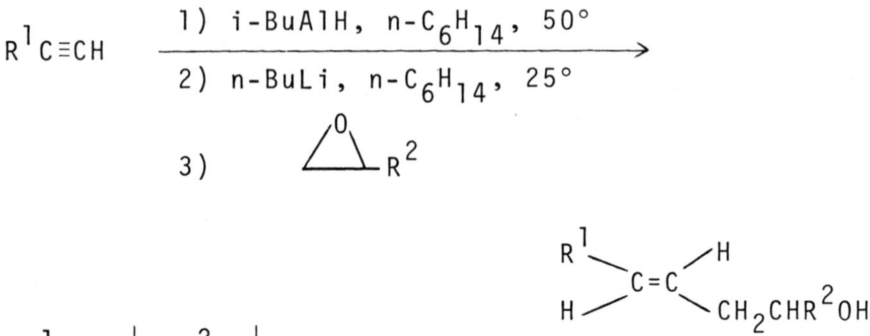

R^1	R^2	% Yield
n-Bu	H	75
n-Bu	Me	74
t-Bu	Me	82

I.G.3-6 S. Baba and E. Negishi, J. Am. Chem. Soc., 98, 6729 (1976).

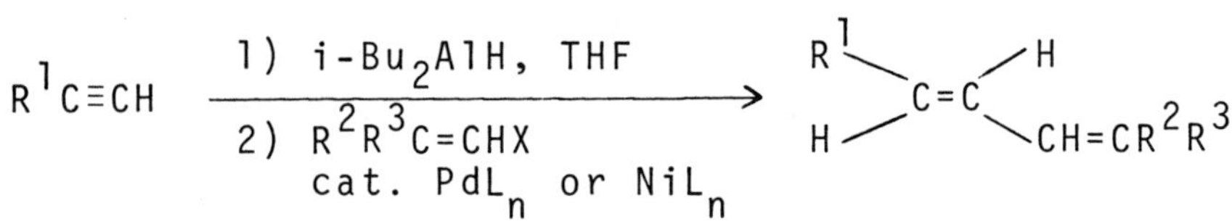

R^1	R^2	R^3	X	% Yield
n-C_5H_{11}	n-Bu	H	I	74
t-Bu	n-Bu	H	I	36
n-Bu	t-Bu	H	I	82
n-Bu	CO_2Me	Me	Br	75

The reaction is stereospecific in terms of both the vinyalane and the vinyl halide.

I.G.3-7 R. A. Coleman, C. M. O'Doherty, H. E. Tweedy, T. V. Harris, and D. W. Thompson, J. Organomet. Chem., 107, C15 (1976).

$$RC{\equiv}C(CH_2)_nOH \xrightarrow[\text{2) } Et_2AlCl,\ CH_2Cl_2,\ -78^\circ]{\text{1) } TiCl_2(C_5H_7O_2)_2,\ py}$$

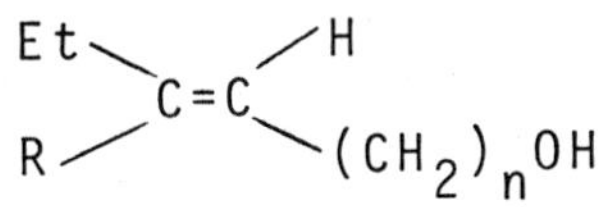

R	n	% Yield
H	2	53
Me	2	53
H	3	20
H	1	0
H	4	0

I.G.3-8 K. Tamao, K. Sumitani, Y. Kiso, M. Zembayashi, A. Fujioka, S. Kodama, I. Nakajima, A. Minato, and M. Kumada, Bull. Chem. Soc., Japan, 49, 1958 (1976).

$$R^1MgX \quad + \quad R^2X' \xrightarrow{L_2NiX_2\ (\text{cat.})} R^1{-}R^2$$

(R = aryl or alkenyl)

Full paper covering the general scope and limitations of the nickel-phosphine complex –catalyzed coupling.

I.G.3-9 P. W. Jennings, D. G. Pillsbury, J. L. Hall, and V. T. Brice, <u>J. Org. Chem.</u>, <u>41</u>, 719 (1976).

$$RX \xrightarrow[\text{Ni(acac)}_2 \text{ or Fe(acac)}_3,\ \text{DMF},\ Ph_3P,\ Et_4NBr]{\text{electrolysis}} R\text{-}R$$

R = $PhCH_2$, 87%

R = Ph, 65%

R = Bu, 28%

I.G.3-10 S. Takahashi, Y. Suzuki, K. Sonogashira, and N. Hagihara, <u>J. Chem. Soc., Chem. Commun.</u>, <u>1976</u>, 839.

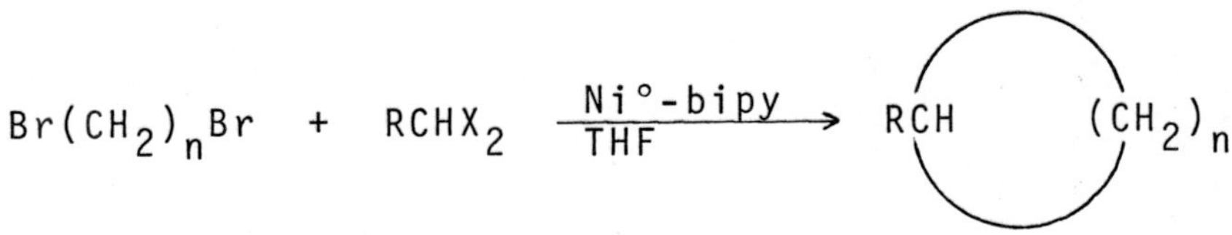

n	$RCHX_2$	% Yield
4	CH_2Br	70
4	CH_2Cl_2	51
4	$MeCHBr_2$	59
4	$CHCl_3$	27
5	CH_2Br_2	37

I.G.3-11 Y. Ito, S. Fujii, T. Konoike, and T. Saegusa, Synth. Commun., 6, 429 (1976).

$$\mathrm{RCHXY} \xrightarrow[\mathrm{DMSO}]{\mathrm{Ag_2O}} \mathrm{RCXYCXYR}$$

R	X	Y	% Yield
$PhCH_2$	CO_2Me	CO_2Me	74
Ph	H	CO_2Me	92
Me	Me	NO_2	40
$-(CH_2)_4CO-$		CO_2Et	90

I.G.3-12 R. C. Larock, J. Org. Chem., 41, 2241 (1976).

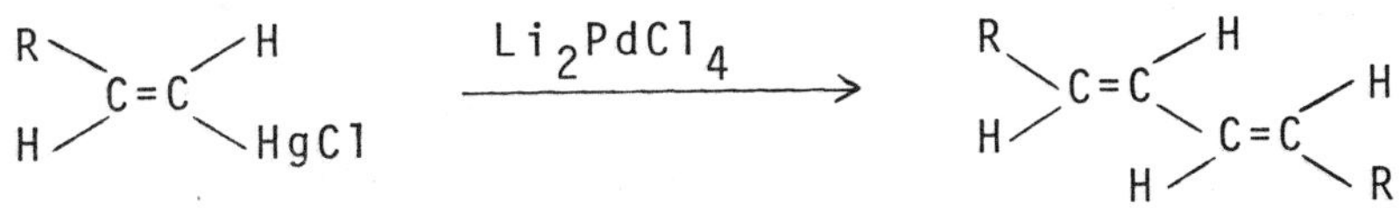

75-100%

I.G.3-13 G. A. Olah and G. K. Surya Prakash, Synthesis, 1976, 607.

$$\mathrm{R^1R^2CHCl} \xrightarrow[\mathrm{THF}]{\mathrm{Ti(II)}} \mathrm{R^1R^2CHCHR^1R^2}$$

R^1	R^2	% Yield
Ph	Ph	85
Ph	H	76
$CH_2{=}CMe$	H	76

I.G.3-14 E. J. Corey, R. L. Danheiser, and S. Chandrasekaran, J. Org. Chem., 41, 260 (1976).

$$R^1COR^2 + R^3COR^4 \xrightarrow[\text{THF}]{TiCl_4\text{-}Mg(Hg)} R^1R^2\underset{|}{\overset{OH}{C}}\text{—}\underset{|}{\overset{OH}{C}}R^3R^4$$

R^1	R^2	R^3	R^4	% Yield
Ph	H	Ph	H	84
$-(CH_2)_5-$		$-(CH_2)_5-$		93
$-(CH_2)_4-$		Me	Me	65
$-(CH_2)_5-$		Me	H	72

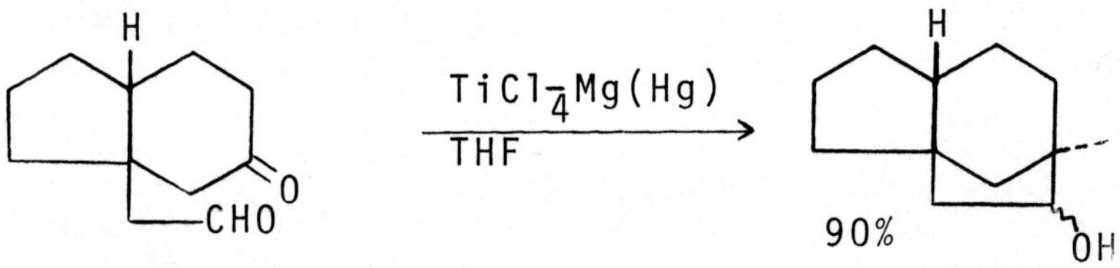

Other titanium chloride species were studied. The $CpTiCl_3$-$LiAlH_4$ reagent was efficiently employed in many cases.

I.G.3-15 N. E. Glushkova and N. P. Kharitonov, J. Gen. Chem. USSR, 45, 1981 (1975).

$$p\text{-}XC_6H_4CHO + Et_3SiH \xrightarrow[110\text{-}140°]{Ni(colloidal)}$$

$p\text{-}XC_6H_4CH(OSiEt_3)CH(OSiEt_3)C_6H_4X\text{-}p$

A

+

$p\text{-}XC_6H_4CH_2OSiEt_3$

B

X	% Yield A	% Yield B
H	59	24
Me	38	43
Me_2N	41	55

I.G.3-16 T. N. Mitchell and B. Kleine, Tetrahedron Lett., 1976, 2173.

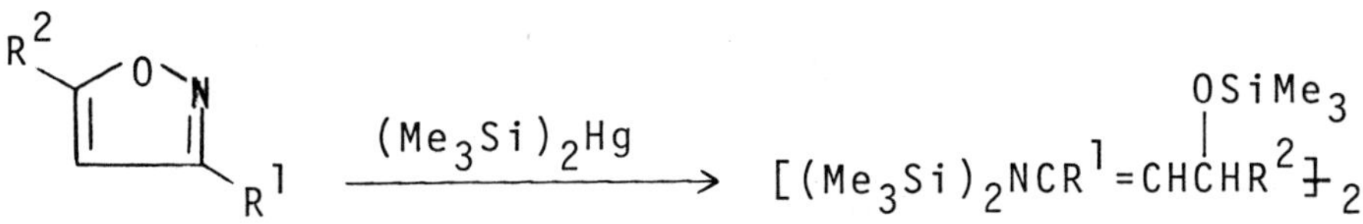

"almost quantitative"

I.G.3-17 G. Cahiez, D. Bernard, and J. F. Normant, J. Organomet. Chem., 113, 99 (1976).

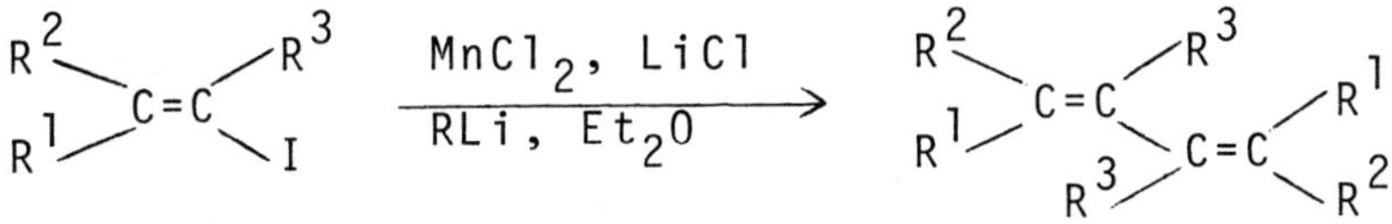

R^1, R^2, R^3, = H or alkyl
70-91%

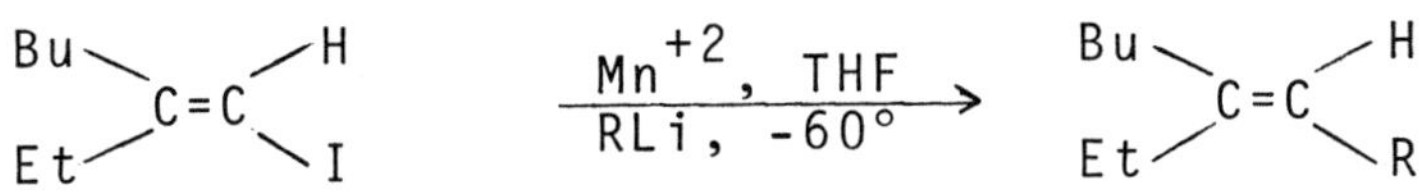

R = Me or n-Bu, 85-90%

I.G.3-18 R. F. Cunico and F. J. Clayton, J. Org. Chem., 41, 1480 (1976).

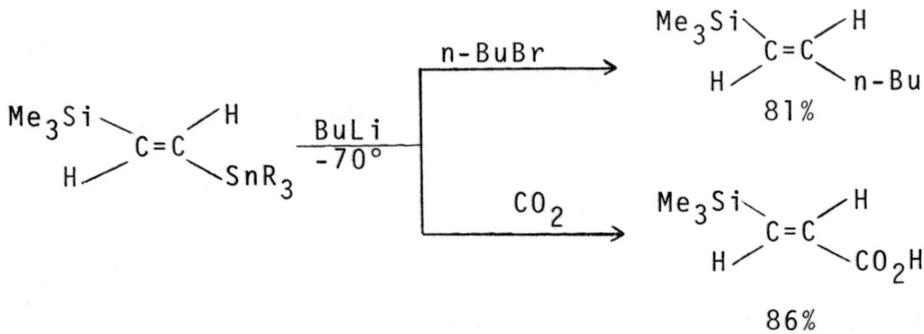

I.G.3-19 W. E. Truce, A. W. Borel, and P. J. Marek, J. Org. Chem., 41, 401 (1976).

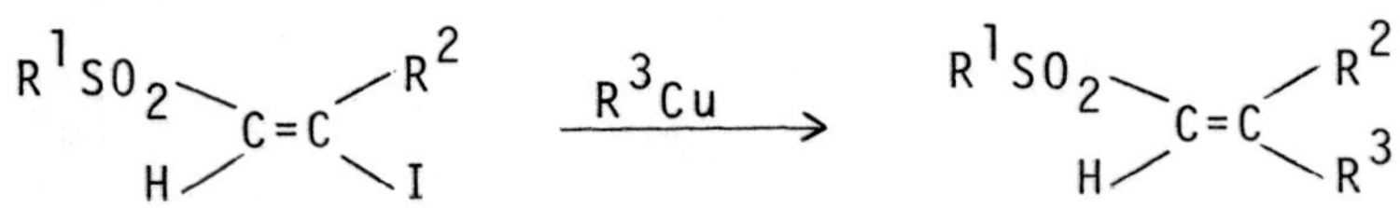

R^1	R^2	R^3	% Yield
Et	n-Bu	Me	96
Et	Ph	n-Bu	90
Et	t-Bu	Me	47
Ph	Ph	Ph	50
Ph	H	PhC≡C	63
Me	Me	PhC≡C	53

I.G.3-20 K. Tamao, M. Zembayashi, and M. Kumada, Chem. Lett., 1976, 1237; see also: *ibid*, 1239.

$$\mathrm{RMgBr} \;+\; \mathrm{BrCH{=}CHOEt} \xrightarrow[\mathrm{Et_2O\ or\ THF}]{\mathrm{Ni(dppp)Cl_2\ (cat.)}} \mathrm{RCH{=}CHOEt}$$

R = n-Bu, 66%

R = Ph, 70%

I.G.3-21 F. Derguini-Boumechal and G. Linstrumelle, Tetrahedron Lett., 1976, 3225.

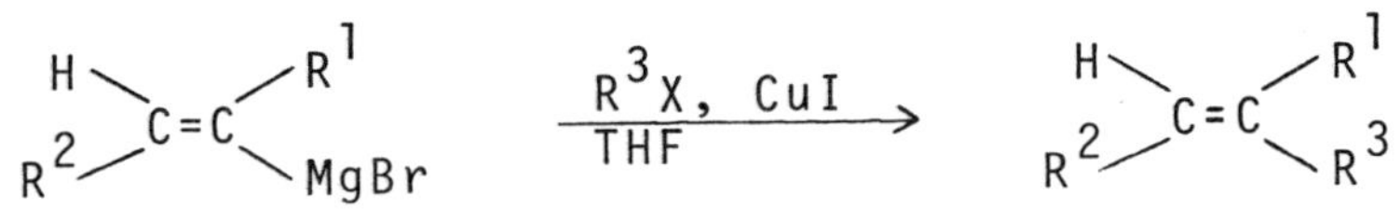

R^1	R^2	R^3X	% Yield
H	H	$n\text{-}C_8H_{17}I$	82
Me	H	$n\text{-}C_8H_{17}OTs$	80
Me	Me	n-PrI	97 (88% Z, 12% E)

I.G.3-22 F. Pochat and E. Levas, Tetrahedron Lett., 1976, 1491.

$$R^1R^2C(SEt)_2 \xrightarrow[\text{MeCN, 60°}]{\text{Hg(CN)}_2\text{, I}_2} R^1R^2C(SEt)CN$$

47-80%

I.G.3-23 I. Ojima, Y. Miyazawa, and M. Kumagai, J. Chem. Soc., Chem. Commun., 1976, 927.

$$MeCOCO_2R \xrightarrow[\text{TiCl}_4\text{, CH}_2\text{Cl}_2]{\text{CH}_2\text{=CHCH}_2\text{SiMe}_3} CH_2{=}CHCH_2CMe(OH)CO_2R$$

82-95%

I.G.3-24 A. Hosomi and H. Sakurai, Tetrahedron Lett., 1976, 1295; see also: G. Deleris, J. Dunoguès, and R. Calas, *ibid*, 2449.

$$Me_3SiCHR^1CH{=}CHR^2 \xrightarrow[\text{TiCl}_4\text{, CH}_2\text{Cl}_2]{R^3COR^4} R^1CH{=}CHCHR^2C(OH)R^3R^4$$

R^1	R^2	R^3	R^4	% Yield
H	H	n-Pr	H	87
H	H	$-(CH_2)_5-$		83
Me	H	Me	Me	72
Ph	H	n-Pr	H	54
H	Me	Me	Me	45
H	Ph	n-Pr	H	87

I.G.3-25 A. Hosomi, M. Endo, and H. Sakurai, Chem. Lett., 1976, 941.

$$R^1R^2C(OMe)_2 \xrightarrow[\substack{TiCl_4,\ CH_2Cl_2 \\ -78^\circ}]{CH_2{=}CHCH_2SiMe_3} CH_2{=}CHCH_2C(OMe)R^1R^2$$

R^1	R^2	% Yield
n-Bu	H	77
n-PrCHBr	H	91
Me	Me	98
$-(CH_2)_5-$		71

1.G.3-26 R. C. Larock and J. C. Bernhardt, Tetrahedron Lett., 1976, 3097.

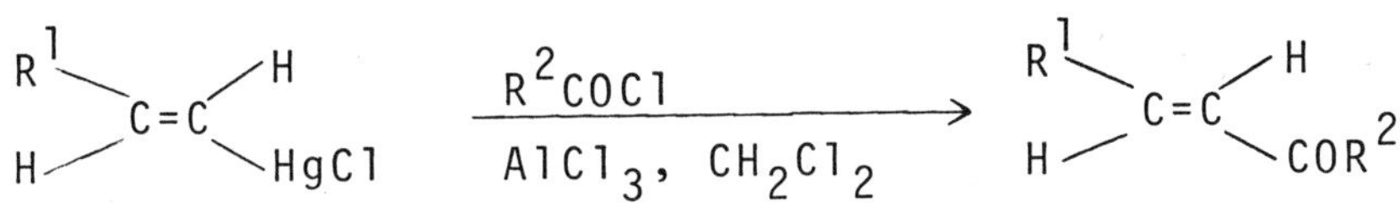

72-100%

I.G.3-27 J. Meijer, H. Westmijze and P. Vermeer, Recl. Trav. Chim. Pays-Bas, **95**, 102 (1976).

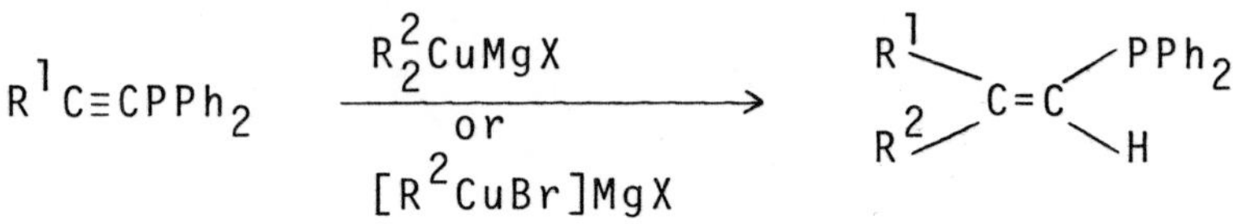

Excellent yields obtained but are dependent on the copper reagent employed.

I.G.3-28 A. Alexakis, A. Commercon, J. Villieras, and J. F. Normant, Tetrahedron Lett., **1976**, 2313.

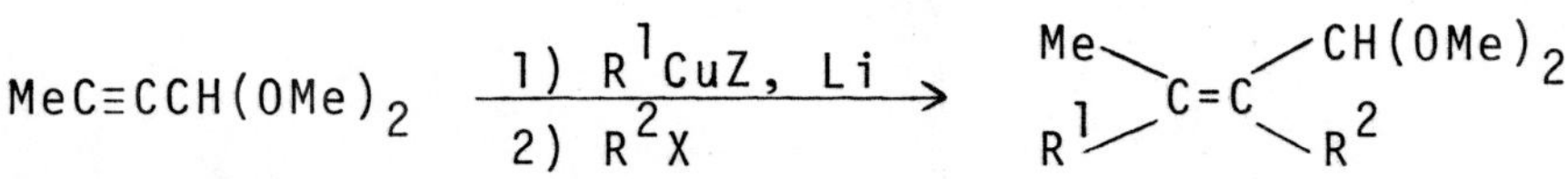

R^1CuZ, Li = n-BuCu(t-BuO)Li, R^2X = Me, 87%

R^1CuZ, Li = Et_2CuLi, R^2X = CH_2=$CHCH_2Br$, 85%

I.G.3-29 K. Kaneda, T. Uchiyama, M. Terasawa, T. Imanaka, and S. Teranishi, Chem. Lett., 1976, 449.

$$CH_2{=}CHCH_2X \;+\; R^1C{\equiv}CR^2 \xrightarrow{\text{Pd-polymer cat.}}$$

$$CH_2{=}CHCH_2CR^2{=}CR^1X \;(A) \;+\; CH_2{=}CHCH_2CR^1{=}CR^2X \;(B)$$

X	R^1	R^2	% Yield A	% Yield B
Cl	Ph	H	61	23
Br	Ph	H	79	13
Cl	n-Pr	H	78	-
Cl	Et	Et	54	
Cl	H	H	75	

I.G.3-30 Y. Frangin and M. Gaudemar, Bull. Soc. Chim. Fr., 1976, 1173.

$$RC{\equiv}CY \xrightarrow[\text{2) } H_3O^+]{\text{1) 2.5 eq. } CH_2{=}CHCH_2ZnBr \text{, THF}} CH_2{=}CHCH_2CR{=}CH_2 \;A \;+\; (CH_2{=}CHCH_2)_2CRMe \;B$$

R	Y	% Yield	A/B
$n\text{-}C_8H_{17}$	H	72	61/39
$n\text{-}C_7H_{15}$	MgBr	64	0/100
Ph	H	73	92/8
Ph	H	80	6/94

Other systems also studied.

I.G.3-31 H. Klein, H. Eijsinga, H. Westmijze, J. Meijer, and P. Vermeer, Tetrahedron Lett., 1976, 947; see also: P.H.M. Schreurs, J. Meijer, P. Vermeer, and L. Brandsma, *ibid*, 2387.

$$H_2C{=}C{=}CHOMe \xrightarrow[\text{THF}]{\text{RCu(Br or R)MgX}} RCH_2CH{=}CHOMe \quad 80\text{-}90\%$$

The E/Z ratio of the product is influenced by the type of organocopper reagent used.

I.G.3-32 L. M. Zubritskii, L. N. Cherkasov, T. N. Fomina, and K. V. Bal'yan, J. Gen. Chem., USSR, 46, 440 (1976).

$$Me_3SiC{\equiv}CCH{=}CH_2 \xrightarrow[\text{2) } H_2O]{\text{1) PhMgBr, } NiCl_2} Me_3SiC{\equiv}CCH_2CH_2Ph \quad 30\text{-}35\%$$

I.G.3-33 H. Matsumoto, T. Nikado, and Y. Nagai, J. Org. Chem., 41, 396 (1976).

$$R^1CH{=}CH_2 + Cl_2CR^2CO_2R^3 \xrightarrow[\text{PhH, 120°}]{RuCl_2(PPh_3)_3\text{ (cat.)}} R^1CHR^2CH_2CCl_2CO_2R^3$$

R^1	R^2	R^3	% Yield
C_6H_{13}	Cl	Me	88
Ph	Cl	Et	82
CN	Cl	Et	71
MeCO	Cl	Et	64
C_6H_{13}	H	Me	89

I.G.3-34 N. Genco, D. Marten, S. Raghu, and M. Rosenblum, J. Am. Chem. Soc., **98**, 848 (1976).

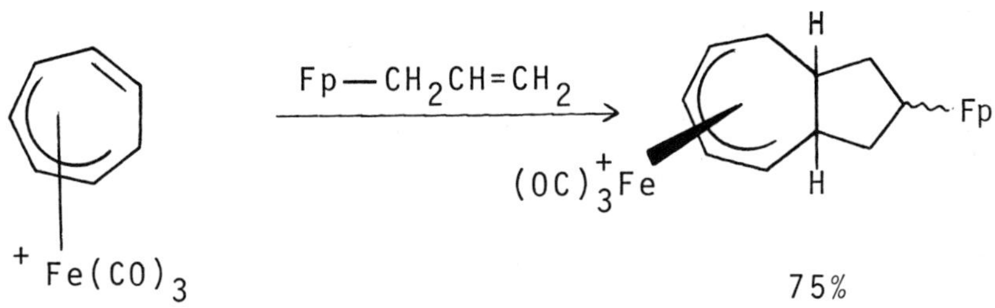

I.G.3-35 A. J. Pearson, Aust. J. Chem., **29**, 1101 (1976).

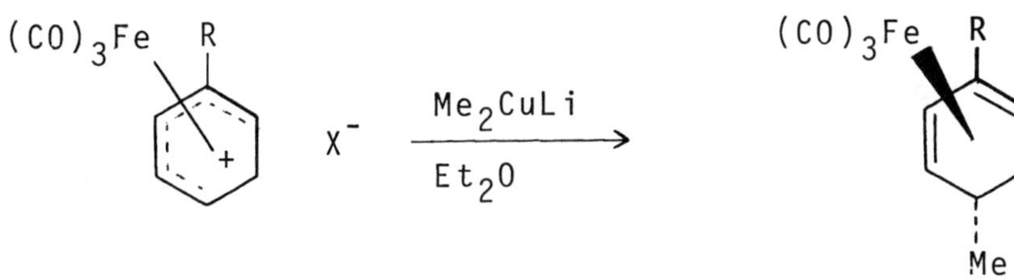

R	X^-	% Yield
H	BF_4^-	80
Me	PF_6^-	60
MeO	BF_4^-	50

I.G.3-36 A. J. Pearson, Aust. J. Chem., 29, 1841 (1976).

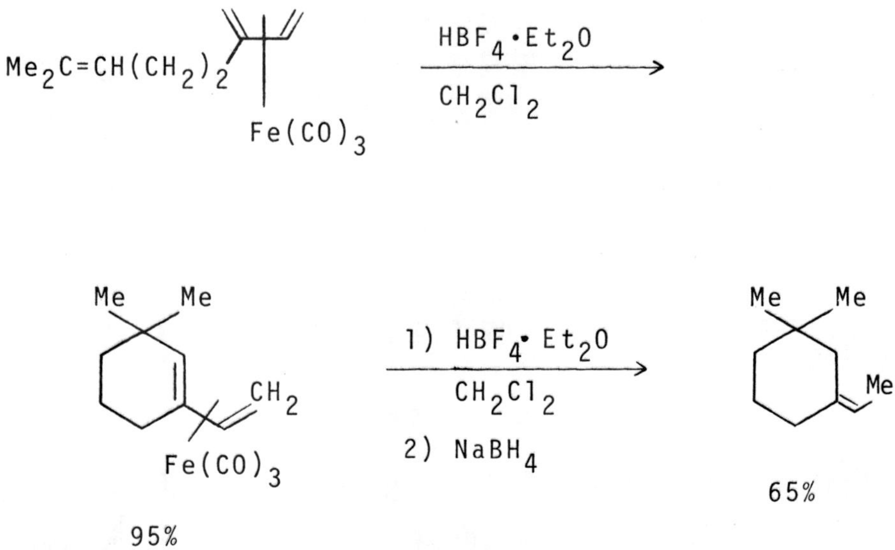

I.G.3-37 A. J. Birch and A. J. Pearson, J. Chem. Soc., Chem. Commun., 1976, 601.

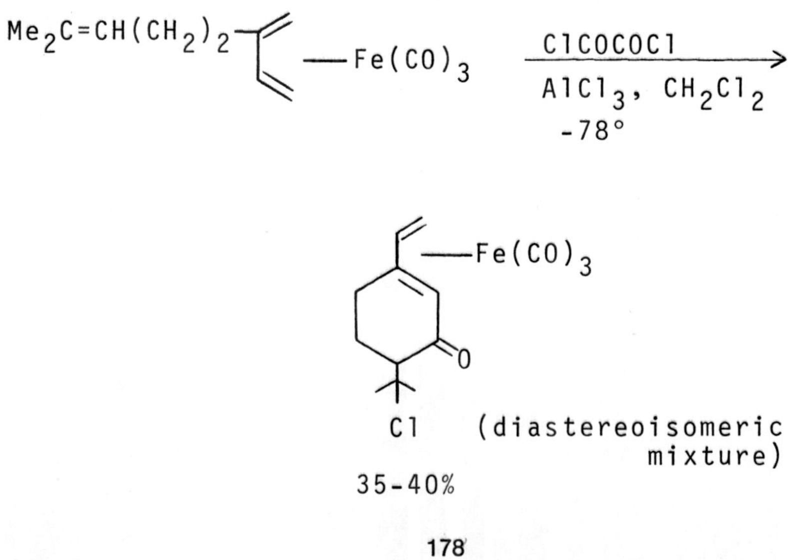

I.G.3-38 D.-J. Collins, W. R. Jackson, and R. N. Timms, Tetrahedron Lett., 1976, 495.

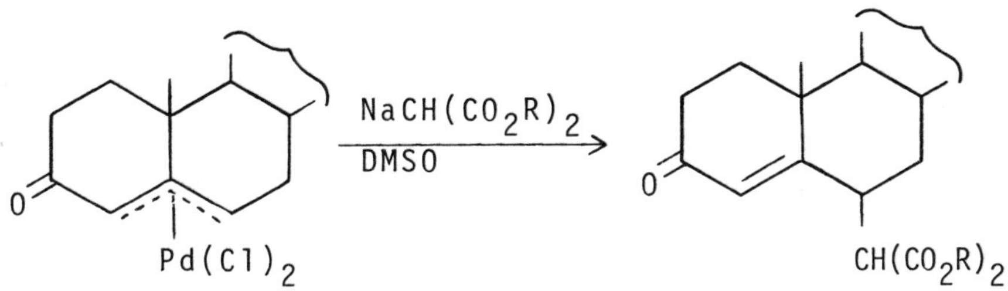

"up to 90% yield"

I.G.3-39 R. C. Larock and M. A. Mitchell, J. Am. Chem. Soc., 98, 6718 (1976).

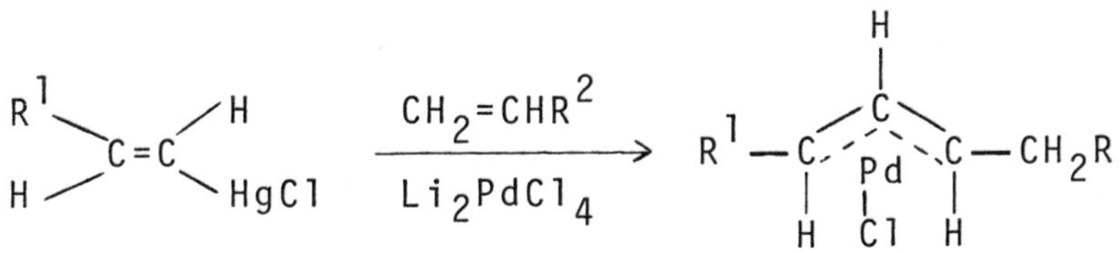

R^1	R^2	% Yield
Ph	CO_2Et	100
t-Bu	CO_2Et	90
t-Bu	H	92
t-Bu	CN	87
t-Bu	Me	59

I.G.3-40 R. L. Funk and K.P.C. Vollhardt, <u>J. Chem. Soc., Chem. Commun.</u>, <u>1976</u>, 833.

$HC{\equiv}CCH_2CH(OTMS)C{\equiv}CH \xrightarrow[CpCo(CO)_2]{Me_3SiC{\equiv}CSiMe_3}$

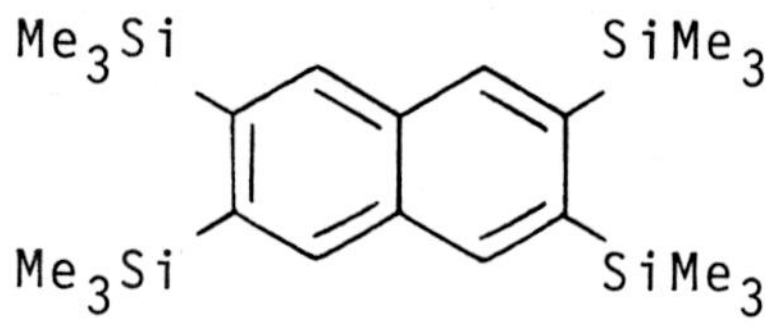

30%

I.G.3-41 R. L. Funk and K.P.C. Vollhardt, <u>J. Am. Chem. Soc.</u>, <u>98</u>, 6755 (1976).

$HC{\equiv}CCHXCH_2YCH_2CH{=}Z$ (with $HC{\equiv}CCH_2$ on the CH) $\xrightarrow[CpCo(CO)_2]{Me_3SiC{\equiv}CSiMe_3}$

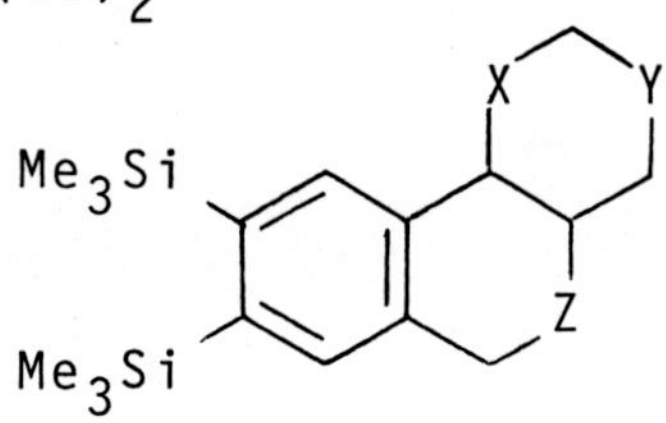

X	Y	Z	% Yield
O	CH_2	CH_2	60
CH_2	O	CH_2	90
CH_2	CH_2	CH_2	80

I.G.3-42 W. Münzenmaier and H. Straub, <u>Synthesis</u>, **1976**, 49.

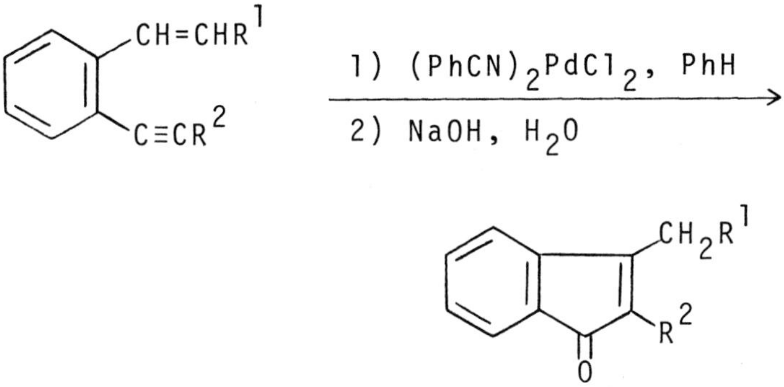

R^1	R^2	% Yield
Ph	Ph	54
t-Bu	Ph	35
Me	Me	30

I.G.3-43 C. F. Lochow and R. G. Miller, <u>J. Am. Chem. Soc</u>., **98**, 1281 (1976).

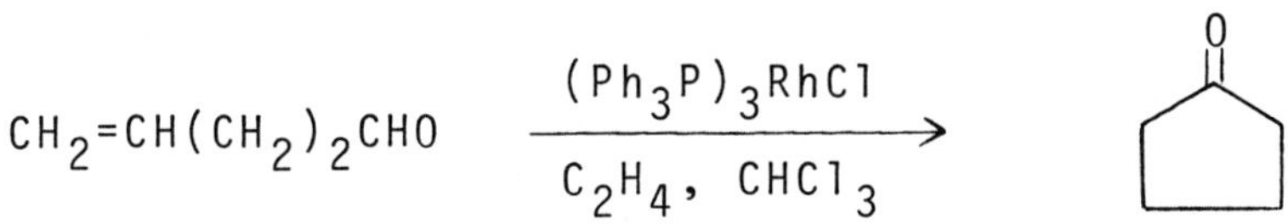

~70%

I.G.3-44 A. Baba, Y. Ohshiro, and T. Agawa, J. Organomet. Chem., 110, 121 (1976).

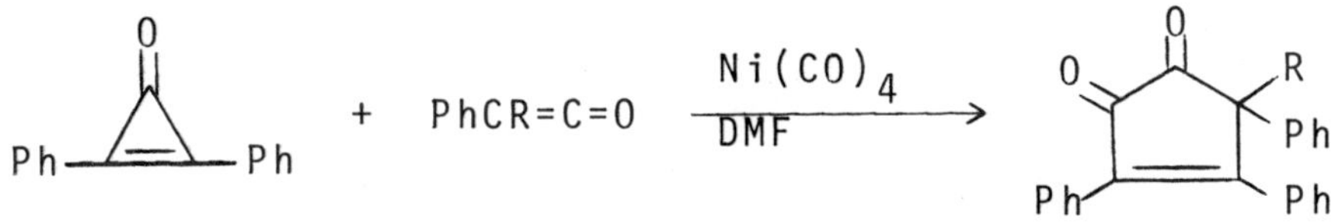

R = Ph or Et, 85%

I.G.e-45 T. Mitsudo, K. Kokuryo, and Y. Takegami, J. Chem. Soc., Chem. Commun., 1976, 722.

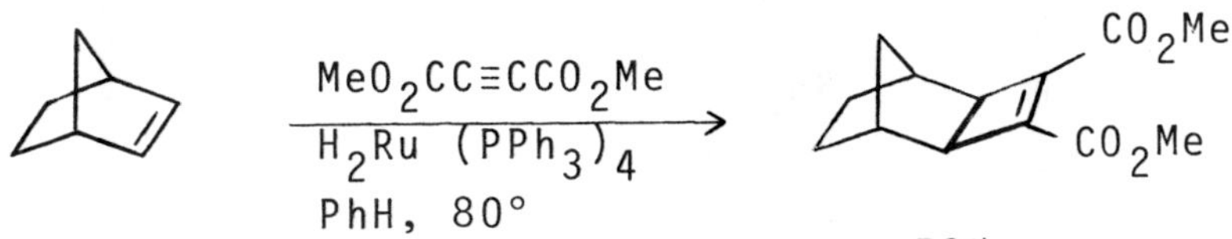

52%

I.G.3-46 R. Baker, M. S. Nobbs, and D. T. Robinson, J. Chem. Soc., Chem. Commun., 1976, 723.

$$CH_2{=}CHCH{=}CH_2 + R^1R^2C{=}NNHPh \xrightarrow[(cod)_2NiPPh_3]{(Ph_3P)_4Pd\ \text{or}}$$

$$CH_2{=}CH(CH_2)_3CH{=}CHCH_2CR^1R^2N{=}NPh \quad +$$

$$CH_2{=}CH(CH_2)_3\underset{\displaystyle CR^1R^2N{=}NPh}{\underset{|}{C}}HCH{=}CH_2$$

Product distribution is dependent on the catalyst used.

I.G.3-47 S. Akutagawa, Bull. Chem. Soc. Japan, **49**, 3646 (1976).

$$CH_2{=}CHCH{=}CH_2 \xrightarrow[NiP(O\text{-}C_6H_4Me)_3\ (cat.)]{RCOMe,\ 70°} CH_2{=}CH(CH_2)_3\overset{H}{C}{=}\overset{H}{C}CH_2C(OH)RMe$$

R	% Yield
Me	62
i-Pr	55
Ph	52

I.G.3-48 P. A. Grieco, M. Nishizawa, N. Marinovic, and W. J. Ehmann, J. Am. Chem. Soc., **98**, 7102 (1976).

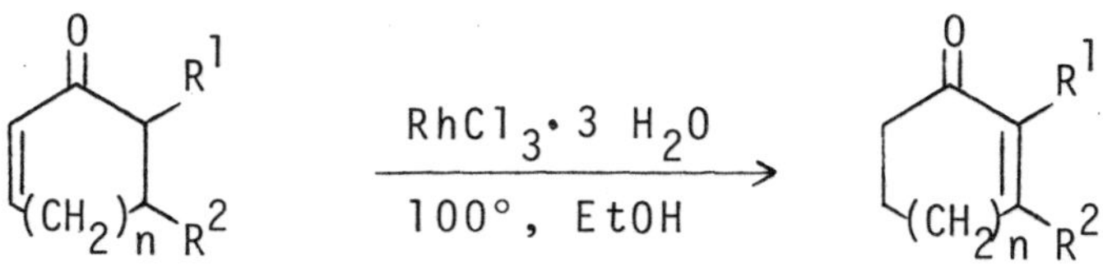

n = 0-3, 62-95%

I.G.3-49 E. C. Taylor, R. L. Robey, K.-T. Liu, B. Favre, H. T. Bozimo, R. A. Conley, C.-S. Chiang, A. McKillop and M. E. Ford, J. Am. Chem. Soc., 98, 3037 (1976).

$$ArCH{=}CR^1COR^2 \xrightarrow[\text{MeOH, HC(OMe)}_3]{\text{Tl(NO}_3)_3} ArCR^1XCH(OMe)_2$$

Ar	R^1	R^2	X	% Yield
Ph	H	H	$CH(OMe)_2$	79
$p\text{-}O_2NC_6H_4$	Me	H	$CH(OMe)_2$	50
H	H	OMe	CO_2Me	96
$p\text{-}ClC_6H_4$	H	OMe	CO_2Me	96

I.G.3-50 J. Levisalles, H. Rudler, and D. Villemin, J. Organomet. Chem., 122, C15 (1976).

$$(RCH_2)_2O \xrightarrow{\text{PhLi, WCl}_6} RPhCHOCH_2R$$

R = Me, 80%

R = $-CH_2CH_2-$, 90%

$$Et_3N \xrightarrow{\text{PhLi, WCl}_6} Et_2NCHPhMe$$

35%

I.G.4-1 H. C. Brown, Pure and Appl. Chem., 47, 49 (1976).

Plenary Lecture: Organoboranes — The Modern Miracle.

I.G.4-2 C. F. Lane and G. W. Kabalka, Tetrahedron, 32, 981 (1976).

Review: Catecholborane. A New Hydroboration Agent.

I.G.4-3 J. Weill-Raynal, Synthesis, 1976, 633.

Review: Formation of Carbon-Carbon Bonds by Using Organoboranes.

I.G.4-4 B. M. Mikhailov, Russ. Chem. Rev., 45, 557 (1976).

Review: Methods of Synthesis and Properties of Allylboranes.

I.G.4-5 E.-I. Negishi, J. Organometal. Chem., **108**, 281 (1976).

Review: Chemistry of Organoborates.

I.G.4-6 S. S. Washburne, J. Organomet. Chem., **123**, 1 (1976).

Review: Silicon-Application to Organic Synthesis. Annual Survey Covering the Year 1974.

I.G.4-7 P. J. Smith, Chem. Ind. (London), **1976**, 1025.

Review: Organic Synthesis *via* Organictin Intermediates – Some Recent Developments.

I.G.4-8 J. Schwartz and J. A. Labinger, Angew. Chem. Int. Ed. Engl., **15**, 333 (1976).

Review: Hydrozirconation: A New Transition Metal Reagent for Organic Synthesis.

I.G.4-9 A. P. Kozikowski and H. F. Wetter, Synthesis, 1976, 561.

Review: Transition Metals in Organic Synthesis.

I.G.4-10 G. Henrici-Olive' and S. Olive', Fortschr. Chem. Forsh., 67, 107 (1976).

Review: Olefin Insertion in Transition Metal Catalysis.

I.G.4-11 K. C. Bishop, III, Chem. Rev., 76, 461 (1976).

Review: Transition Metal Catalyzed Rearrangements of Small Ring Organic Molecules.

I.G.4-12 I. P. Beletskaya, G. A. Artamkina, and O. A. Reutov, Russ. Chem. Rev., 45, 330 (1976).

Review: The Interaction of Organimetallic Derivatives with Organic Halides.

II.A.1-1 G. A. Lee and H. H. Freedman, Tetrahedron Lett., 1641 (1976).

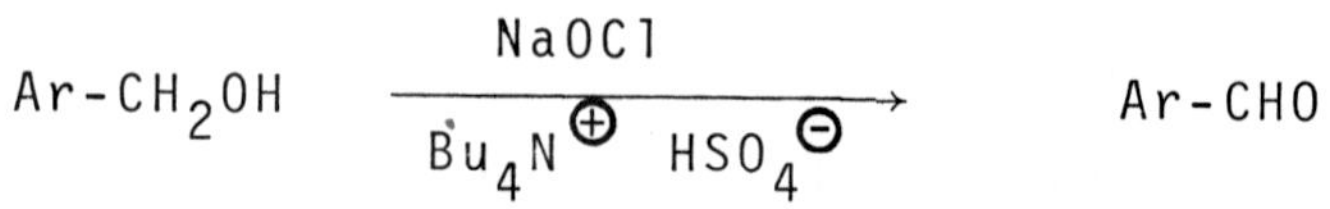

47-100%

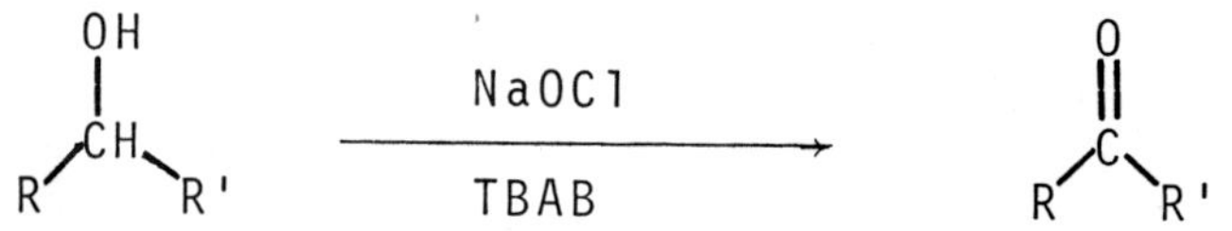

R,R' = alkyl, aryl

II.A.1-2 K. Omura, A. K. Sharma, and D. Swern, J. Org. Chem., 41, 957 (1976).

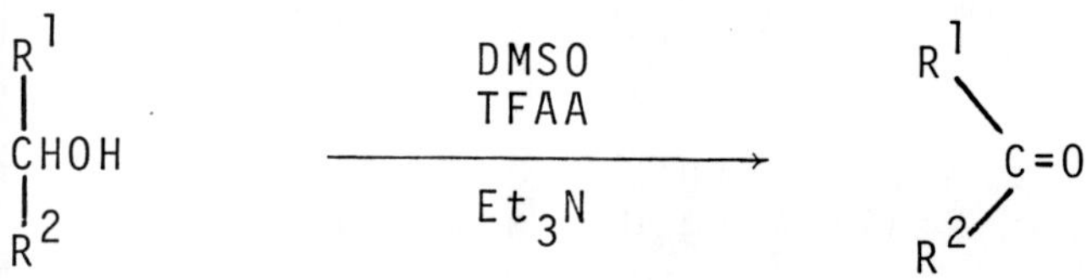

R^1,R^2 = alkyl, aryl, H 44-98%

Eight examples given.

II.A.1-3 R. W. Binkley, J. Org. Chem., 41, 3030 (1976).

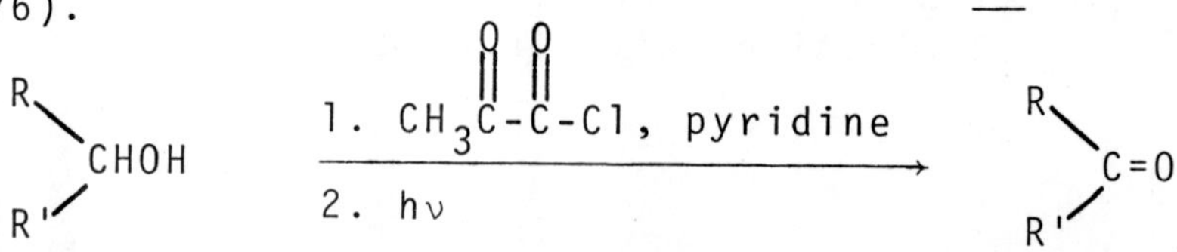

R,R' = H, alkyl, aryl, cyclic 76-100%

II.A.1-4 S. L. Huang, K. Omura, and D. Swern, J. Org. Chem., 41, 3329 (1976).

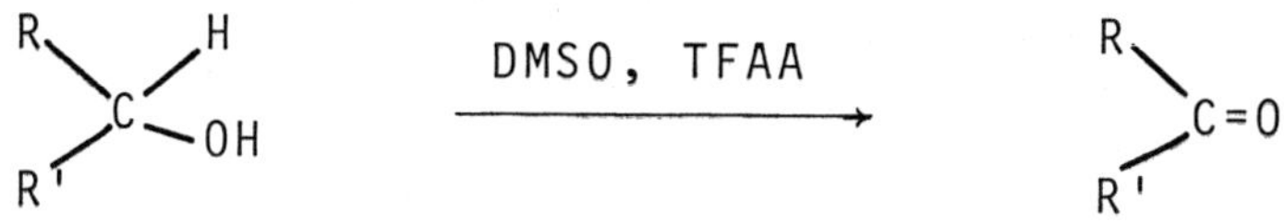

R,R' = alkyl, aryl, cyclic, H

Yields generally 70-95%

Especially useful for very hindered alcohols.

II.A.1-5 G. Cardillo, M. Orena, and S. Sandri, Synthesis, 394 (1976).

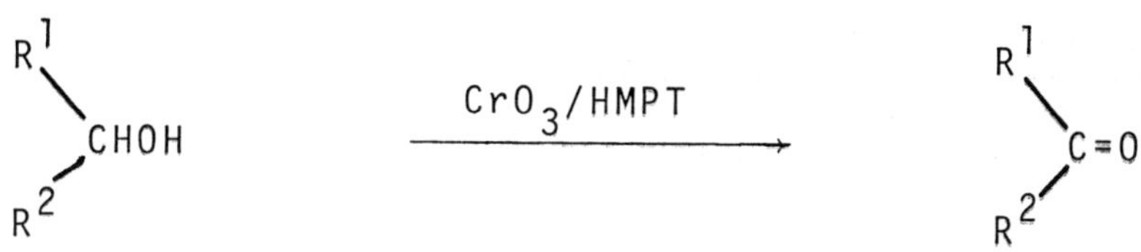

R^1,R^2 = H, alkyl, aryl, cyclic

80-100% for aldehydes, aryl ketones
40-49% for dialkyl ketones

II.A.1-6 G. Cainelli et al., J. Am. Chem. Soc., 98, 6737 (1976).

Use of CrO_3 bound to an ion-exchange resin to effect alcohol oxidations:

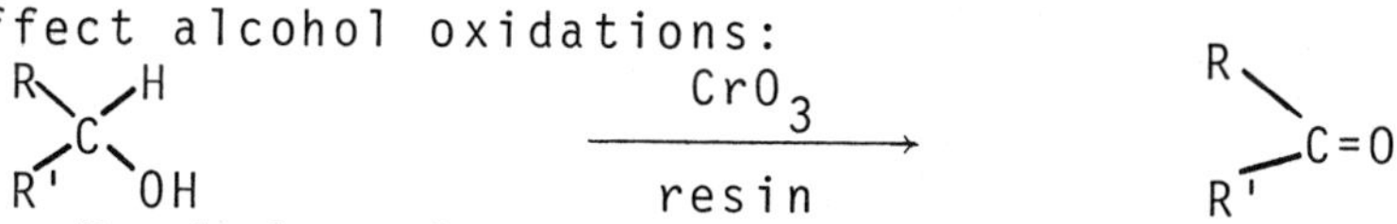

R,R' = H, alkyl, aryl

Yields generally 75-95%

II.A.1-7 G. H. Posner, R. B. Perfetti, and A. W. Runquist, Tetrahedron Lett., 3499 (1976).

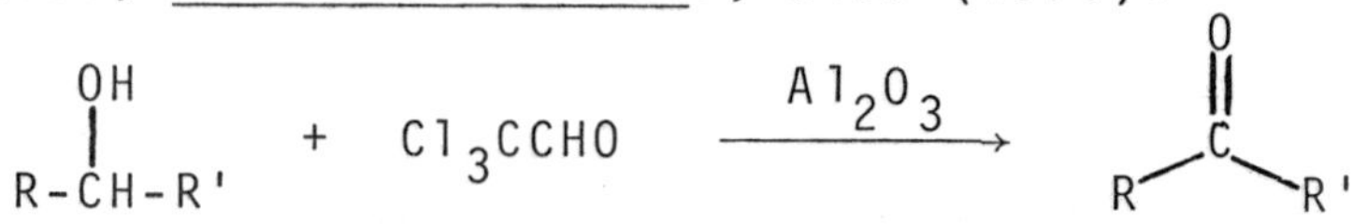

R,R' = alkyl, aryl, cyclic 63-90%

II.A.1-8 M. Fetizon, F. Gomez-Parra, and J.-M. Louis, J. Het. Chem., 13, 525 (1976).

$$R\text{-}CH_2OH \xrightarrow[\text{celite}]{Ag_2CO_3} R\text{-}CHO$$

R = heterocyclic 50-97%

II.A.1-9 W. L. Waters et al., J. Org. Chem., 41, 889 (1976).

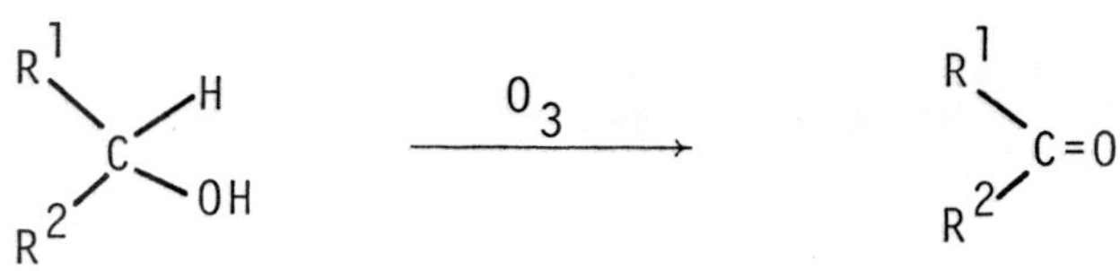

R^1,R^2 = alkyl, cyclic 57-83% (VPC)

II.A.1-10 T. Ogawa and M. Matsui, J. Am. Chem. Soc., 98, 1629 (1976).

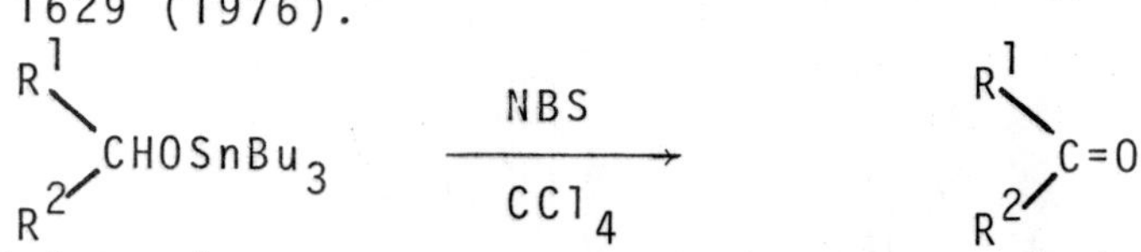

R's = H, alkyl, aryl 60-90%

II.A.1-11 K. Saigo, A. Morikawa, and T. Mukaiyama, Bull. Chem. Soc. Japan, **49**, 1656 (1976).

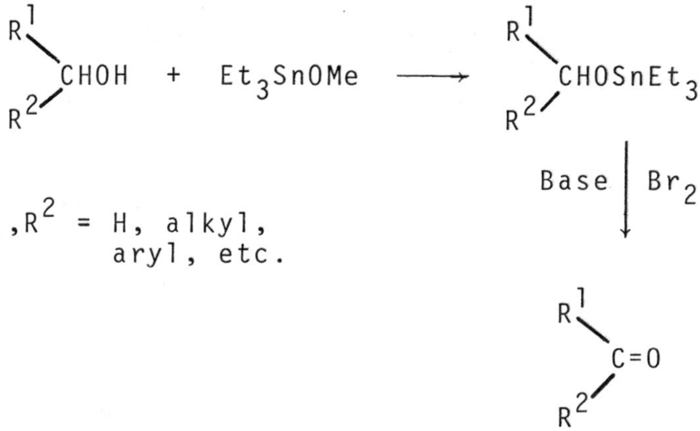

R^1,R^2 = H, alkyl, aryl, etc.

generally 60-80%

II.A.1-12 Y. Ueno and M. Okawara, Tetrahedron Lett., 4597 (1976).

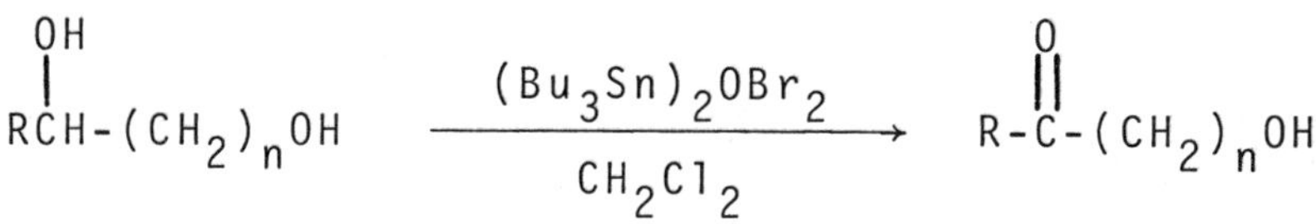

R = Me, Et, Ph

n = 0, 1

66-86%

II.A.1-13 G. A. Olah, J. Welch, and T.-L. Ho, J. Am. Chem. Soc., **98**, 6717 (1976).

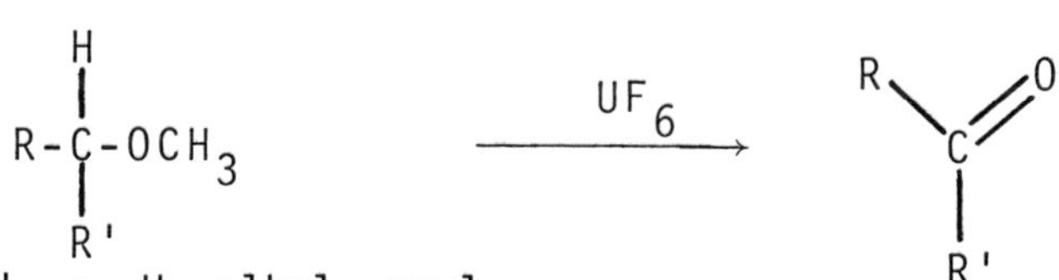

57-86%

R,R' = H, alkyl, aryl

II.A.1-14 M. E. Jung, J. Org. Chem., 41, 1479 (1976).

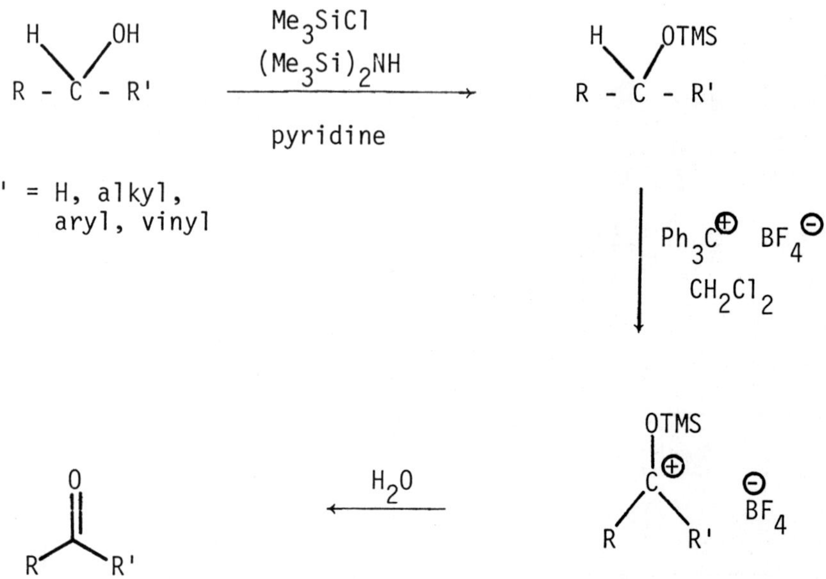

R,R' = H, alkyl, aryl, vinyl

38-100% (VPC)

II.A.1-15 G. A. Olah and T.-L. Ho, Synthesis, 609 (1976).

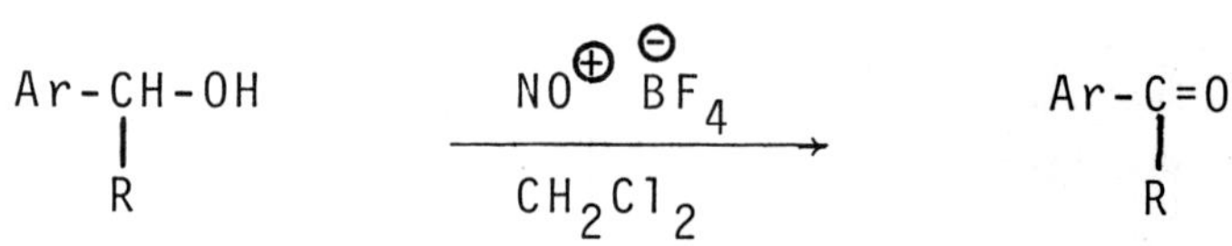

R = H, Me, Ph

60-78%

II.A.1-16 J. J. Kaminski and N. Bodor, <u>Tetrahedron</u>, <u>32</u>, 1097 (1976).

Use of <u>I</u> as a brominating and oxidizing agent, similar to NBS.

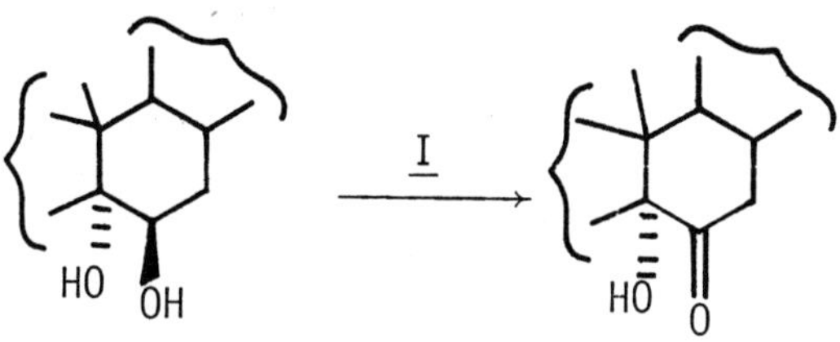

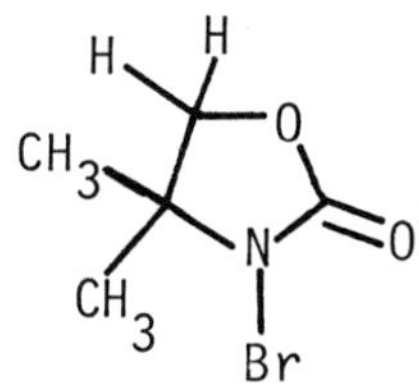

83% <u>I</u>

II.A.1-17 M. Al Neirabeyeh <u>et al</u>., <u>Synthesis</u>, 811 (1976).

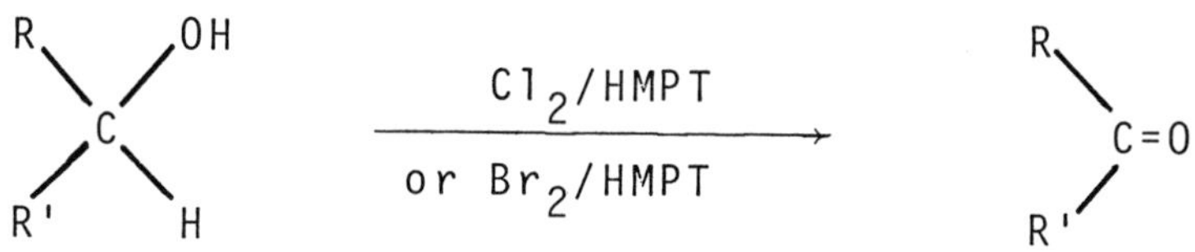

R,R' = H, alky, aryl

Oxidizes 2° alcohols more rapidly than 1°, e.g.:

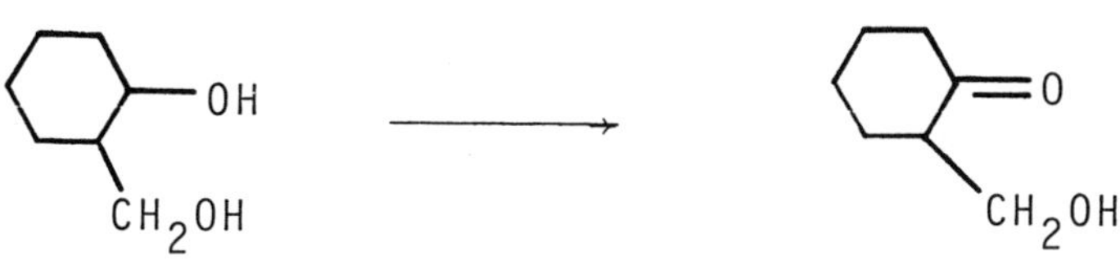

93%

II.A.1-18 M. E. Jung and L. M. Speltz, J. Am. Chem. Soc., 98, 7882 (1976).

A method for the selective oxidation of secondary alcohols, e.g.:

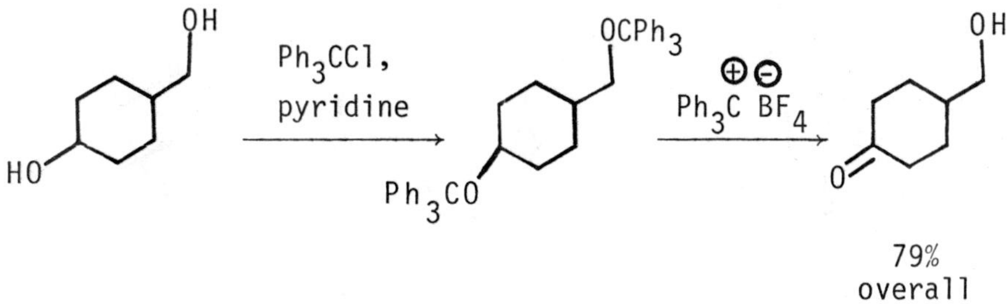

II.A.1-19 T. C. Sharma and V. Saksena, Indian J. Chem., 14B, 143 (1976).

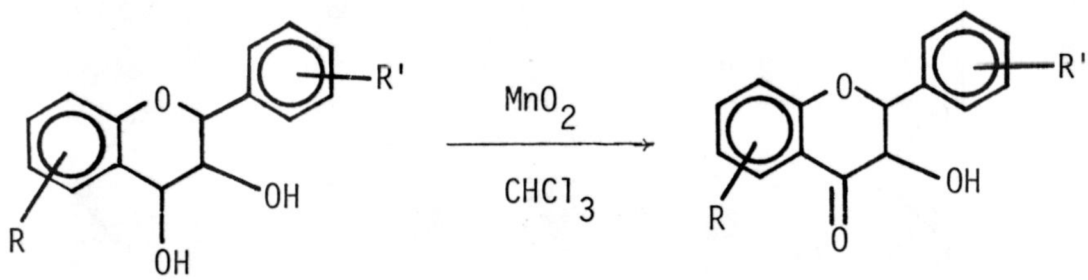

No yields given.

II.A.2-1 S. A. Maslov and E. A. Blyumberg, Russ. Chem. Rev., 45, 155 (1976).

Review: "Liquid-Phase Oxidation of Aldehydes"

II.B.1-1 Z. Cohen et al., J. Org. Chem., 41, 2651 (1976).

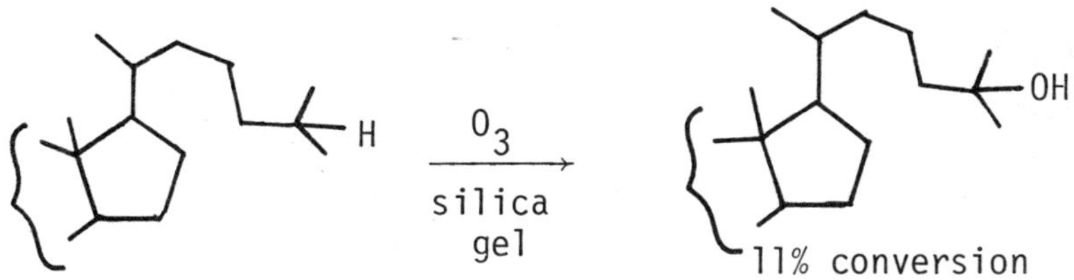

51% yield

II.B.1-2 W. Adam, O. Cueto, and V. Ehrig, J. Org. Chem., 41, 370 (1976).

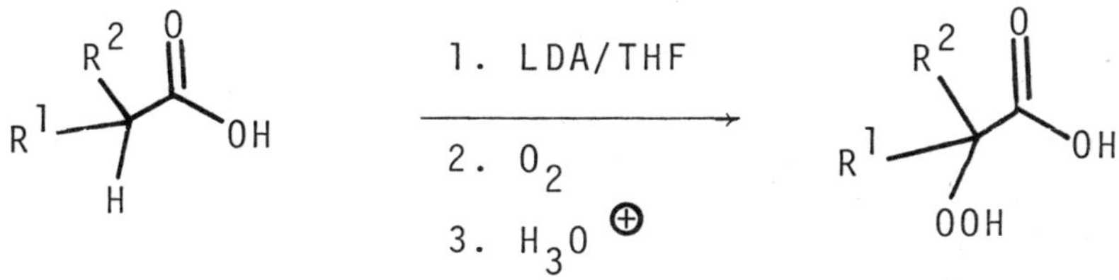

Yields >70%

II.B.1-3 E. J. Corey and S. Knapp, Tetrahedron Lett., 4687 (1976)

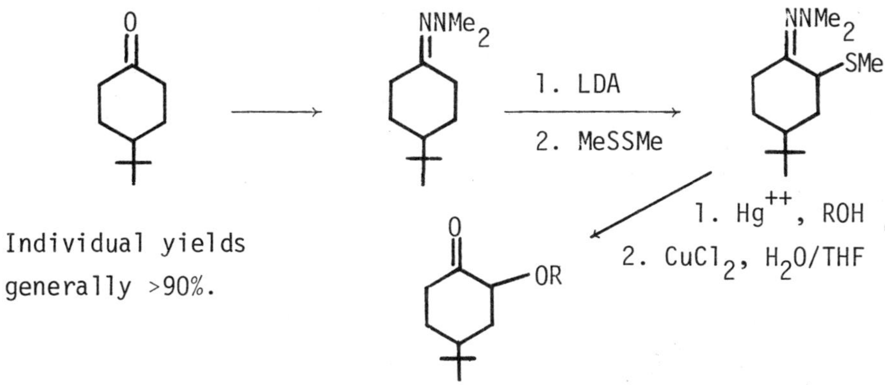

Individual yields generally >90%.

II.B.1-4 H. H. Wasserman and J. L. Ives, J. Am. Chem. Soc., 98, 7868 (1976).

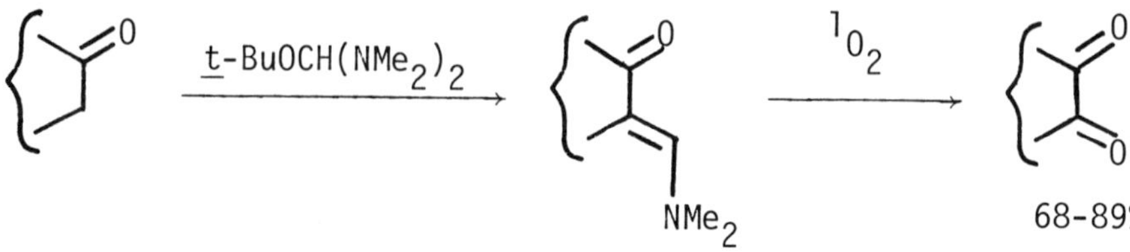

68-89%
overall

II.B.1-5 E. Vedejs and J. E. Telschow, J. Org. Chem., 41, 740 (1976).

$R\text{-}CH_2\text{-}CN \xrightarrow[\text{2. } MoO_5\cdot Py\cdot HMPA]{\text{1. LDA}} R\text{-}CH(OH)\text{-}CN$

R = alkyl, aryl

35-57%

II.B.1-6 G. Ortar, M. P. Arpiani, and A. Romeo, Steroids, 27, 197 (1976).

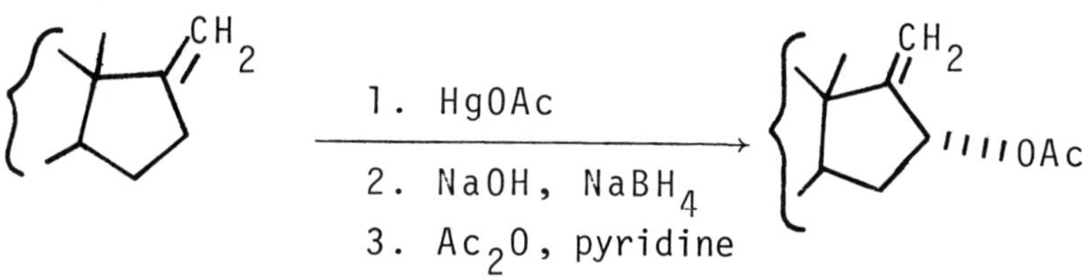

57%

II.B.1-7 D. S. Fullerton and C.-M. Chen, Synth. Commun., 6, 217 (1976).

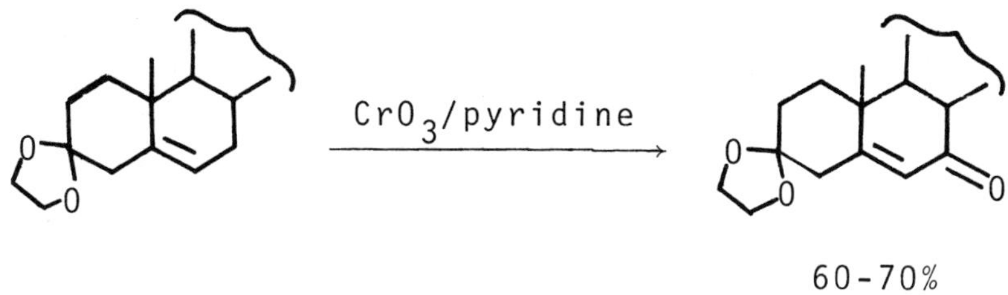

II.B.1-8 G. Stubenrauch and R. Knuppen, Steroids, 28, 733 (1976).

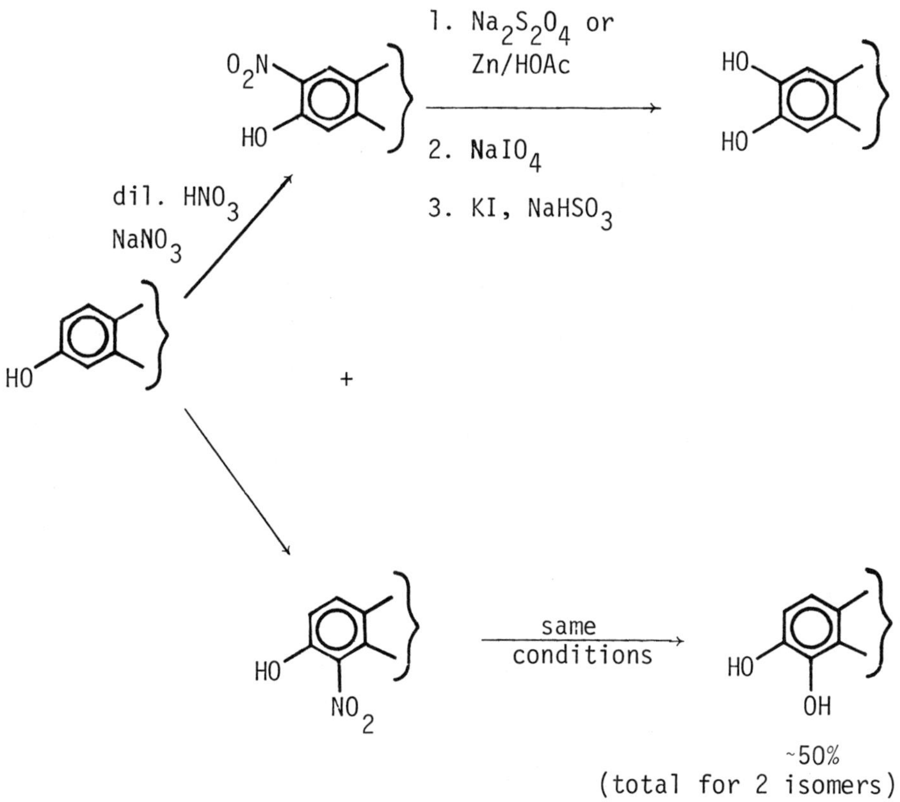

II.B.1-9 J. Buddrus and H. Plettenberg, <u>Angew. Chem. Int. Ed.</u>, <u>15</u>, 436 (1976).

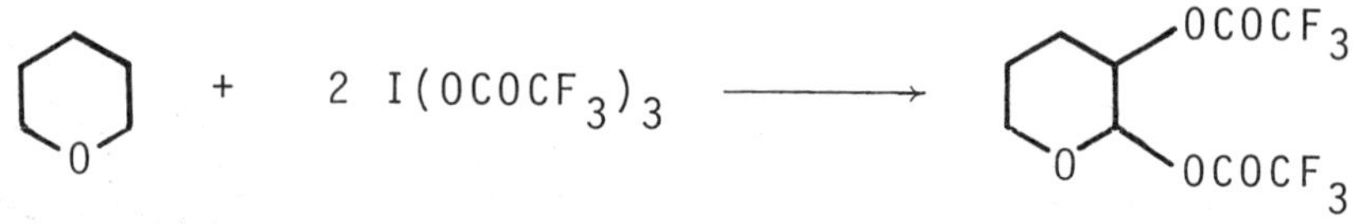

42%

II.B.1-10 R. W. Hoffman and N. Maak, <u>Tetrahedron Lett.</u>, 2237 (1976).

Formation of allyl alcohols from α,β-unsaturated sulfoxides, e.g.:

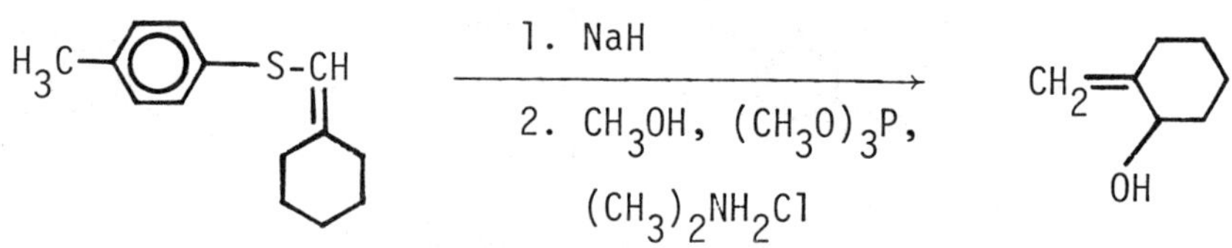

75% yield

(Up to 60% optical purity can be obtained using optically active starting materials.)

II.B.1-11 P.R.O. deMontellano and C. K. Hsu, <u>Tetrahedron Lett.</u>, 4215 (1976).

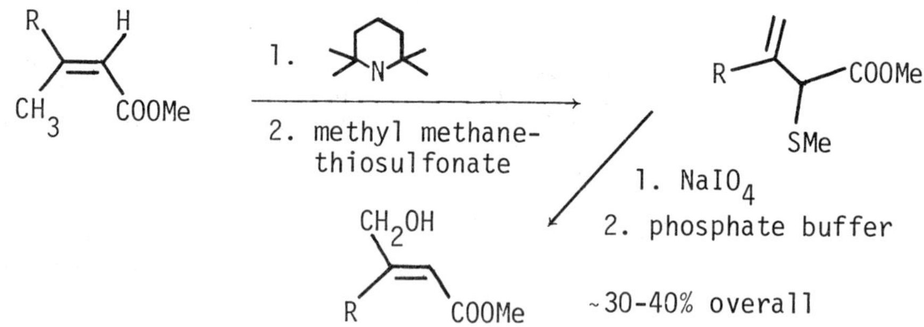

II.B.2-1 N. Inukai *et al.*, *Chem. Pharm. Bull.*, **24**, 820 (1976).

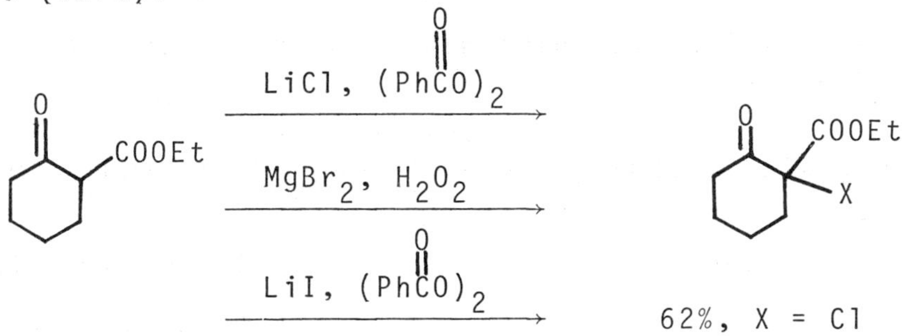

62%, X = Cl
100%, X = Br
72%, X = I

II.B.2-2 A. Rahman, A. Basha, and V. U. Ahmad, *Experientia*, **32**, 1491 (1976).

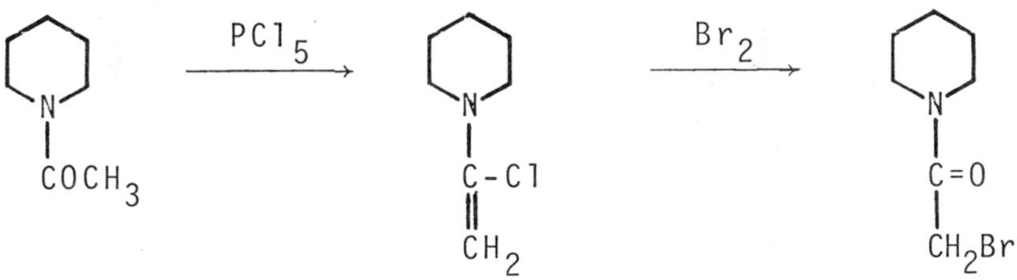

80%

II.B.2-3 L. Blanco, P. Amice, and J. M. Conia, *Synthesis*, 194, 196 (1976).

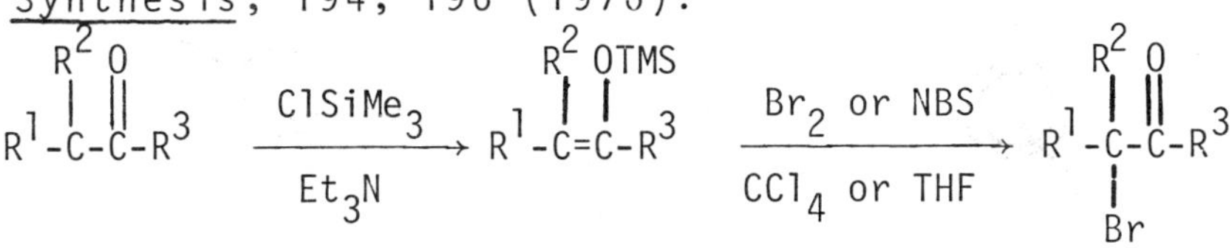

R's = H, alkyl, cyclic 60-95%

II.B.2-4 F. Kamper, H. J. Schöfer, and H. Luftmann, Angew. Chem. Int. Ed., 15, 306 (1976).

Regioselective chlorination of centered methylene groups:

$$CH_2\!-\!\!\left[(CH_2)_nCOOH\right]_2 \xrightarrow[H_2SO_4]{i\text{-}Pr_2NCl} CHCl\!-\!\!\left[(CH_2)_nCOOH\right]_2$$

n = 2-4

58-96% specificity

2-96% yields

II.B.2-5 M.F.G.-Loustalot et al., Synthesis, 33 (1976).

R's = H, alkyl

50-75%

II.B.2-6 R. C. Cambie et al., J. Chem. Soc. Perkin I, 1161 (1976).

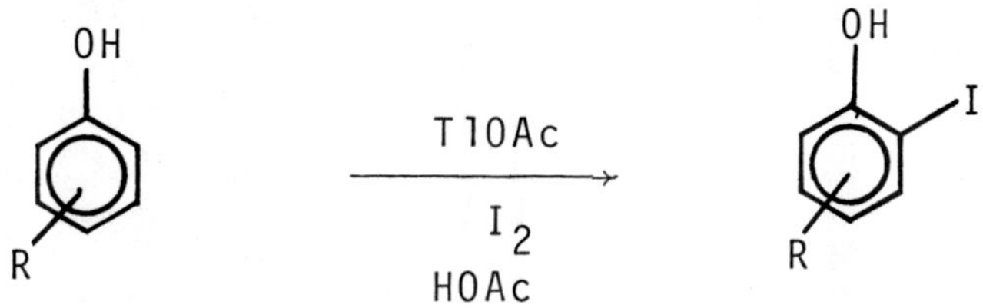

R = H, alkyl, aryl

yields generally 50-80%

II.B.2-7 L. L. Miller and B. F. Watkins, J. Am. Chem. Soc., 98, 1515 (1976).

$$\text{Ar-H} \xrightarrow[\text{Iodine(I)}]{\text{Electrogenerated}} \text{Ar-I}$$

Yields generally 50-90%

II.B.2-8 I. Agranat et al., Experientia, 15, 417 (1976).

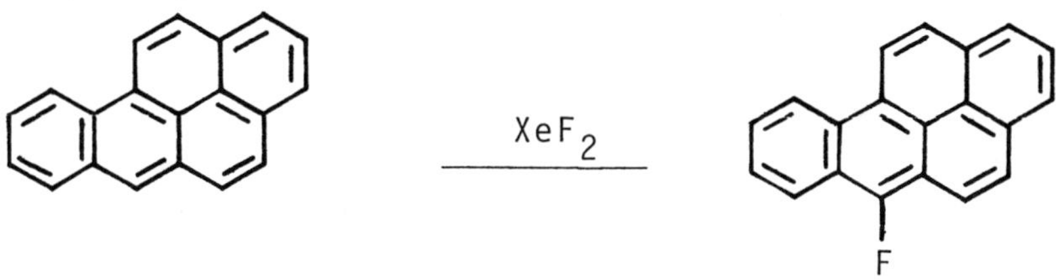

26%

II.B.2-9 D. E. Pearson et al., Synthesis, 621 (1976).

Use of trimethyl phosphate as a reaction medium for aromatic halogenation of substrates sensitive to HX. No HX is generated because of a fast reaction to form methyl halide.

II.B.3-1 P. A. Zoretic and P. Soja, J. Org. Chem., 41, 3587 (1976).

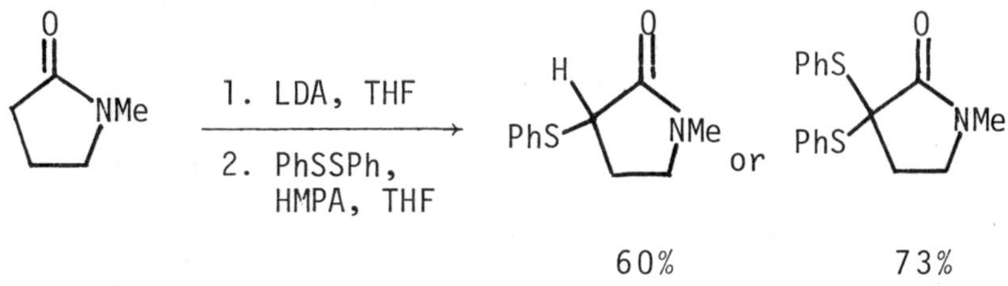

depending on conditions

II.C-1 G. A. Lee and H. H. Freedman, Tetrahedron Lett., 1641 (1976).

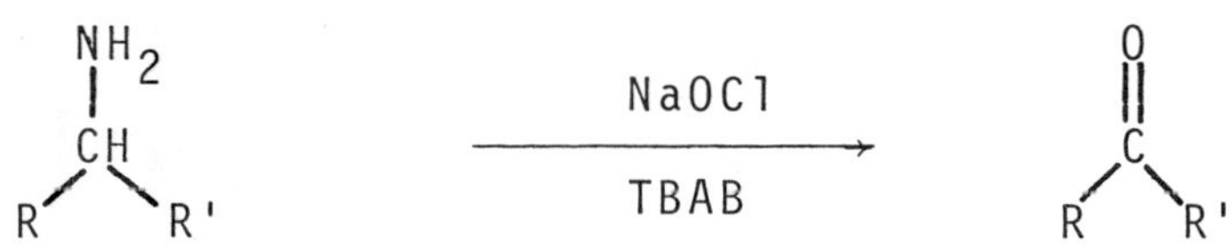

R,R' = alkyl, aryl 84-98%

II.C-2 R. Kirchoff, Tetrahedron Lett., 2533 (1976).

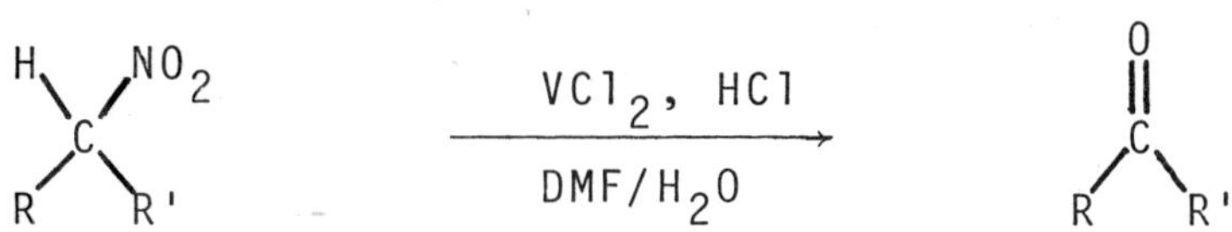

R = H, alkyl 24% (heptaldehyde)

R' = alkyl, benzyl 53-71% (ketones)

II.D-1   M. R. Czarny, _Synth. Commun._, **6**, 285 (1976).

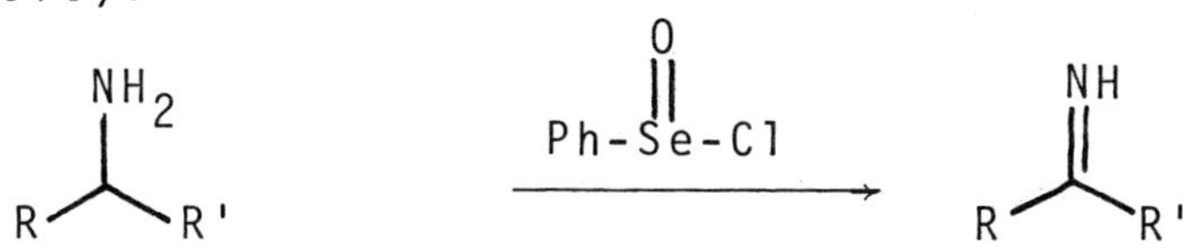

R,R' = Ph; also 2-adamantylamine   97%

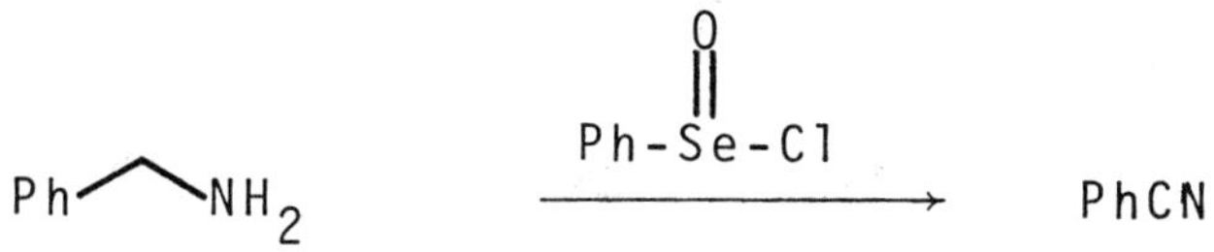

85%

II.D-2   K. J. Chapman and L. K. Dyall, _Aust. J. Chem._, **29**, 367 (1976).

$$ArNH_2 + Cl_2O \longrightarrow ArNHCl$$

"high yields"

II.E-1   T. Aida _et al._, _Bull. Chem. Soc. Japan_, **49**, 1441 (1976).

$$2RSH \xrightarrow[DMSO]{I_2,\ HI} R\text{-}S\text{-}S\text{-}R$$

R = alkyl, aryl   generally >90%

II.E-2 M. Hojo and R. Masuda, <u>Tetrahedron Lett</u>., 613 (1976).

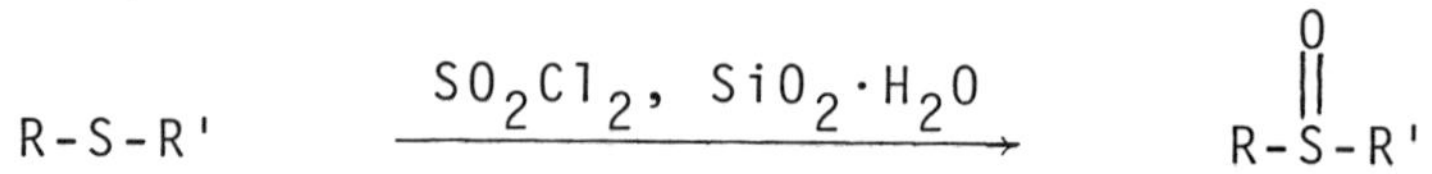

R,R' = alkyl, aryl, vinyl, CH_2Cl 82-100% (crude)

II.E-3 R. Louw <u>et al</u>., <u>J. Chem. Soc., Chem. Commun</u>., 496 (1976).

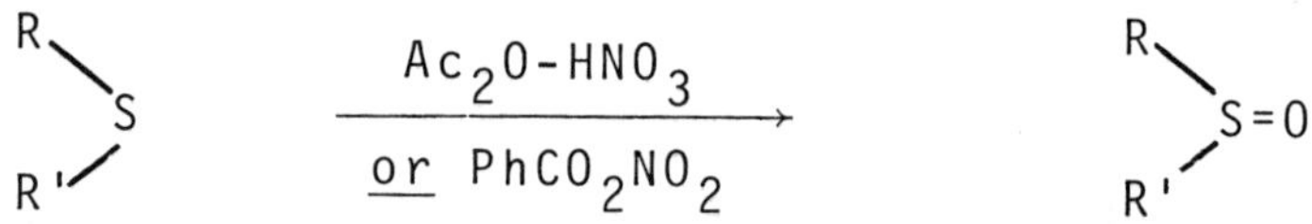

R,R' = 1° alkyl, aryl 83-100%

II.E-4 C. R. Harrison and P. Hodge, <u>J. Chem. Soc. Perkin I</u>, 2252 (1976).

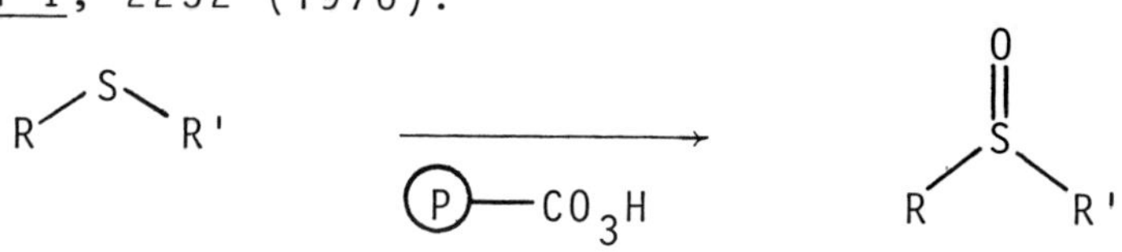

Wide variety of substrates including penicillins, etc. ~70-100%

II.E-5 O. G. Lowe, <u>J. Org. Chem</u>., <u>41</u>, 2061 (1976).

$$\text{RSH or RSSR} \xrightarrow[H_2O,\ Br_2]{Me_2SO} RSO_3H$$

II.F.1-1 M. Parrilli *et al.*, *Tetrahedron Lett.*, 207 (1976).

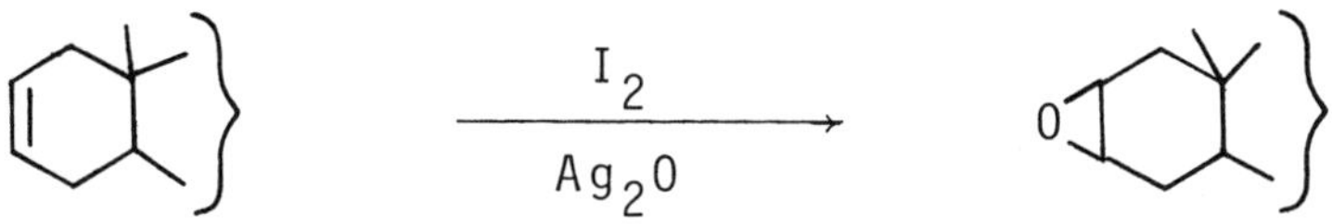

70-86%

Mixture of α and β.

II.F.1-2 N. Shimiza and P. D. Bartlett, *J. Am. Chem. Soc.*, **98**, 4193 (1976).

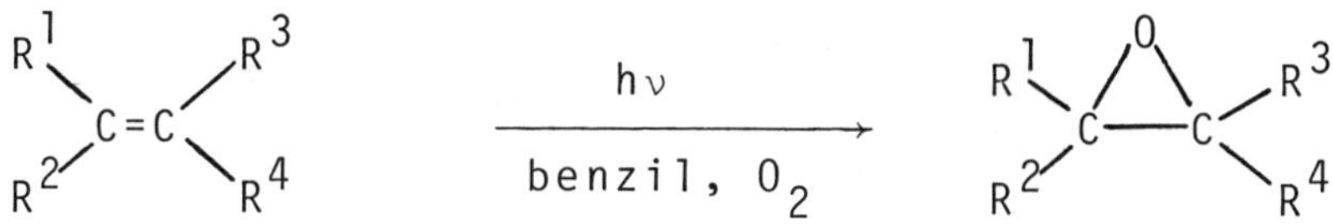

R's = alkyl, aryl, H

5-98%

trans product preferred

II.F.1-3 C. R. Harrison and P. Hodge, *J. Chem. Soc. Perkin I*, 605 (1976).

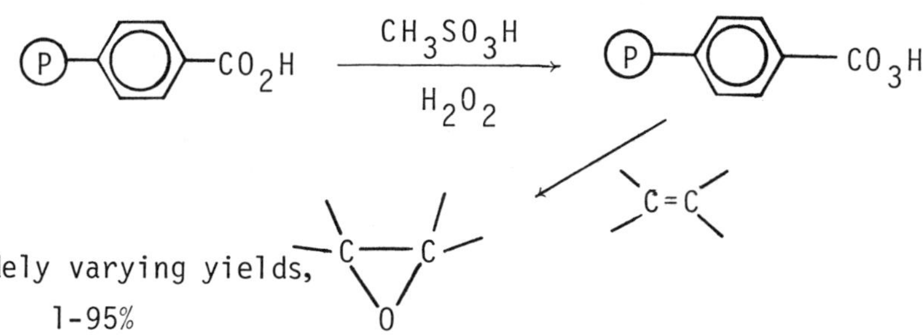

widely varying yields, 1-95%

II.F.2-1 V. G. Atabekyan, M. V. Mukhina, and A. S. Sopova, J. Org. Chem. (USSR), 12, 1234 (1976).

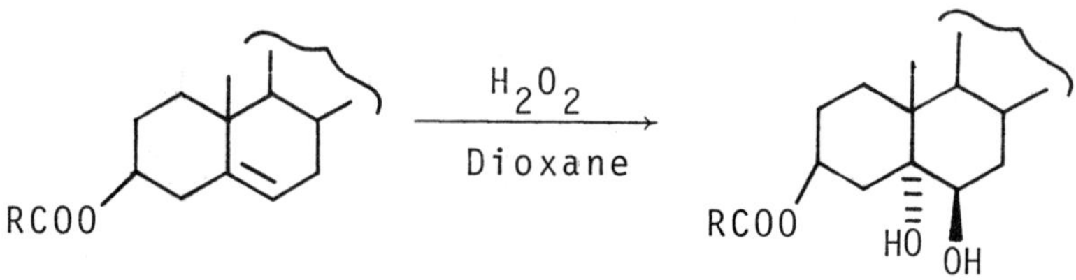

40-75%

II.F.2-2 E. Glotter and A. Schwartz, J. Chem. Soc. Perkin I, 1660 (1976).

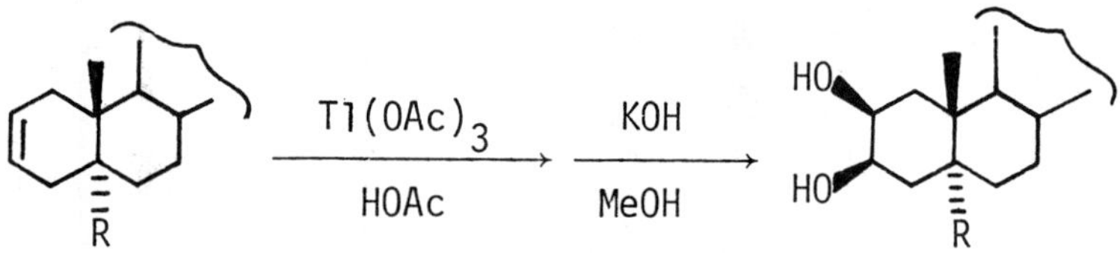

R = H, OH

fair yields

II.F.2-3 V. Van. Rheenen, R. C. Kelly, and D. Y. Cha, Tetrahedron Lett., 1973 (1976).

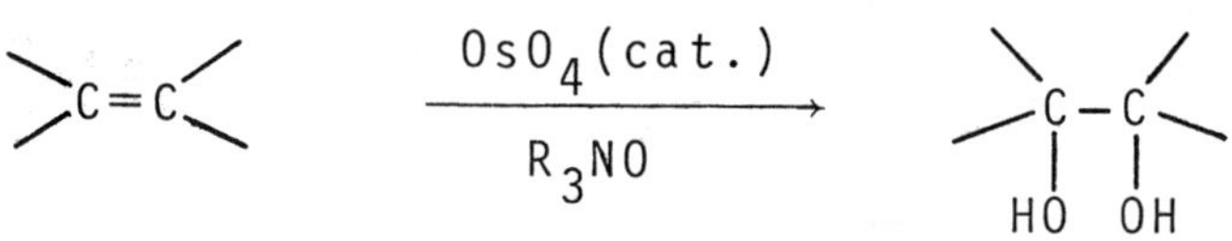

Use of tertiary amine oxides as the oxidants in catalytic oxidations of olefins to cis-1, 2-glycols. Yields are comparable to those obtained using a full mole of O_5O_4.

II.F.3-1 V. F. Mel'nikov, V. I. Koshutin, and V. A. Smirnov, J. Gen. Chem. (USSR), 45, 2072 (1975).

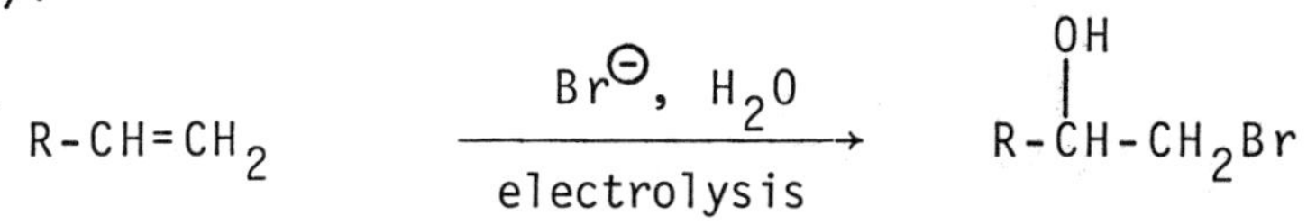

R = 1° alkyl

~60%
few details

II.F.3-2 A. Zwierzak and K. Osowska, Angew. Chem. Int. Ed., 15, 302 (1976).

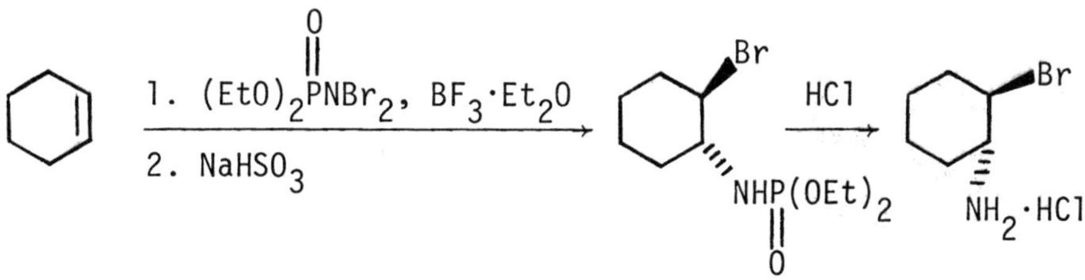

no yield 72%

II.F.3-3 M. Adinolfi et al., Tetrahedron Lett., 3661 (1976).

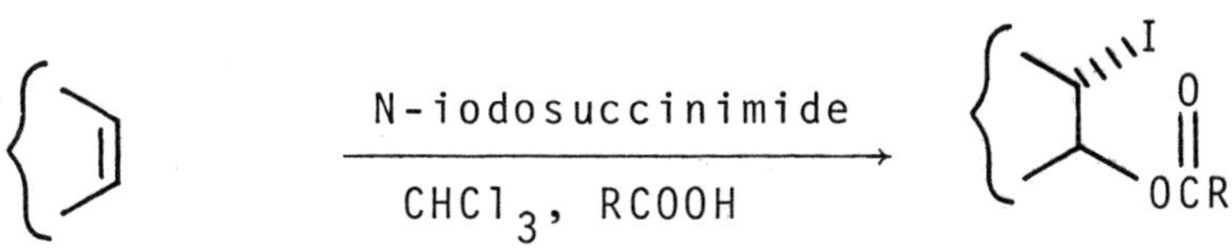

R = CH_3, Ar, CF_3, H

70-95%

II.F.3-4 K. B. Sharpless and S. P. Singer, J. Org. Chem., 41, 2504 (1976).

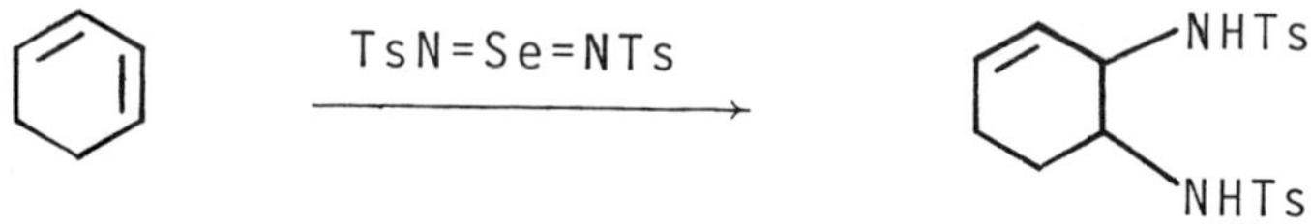

37%

II.F.3-5 K. B. Sharpless, A. O. Chong, and K. Oshima, J. Org. Chem., 41, 177 (1976).

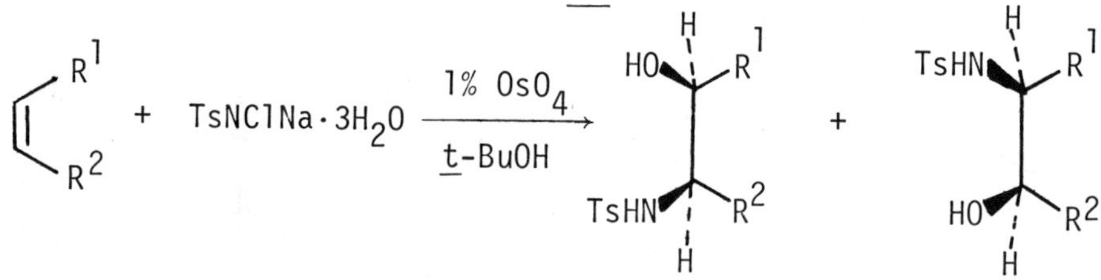

R^1, R^2 = alkyl, aryl, cyclic Total yields 34-87%

II.F.3-6 J. Kalvoda and H. Kaufmann, J. Chem. Soc., Chem. Commun., 209, 210 (1976).

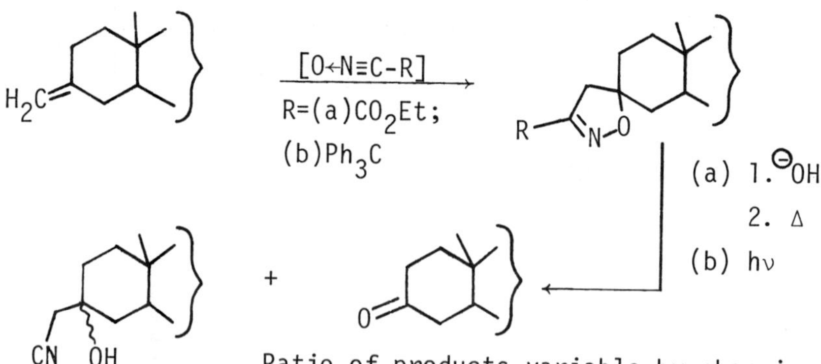

Ratio of products variable by changing decarboxylation conditions.

II.F.3-7 G. M. Rubottom, J. M. Gruber, and G. M. Mong, J. Org. Chem., 41, 1673 (1976).

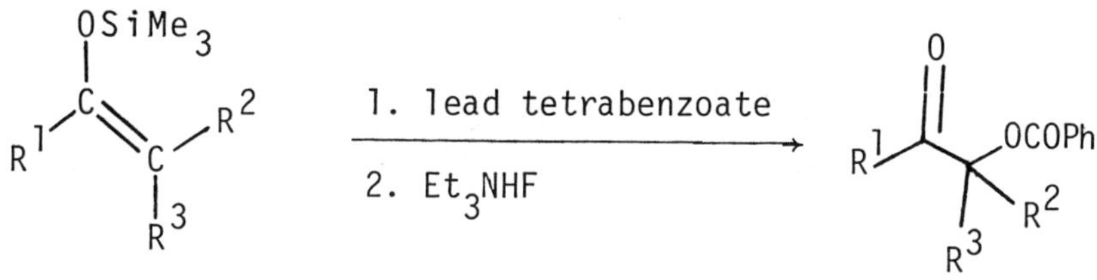

R's = H, alkyl, aryl

77-97% (isolated)

II.G-1 P. Jacob III et al., J. Org. Chem., 41, 3627 (1976).

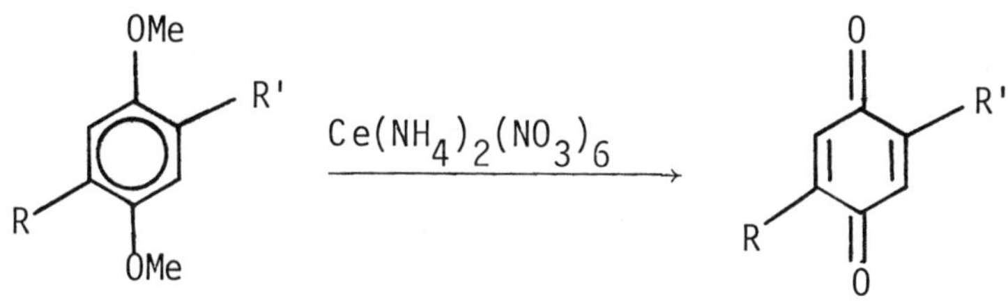

R = H, Me, OMe

~60-90%

R' = H, 1° alkyl, vinyl, CH_2OH

II.G-2 A. McKillop, D. H. Perry, and M. Edwards, J. Org. Chem., 41, 282 (1976).

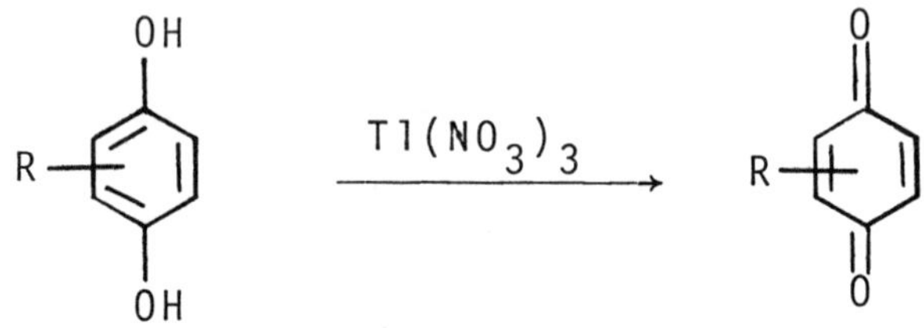

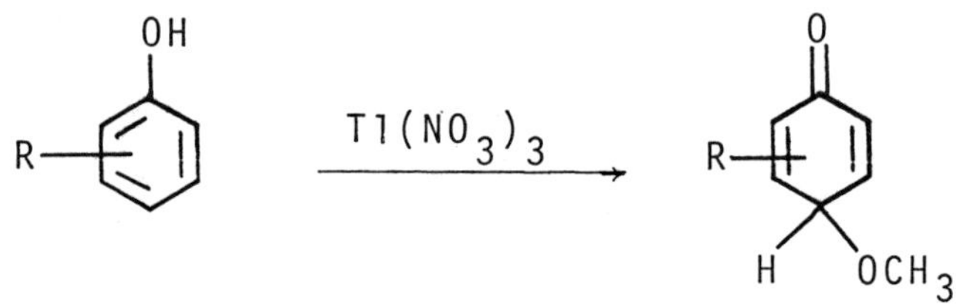

yields ~70-98%

II.H-1 H.-J. Liu, Can. J. Chem., 54, 3113 (1976).

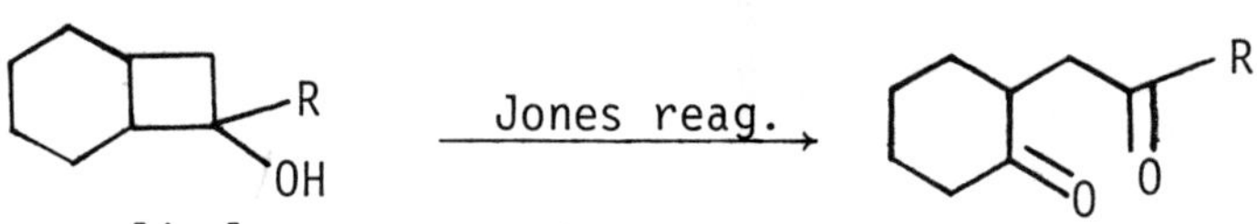

R = alkyl

50-85%

II.H-2 B. Danieli and G. Palmisano, Chem. and Ind., 565 (1976).

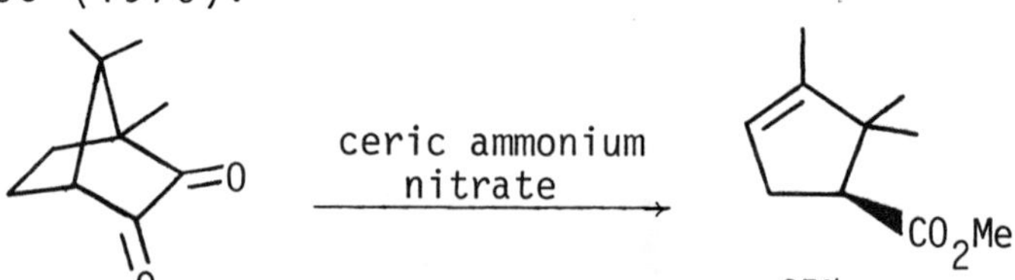

~65%

II.H-3/II.I-1   I. Saito *et al.*, *Synthesis*, 255 (1976).

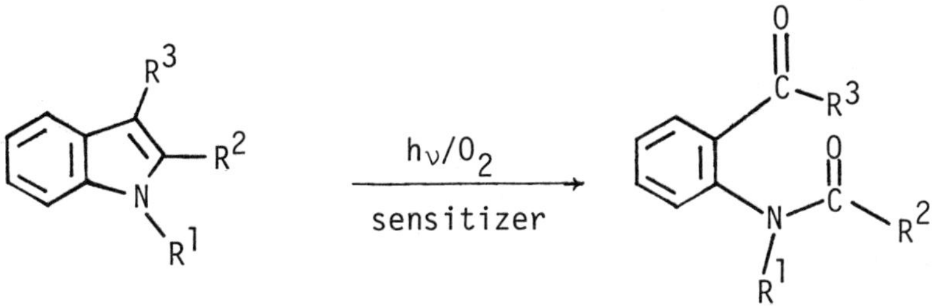

R^1 = Me, H
R^2 = H, Me, Ph
R^3 = Me, -COR, $-CH_2COOH$

generally 70-90%

II.H-4/II.I-2   R. W. Binkley, *Synth. Commun.*, **6**, 281 (1976).

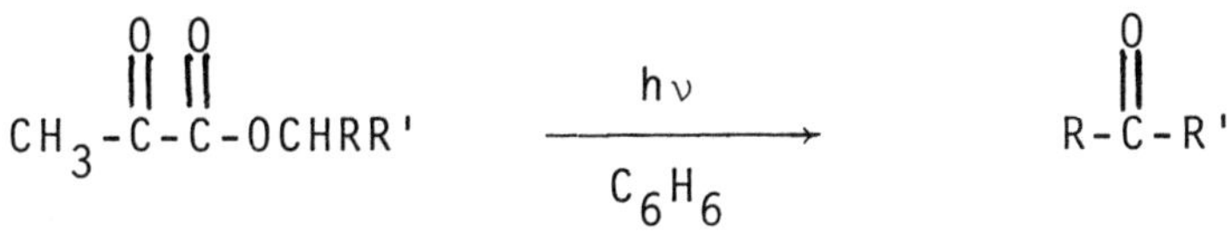

R,R' = H, alkyl, aryl   76-100%

II.I-3   N. Shimizer and P. D. Bartlett, *J. Am. Chem. Soc.*, **98**, 4193 (1976).

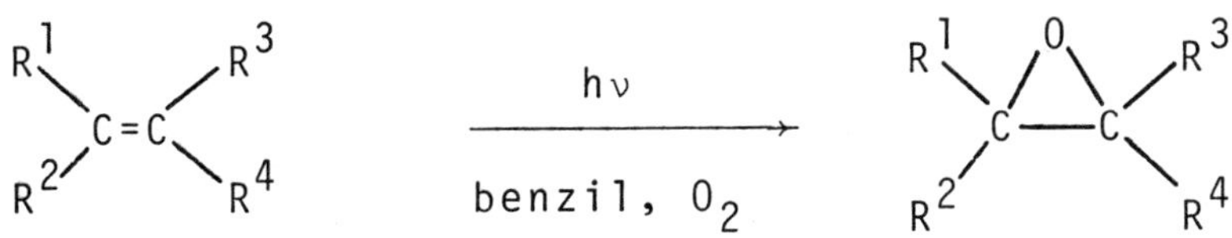

R's = alkyl, aryl, H

5-98%
trans product preferred

II.J-1 M. R. Czarny, Synth. Commun., 6, 285 (1976).

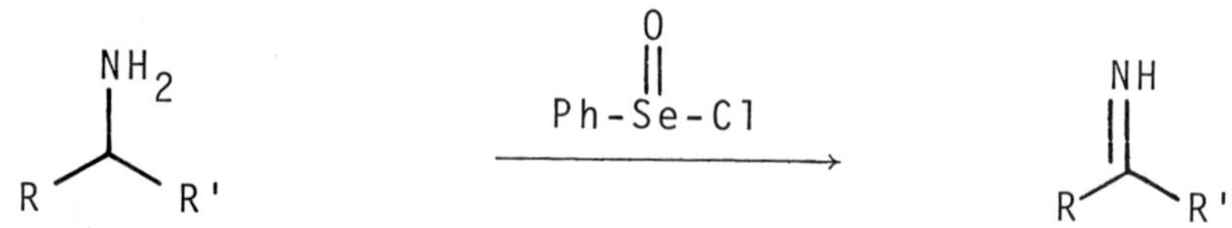

R,R' = Ph; also works for 2-adamantylamine

97%

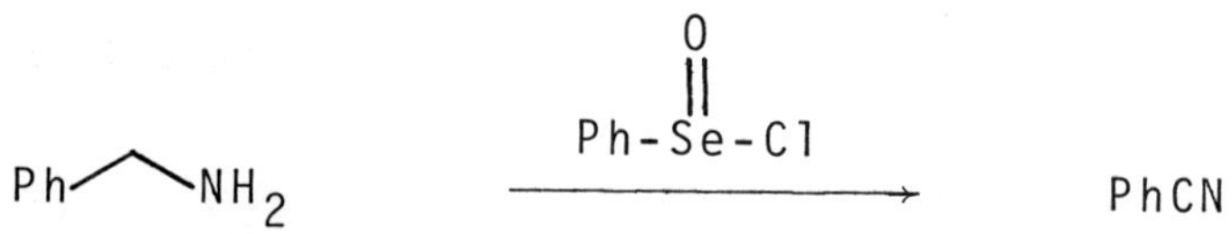

85%

II.K-1 A. H. Haines, Chem. and Ind., 883 (1976).

Review: "Some Recent Developments in Oxidation in Organic Chemistry"

II.K-2 A. J. Fatiadi, Synthesis, 65, 133 (1976).

Review: "Active Manganese Dioxide Oxidation in Organic Chemistry"

II.K-3 R. N. Butler, Chem. and Ind., 499 (1976).

Review: "Lead Tetraacetate Acetoxylation of the Heteroallylic System (>C=X-YH)"

III.A-1 G. Brieger and T.-H. Fu, *J. Chem. Soc., Chem. Commun.*, 757 (1976).

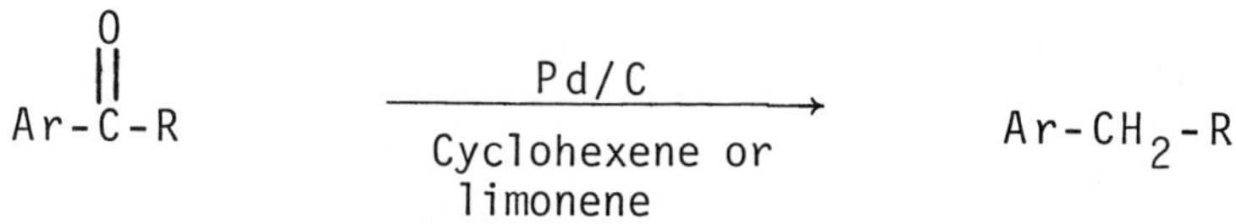

R = H, Me

70-90%

Reaction fails when Ar is phenolic.

III.A-2 Y. Kikugawa, *Chem. Pharm. Bull.*, **24**, 1059 (1976).

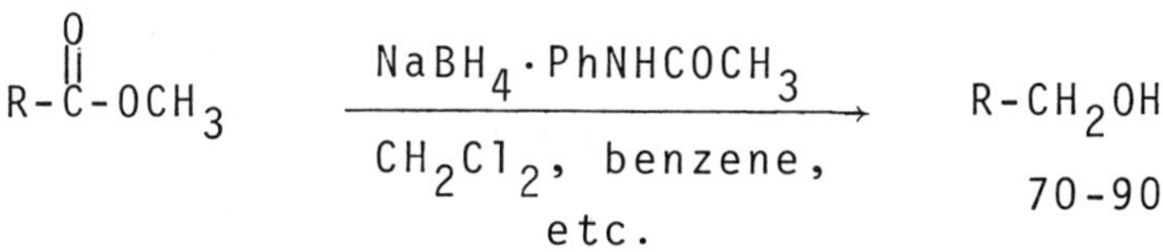

70-90%

R = alkyl, aryl, cyclic

Also reduces ketones, aldehydes.

III.A-3 T. Fujisawa, K. Sugimoto, and H. Ohta, *J. Org. Chem.*, **41**, 1667 (1976).

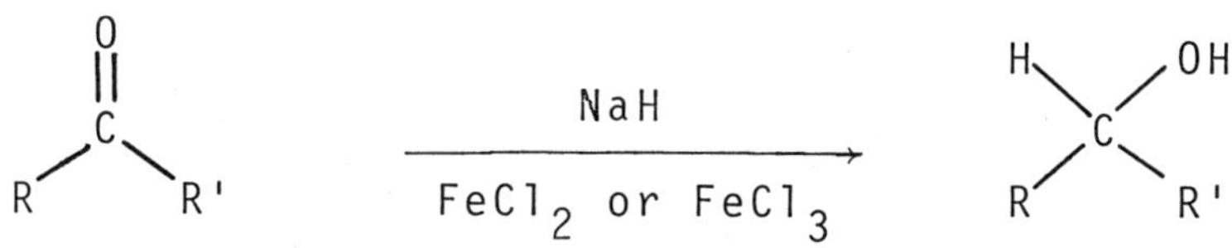

R,R' = H, Ph, alkyl

75-82% (VPC)

Reaction fails with benzophenone.

III.A-4 R. Grandi *et al.*, *J. Org. Chem.*, **41**, 1755 (1976).

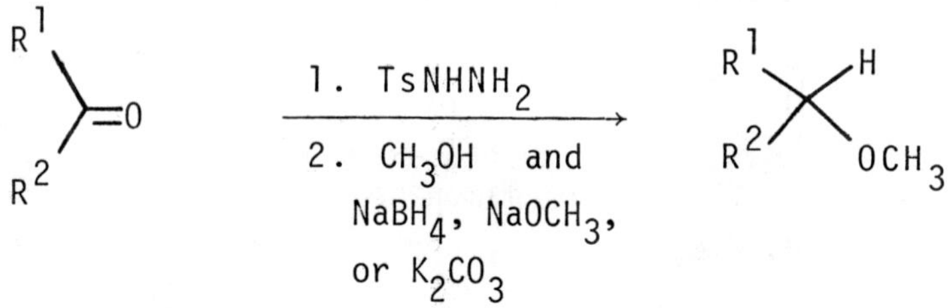

R^1, R^2 = H, alkyl, aryl

Widely varying yields.

III.A-5 M. Sevrin, D. Van Ende and A. Krief, *Tetrahedron Lett.*, 2643 (1976).

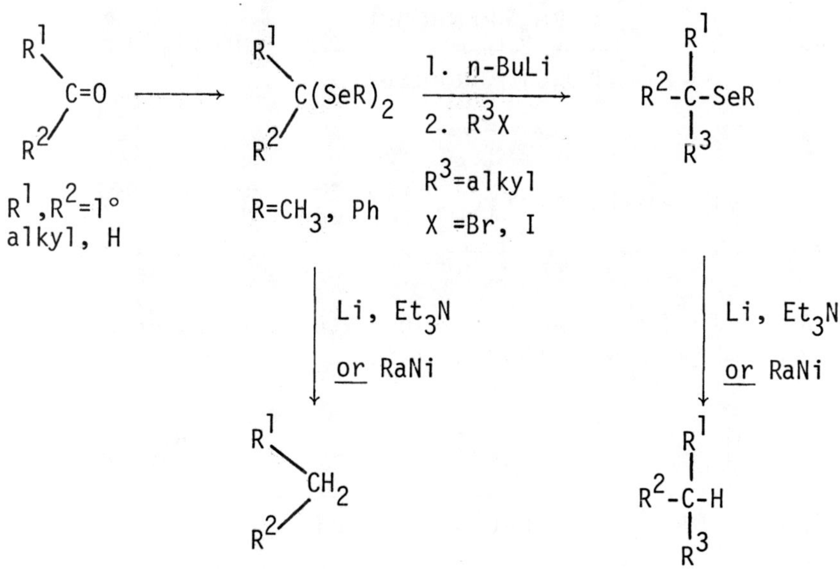

85-92%

III.A-6 T. Fujisawa, K. Sugimoto, and H. Ohta, Chem. Lett., 581 (1976).

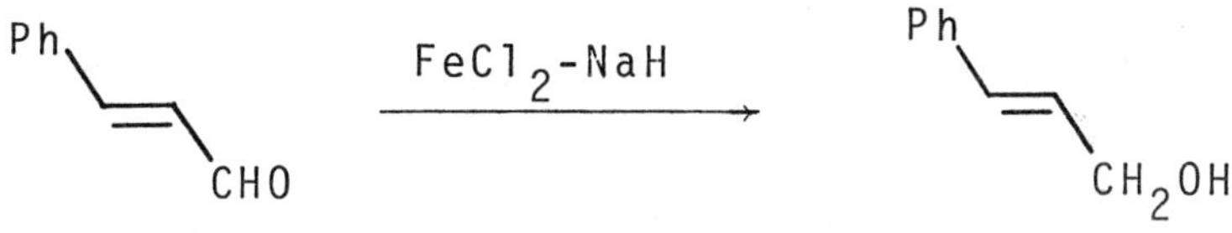

42%

III.A-7 P. S. Gradeff and G. Formica, Tetrahedron Lett., 4681 (1976).

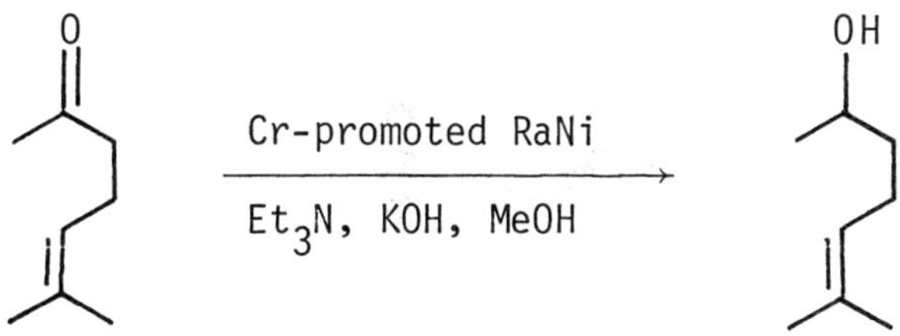

88%

III.A-8 S. Krishnamurthy and H. C. Brown, J. Am. Chem. Soc., 98, 3383 (1976).

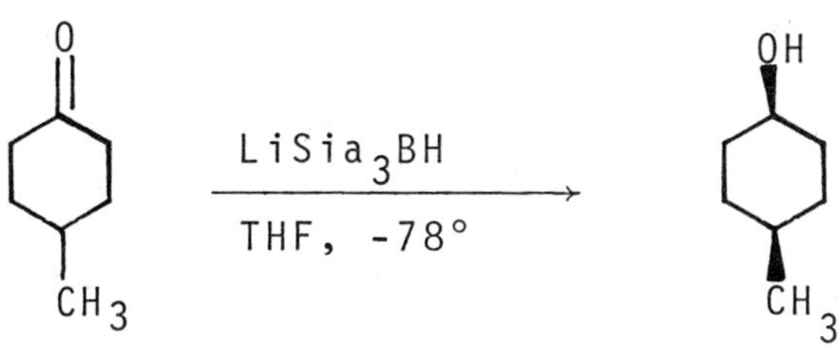

99%

Gives the less stable isomer, generally in about 99% yield. Nine examples cited.

III.A-9 Y. Yamamoto *et al.*, *J. Am. Chem. Soc.*, **98**, 1965 (1976).

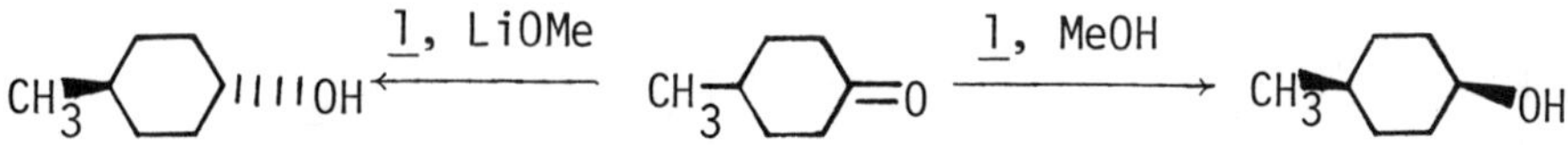

90% 84%

Several additional examples cited.

1 = 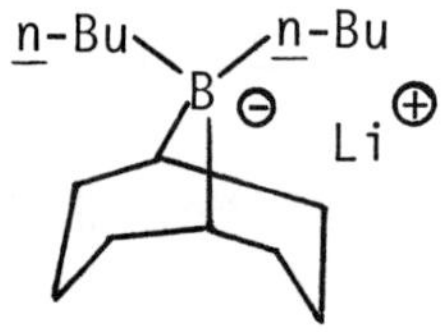

III.A-10 D. Nasipuri *et al.*, *J. Chem. Soc. Perkin I*, 321 (1976).

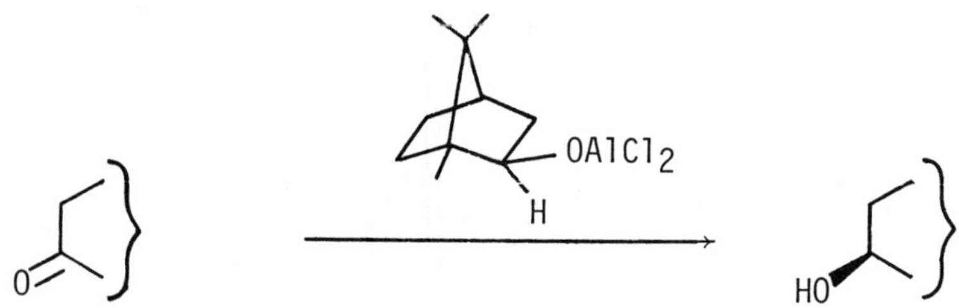

various 3-oxo triterpenoids

~70% chemical yield

75-100% axial alcohol

III.A-11 I. Ojima *et al.*, *J. Organometal. Chem.*, **122**, 83 (1976).

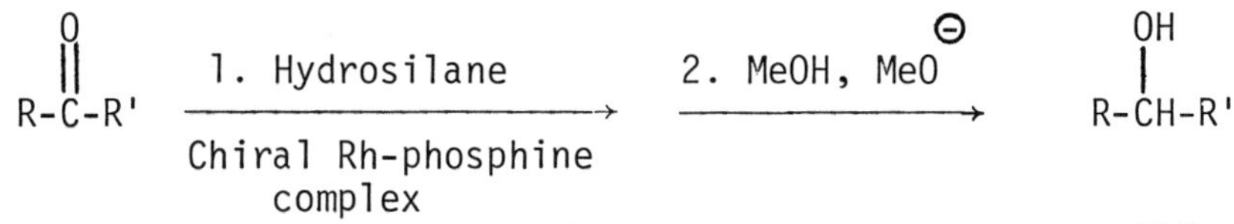

R = alkyl

R' = alkyl, aryl

>90%

4-58% ee

III.A-12 T. Hayashi, T. Mise, and M. Kumada, *Tetrahedron Lett.*, 4351 (1976).

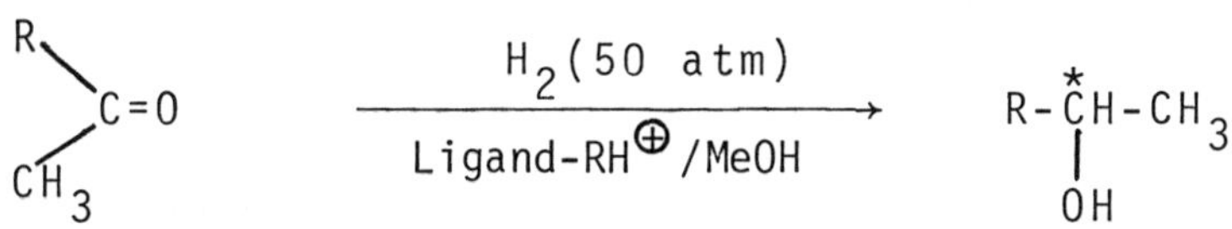

R = Ph, *t*-Bu, COOH

generally >90%
up to 83% ee

Ligand =

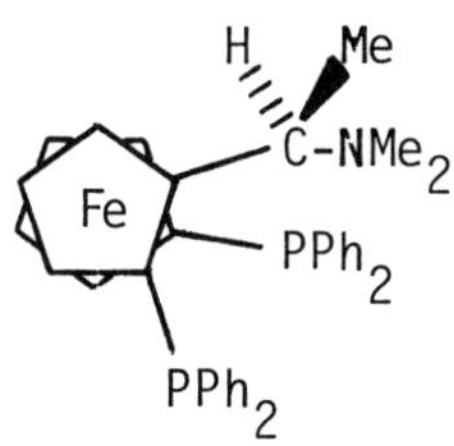

III.A-13 E. C. Ashby and J. R. Boone, *J. Org. Chem.*, **41**, 2890 (1976).

A study of the stereochemistry of reduction of ketones by a variety of simple and complex metal hydrides of the main group elements.

III.A-14 S. Yamada, M. Kitamoto, and S. Terashima, Tetrahedron Lett., 3165 (1976).

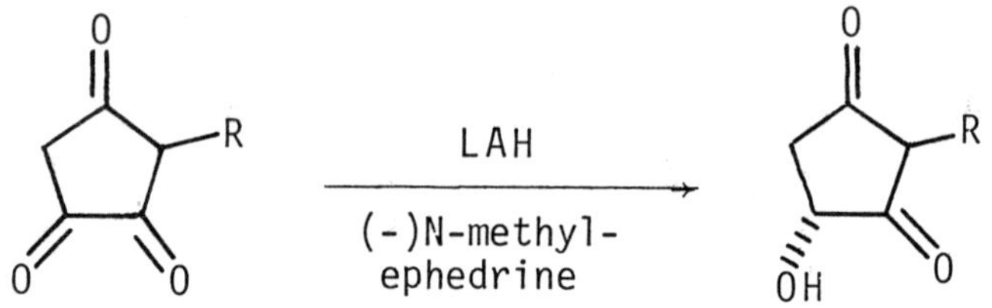

R = alkyl, ester

yields ~40%, 55% ee

III.A-15 B. S. Deol, D. D. Ridley and G. W. Simpson, Aust. J. Chem., 29, 2459 (1976).

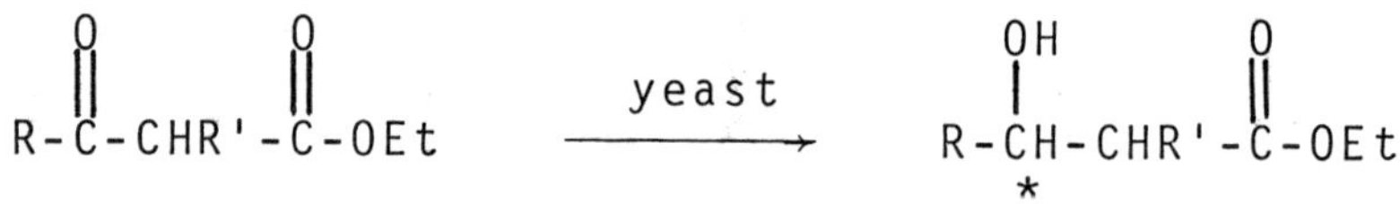

R,R' = alkyl, aryl, cyclic

~70-80%
high optical purity

Also works with

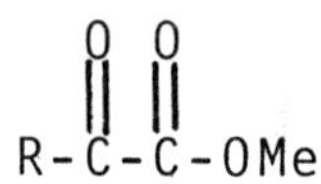

and

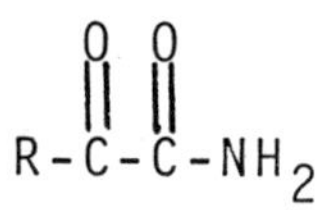

III.A-16 P. L. Russ and E. A. Caress, J. Org. Chem., 41, 149 (1976).

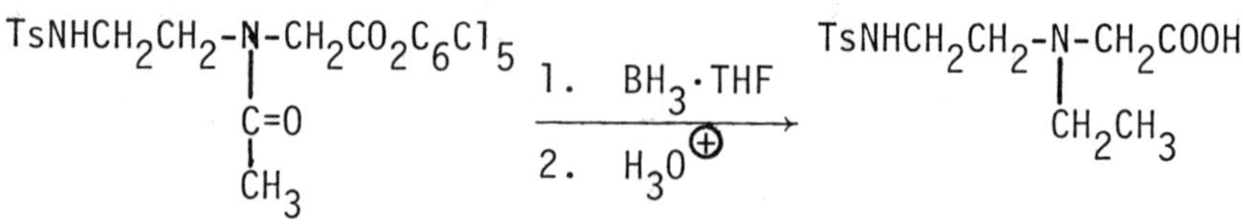

III.A-17 M. F. Grundon, D. G. McCleery, and J. W. Wilson, Tetrahedron Lett., 295 (1976).

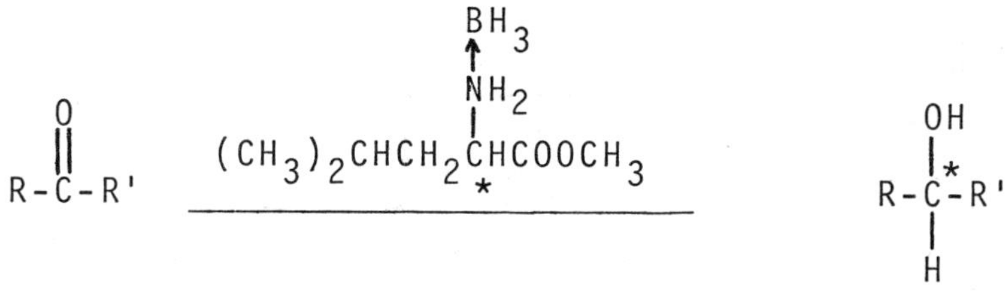

15-22% ee

III.A-18 T.-L. Ho and G. A. Olah, Synthesis, 815 (1976).

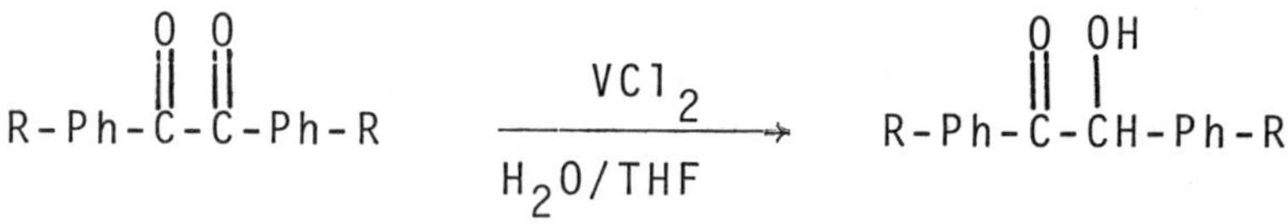

R = H, Me, Ph, OMe

82-92%

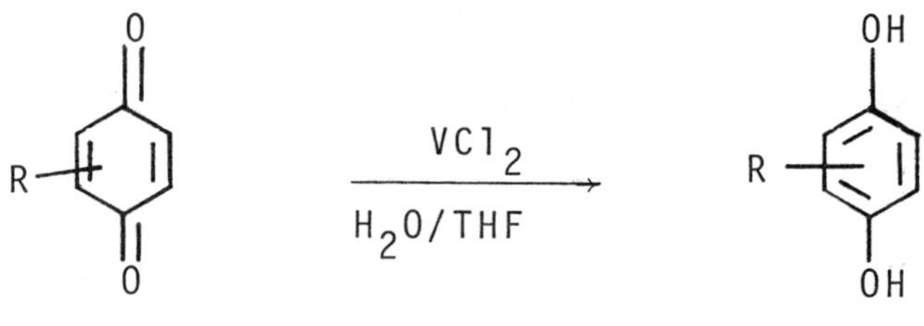

R = Me, Cl, benzo-quinone, etc.

~90%

III.A-19 K. Nishiyama *et al.*, *J. Chem. Soc., Chem. Commun.*, 101 (1976).

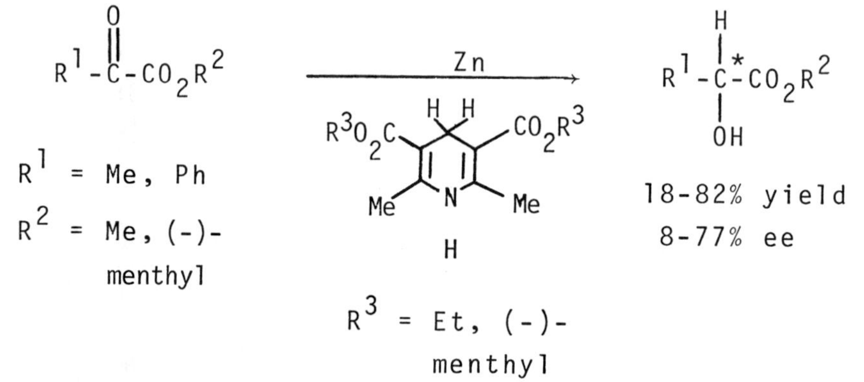

III.A-20 R. Kanazawa and T. Tokoroyama, *Synthesis*, 526 (1976).

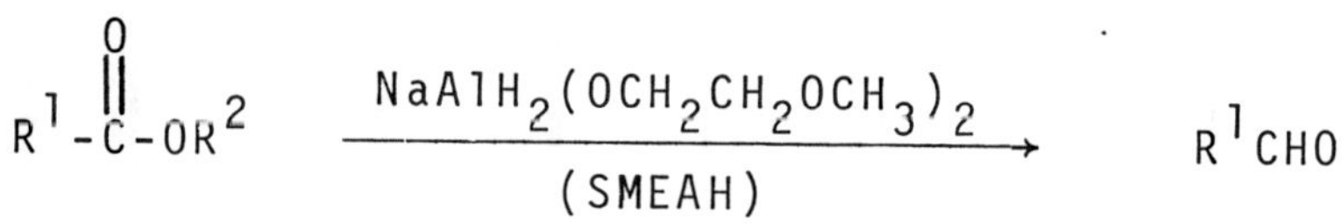

R^1 = alkyl, aryl, cinnamyl

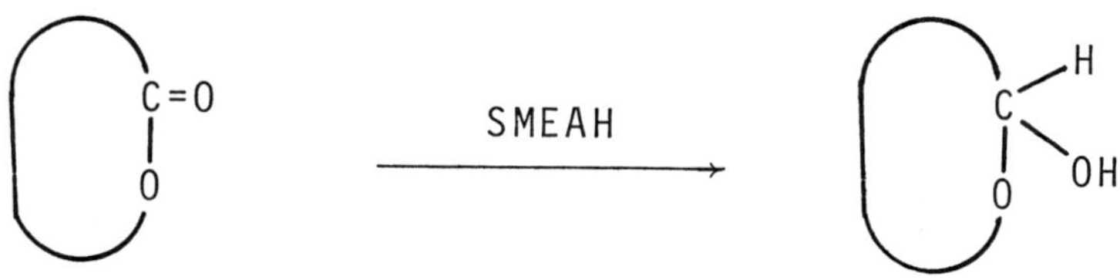

5 or 6-membered ring

60-90%

III.A-21 N. Chatterjie, J. G. Umans, and C. E. Inturrisi, J. Org. Chem., 41, 3624 (1976). See also J. Med. Chem., 18, 490 (1975).

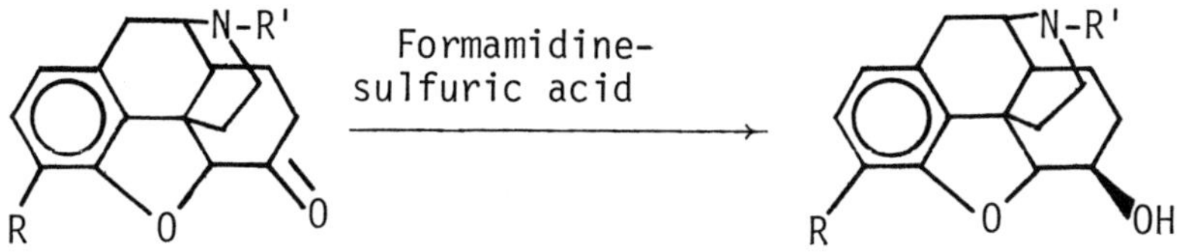

R = OH, OCH_3 63%

R' = 1° alkyl

III.A-22 Y. Maki et al., Chem. and Ind., 322 (1976).

$$R\text{-}\overset{O}{\overset{\|}{C}}\text{-}NH_2 \xrightarrow[HSCH_2CH_2SH]{NaBH_4,\ THF} RCH_2NH_2$$

R = pentyl, phenyl ~90%

III.A-23 N. Umino, T. Iwakuma, and N. Itoh, Tetrahedron Lett., 763 (1976).

$$R\text{-}\overset{O}{\overset{\|}{C}}\text{-}NHR' \xrightarrow[R''\text{-}COOH]{NaBH_4} R\text{-}CH_2NHR'$$

R = alkyl, aryl

R' - H, alkyl

yields generally 60-90%

R" = CH_3, CF_3, Ph

III.A-24 A. Rahman *et al.*, *Tetrahedron Lett.*, 219 (1976).

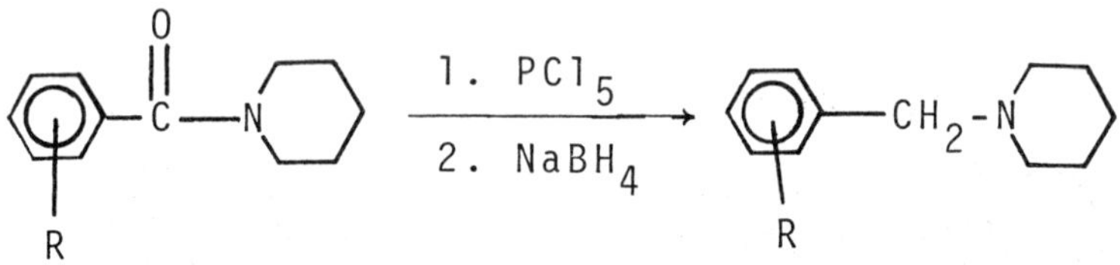

75-90%

III.B-1 N. Umino, T. Iwakuma, and N. Itoh, *Tetrahedron Lett.*, 2875 (1976).

$$R\text{-}CN \xrightarrow[\text{THF}]{NaBH_3(OCOCF_3)} RCH_2NH_2$$

R = alkyl, aryl 70-89%

III.C-1 T. Fujisawa and H. Ohta, *Bull. Chem. Soc. Japan*, **49**, 2341 (1976).

$$RSSR + PhI \xrightarrow{h\nu} RSPh$$

R = alkyl, aryl 20-100%

III.C-2 J. Drabowicz and M. Mikolajczyk, *Synthesis*, 527 (1976).

$$R^1\text{-}\overset{\overset{\displaystyle O}{\|}}{S}\text{-}R^2 \xrightarrow{LiAlH_4,\ TiCl_4,\ \text{ether}} R^1\text{-}S\text{-}R^2$$

R's = aryl, 1°, 2°, 3° alkyl

III.C-3 G. A. Olah, G.K. SuryaPrakash and T.-L. Ho, <u>Synthesis</u>, 810 (1976).

$$R_2S{=}O \xrightarrow[VCl_2,\ H_2O/THF]{MoOCl_3,\ Zn,\ THF} R_2S$$

R = Ph, Bz, 74-91%

<u>n</u>-Bu, $-(CH_2)_4-$

III.C-4 D. W. Chasar and T. M. Pratt, <u>Synthesis</u>, 262 (1976).

R,R' = alkyl, aryl 72-100%

III.D-1 J. O. Morley, <u>Synthesis</u>, 528 (1976).

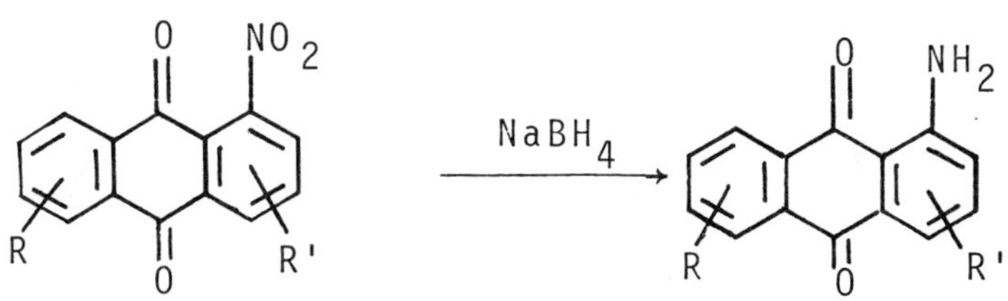

R,R' = alkyl, COOH, SO_3Na, NH_2 ~80-90%

III.E.1-1 T. Fujisawa, K. Sugimoto, and H. Ohta, <u>Chem. Lett.</u>, 581 (1976).

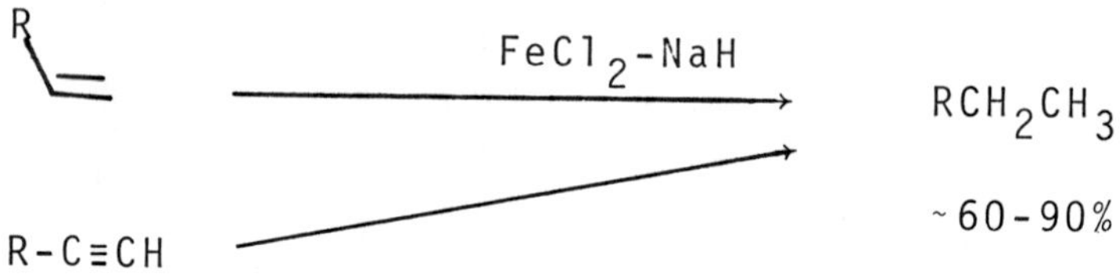

R = alkyl, Ph

III.E.1-2 N. J. Cusack *et al.*, *Tetrahedron*, **32**, 2157 (1976).

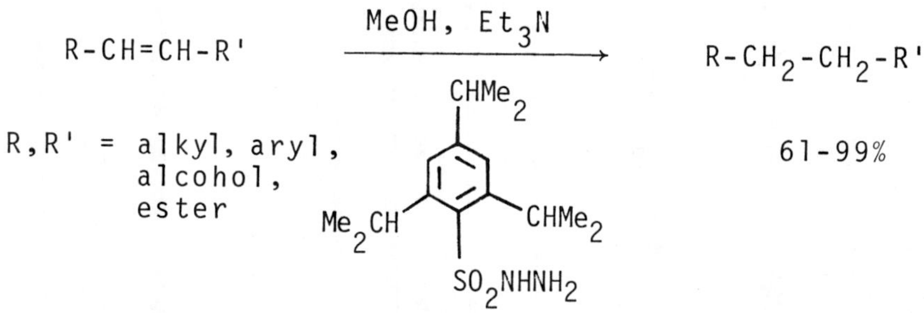

A convenient precursor for diimide reductions.

III.E.1-3 Y. Nakamura and H. Hirai, <u>Chem. Lett.</u>, 165 (1976).

Selective catalyst for reduction of conjugated diolefins to monoolefins at 30° and 1 atm. H_2.

III.E.1-4 E. C. Ashby, J. J. Lin, and R. Kovar, J. Org. Chem., 41, 1939 (1976).

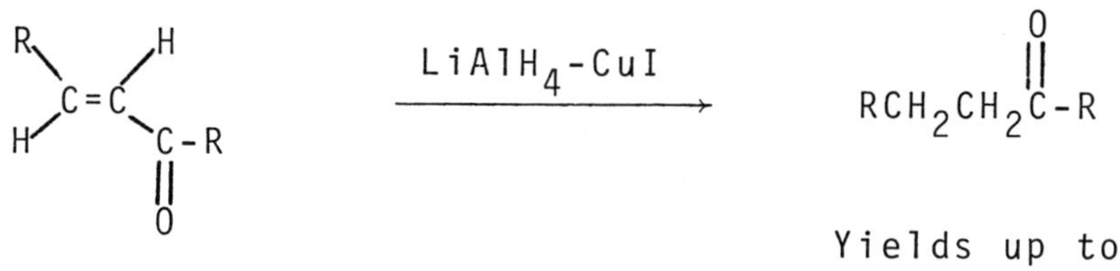

Yields up to 101% (VPC)

R = CH_3, t-Bu

III.E.1-5 G. Descotes and D. Sinou, Tetrahedron Lett., 4083 (1976).

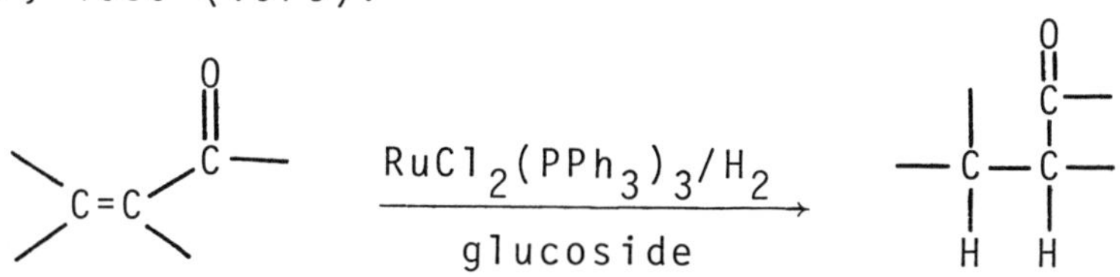

Widely varying yields, up to 34% ee.

III.E.1-6 G. P. Boldrini and A. Umani-Ronchi, Synthesis, 596 (1976).

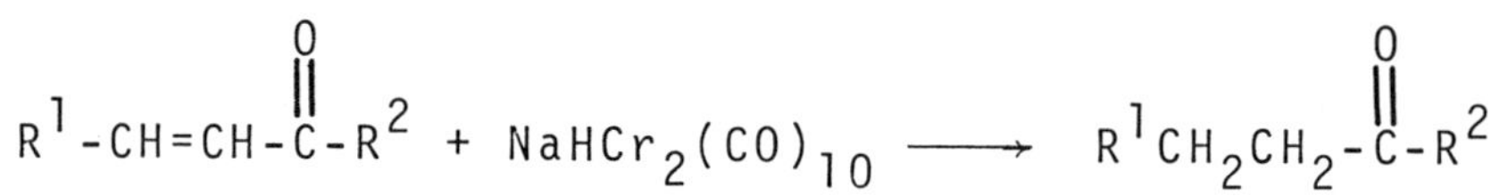

R^1 = alkyl, aryl

R^2 = H, alkyl, aryl

40-80%

III.E.1-7 J. M. Fortunato and B. Ganem, J. Org. Chem., 41, 2194 (1976).

A detailed survey of the use of lithium and potassium tri-sec-butylborohydrides (L- and K-Selectrides) for the conjugate reduction and reductive alkylation of unsaturated ketones and esters.

III.E.1-8 F. Toda and M. Kanno, Bull. Chem. Soc. Japan, 49, 2643 (1976).

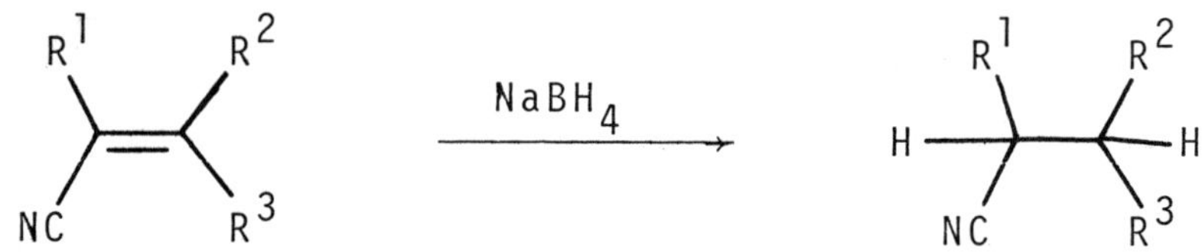

R^1 = aryl, n-alkyl 56-100%
R^2 = H, Ph, CN, NO_2, OMe
R^3 = H, Ph

III.E.1-9 R. O. Hutchins, et al., J. Org. Chem., 41, 3328 (1976).

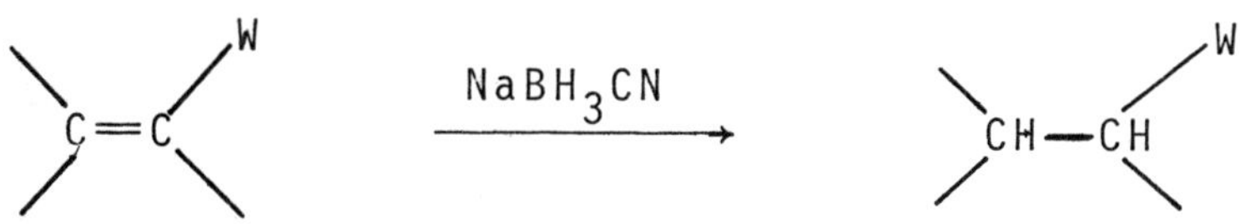

W = COOR, CN, NO_2 Yields generally 60-90%

III.E.1-10 F. Toda and K. Iida, Chem. Lett., 695 (1976).

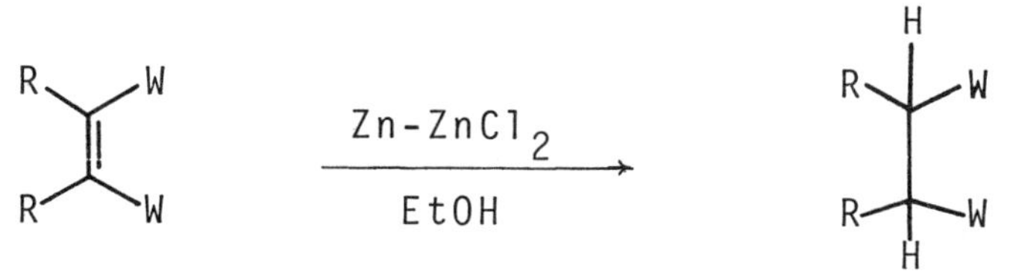

R = alkyl, aryl, vinyl ~70-90%

W = COR, COOR, CN, etc.

III.E.1-11 K. Junghaus, Chem. Ber., 109, 395 (1976).

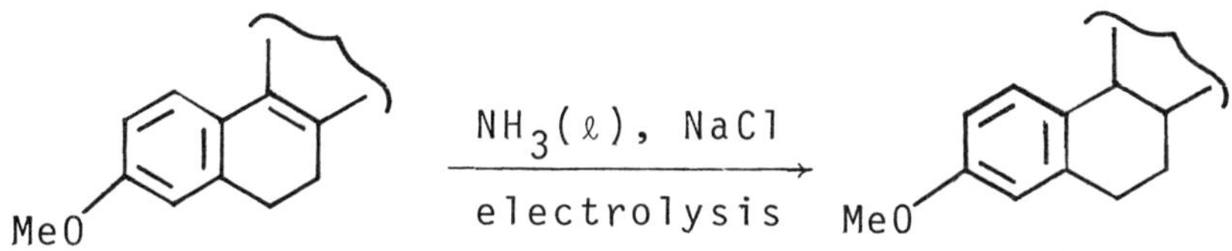

Widely varying yields.

III.E.1-12 T. Hayashi et al., Tetrahedron Lett., 1133 (1976).

Use of I as a chiral hydrogenation catalyst:

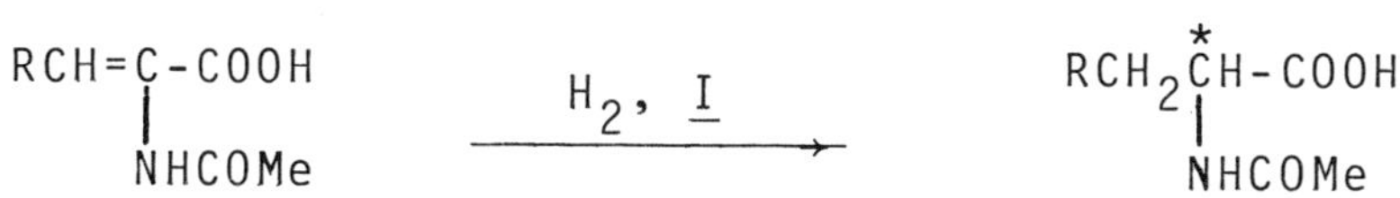

R = aryl

I = 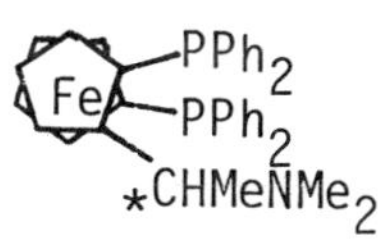

predominantly (S)

86-94% chemical yield

52-89% ee

III.E.1-13 K. M. Minachev, Y. S. Khodakov, and V. S. Nakhshunov, Russ. Chem. Rev., **45**, 142 (1976).

Review: "Hydrogenation of Alkenes on Oxide Catalysts"

III.E.1-14/III.E.2-1 A. Sisak and F. Ungvary, Chem. Ber., **109**, 531 (1976).

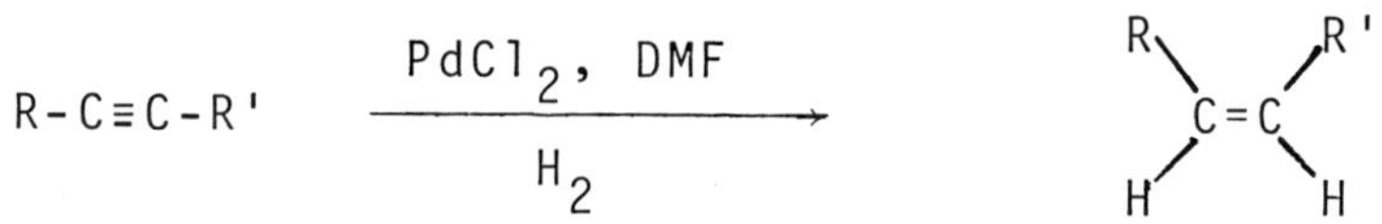

Nearly quantitative for 1-heptyne, 2-pentyne

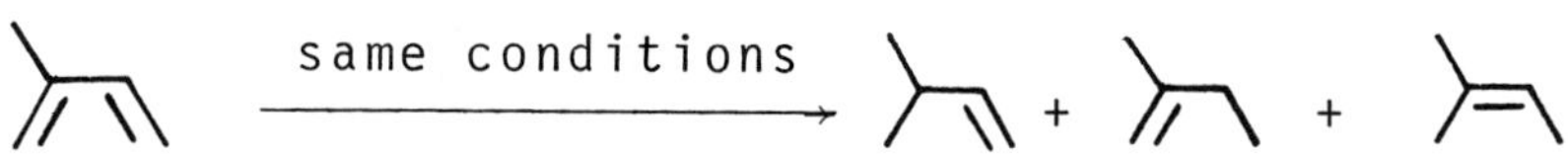

29% 25% 46%

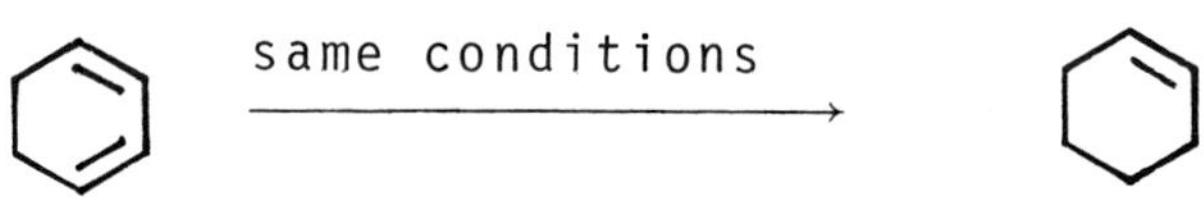

100%

III.E.1-15/III.E.2-2 P. W. Chum and S. E. Wilson, Tetrahedron Lett., 15 (1976).

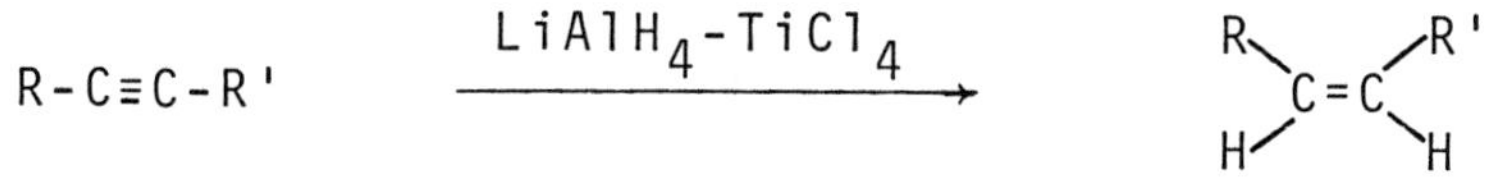

1-octyne or 4-octyne → 1-octene, 81%; *cis*-4-octene, 73%

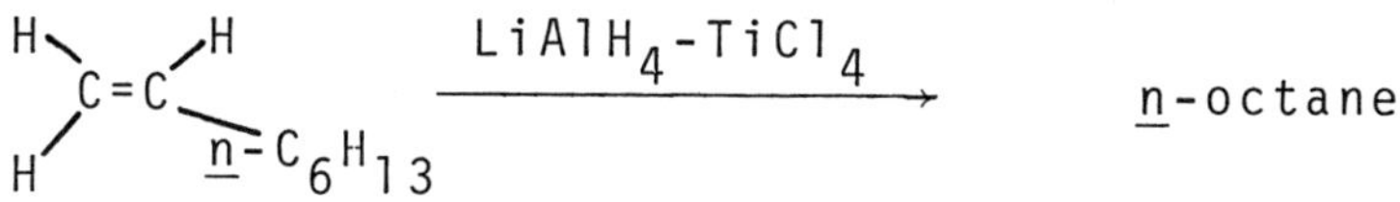

92%

III.E.2-3 J. K. Crandall and F. Collonges, J. Org. Chem., 41, 4089 (1976).

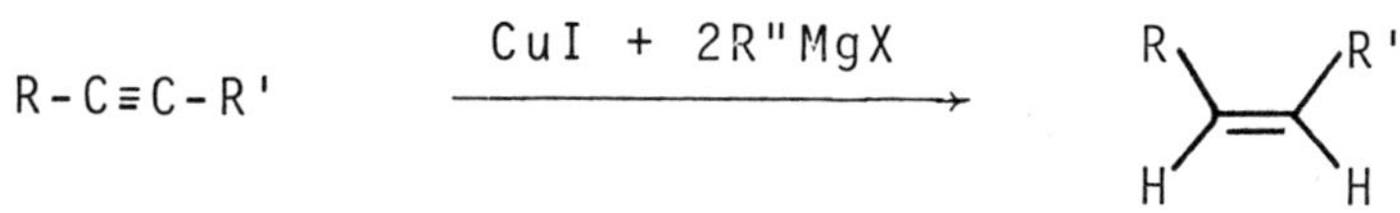

R,R' = Me, *n*-C_5H_{11}, Ph

R" = *n*-Bu, Et

65-100% for relatively unhindered alkynes

III.E.2-4 R. R. Schrock and J. A. Osborn, J. Am. Chem. Soc., 98, 2143 (1976).

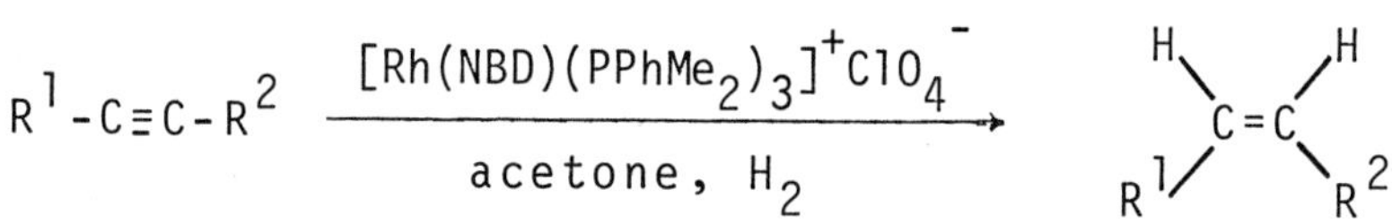

R^1,R^2 = alkyl

>95%

III.E.2-5 E. L. Muetterties, Bull. Soc. Chim. Belges, 85, 451 (1976).

Review: "Metal Clusters in Catalysis-Reduction of Triple Bonds"

III.F.1-1 F. Bohlmann, J. Staffeldt, and W. Skuballa, Chem. Ber., 109, 1586 (1976).

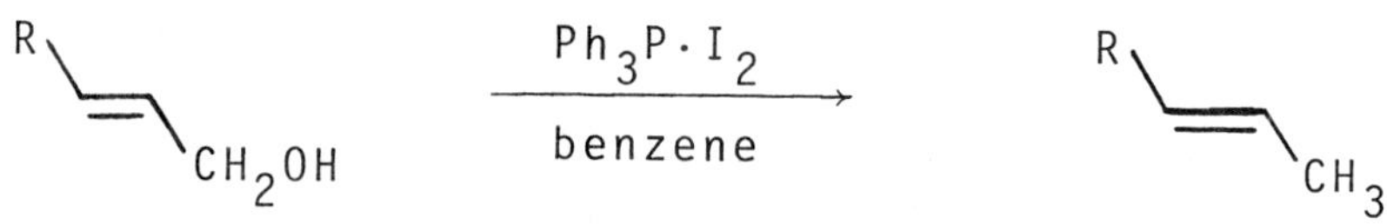

R = alkyl, aryl, acetylenic

24-60%

III.F.1-2 M. G. Adlington, M. Orfanopoulas, and J L. Fry, Tetrahedron Lett., 2955 (1976).

$$\text{R-OH} \xrightarrow[\text{Et}_3\text{SiH}]{\text{BF}_3} \text{R-H}$$

R = 2°, 3°

poor to fair yields

III.F.1-3 Y. Fujimoto and N. Ikekawa, Chem. Pharm. Bull., 24, 825 (1976).

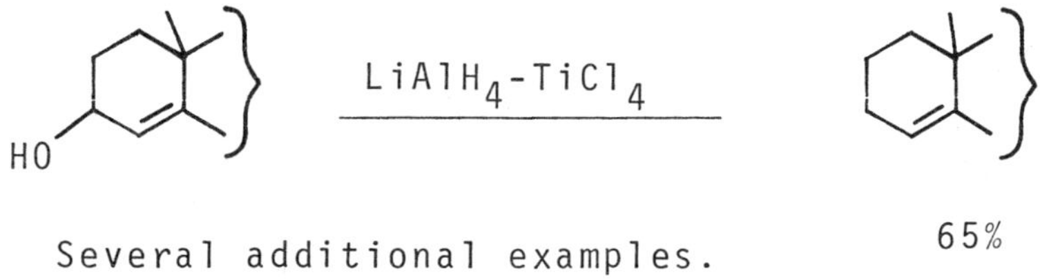

Several additional examples.

III.F.1-4 H. C. Beyerman et al., Rec. Trav. Chim. Pays-Bas, 95 43 (1976).

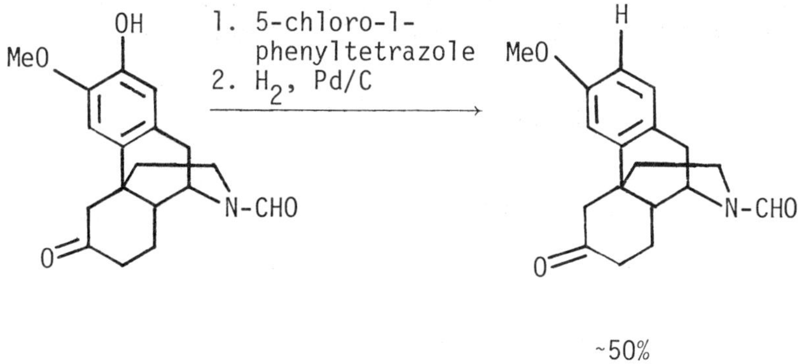

III.F.1-5 S. Krishnamurthy and H. C. Brown, J. Org. Chem., 41, 3064 (1976).

R-OTs $\xrightarrow[\text{THF}]{LiEt_3BH}$ R-H

R = 1°,2° alkyl, cyclic — 81-100%

e.g.

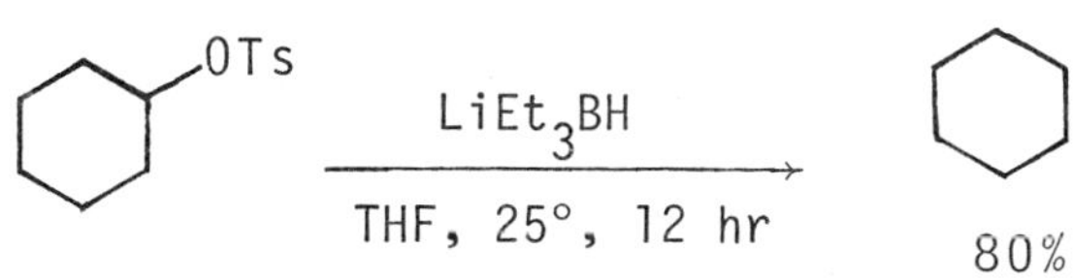

III.F.1-6 Y. Fujimoto and T. Tatsuno, Tetrahedron Lett., 3325 (1976).

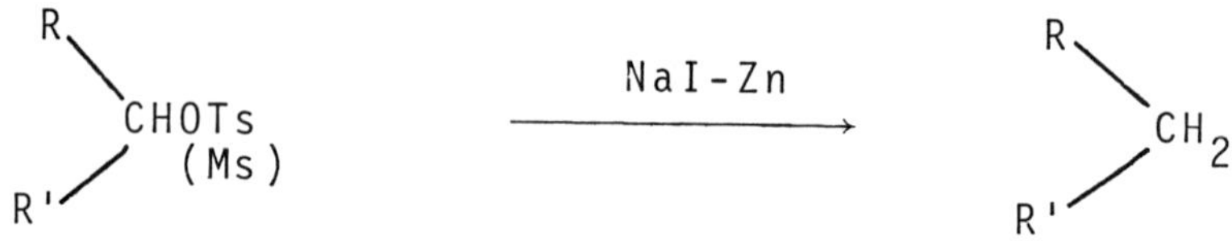

R = alkyl
R' = H, alkyl

yields generally 70-90%

III.F.2-1 M. P. Doyle, C. C. McOsker, and C. T. West, J. Org. Chem., 41, 1393 (1976).

$$RX + Et_3SiH \xrightarrow{AlCl_3} RH + Et_3SiX$$

R = 1°,2°,3° alkyl

X = Br, Cl

26-100% (VPC)

III.F.2-2 W. E. Parham, L. D. Jones and Y. A. Sayed, J. Org. Chem., 41, 1184, 1187 (1976).

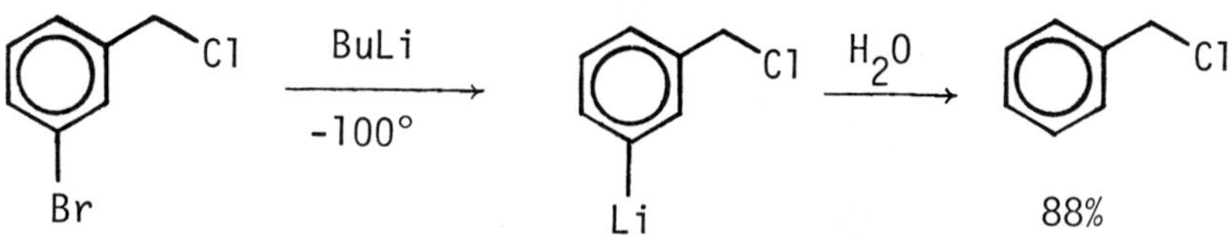

88%

Other electrophiles may also be added.

III.F.2-3 T.-L. Ho and G. Olah, Synthesis, 807 (1976).

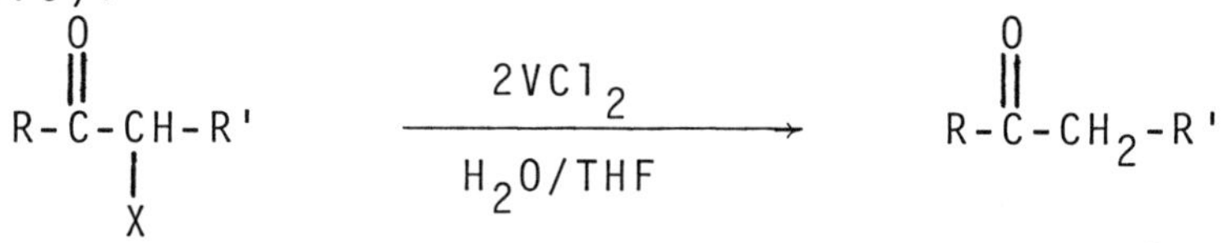

X = Br, Cl

R = alkyl, aryl

R' = H, alkyl

80-96%

III.F.4-1 A. J. Boulton, et al., Tetrahedron Lett., 2689 (1976).

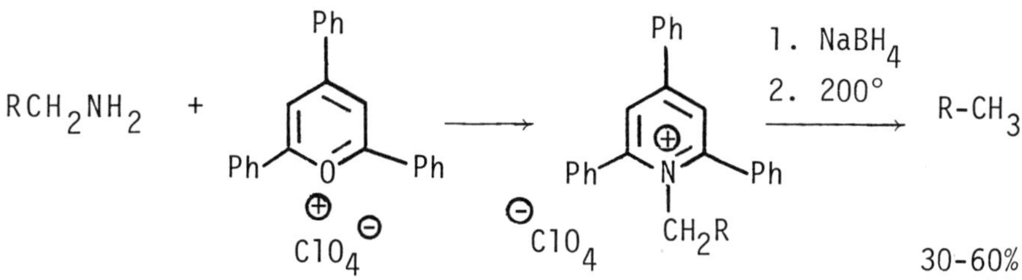

30-60% overall

R = vinyl, phenyl, furyl, pyridyl

III.F.4-2 A. Rahman and A. Basha, J. Chem. Soc., Chem. Commun., 594 (1976).

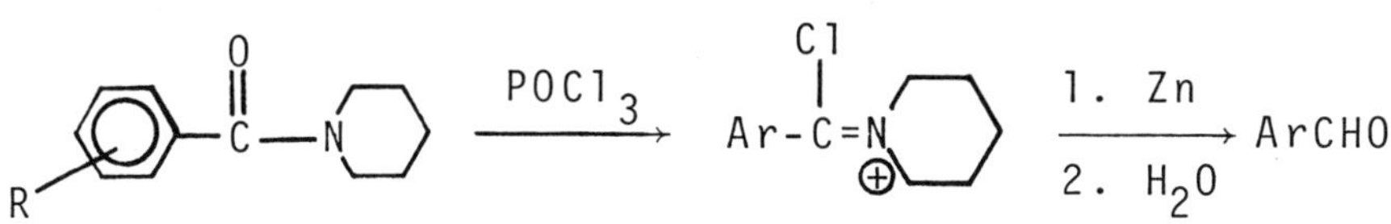

R = H, Cl, Br

90-95%

III.G-1 P. B. Dervan and M. A. Shippey, *J. Am. Chem. Soc.*, **98**, 1265 (1976).

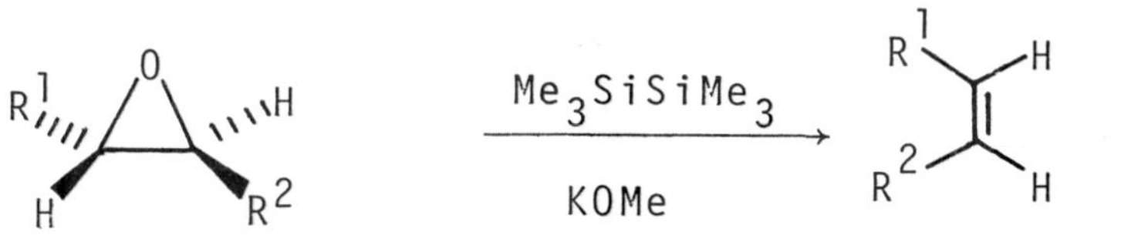

R^1,R^2 = alkyl

75-99% (VPC)

III.G-2 M. T. Reetz and M. Plachky, *Synthesis*, 199 (1976).

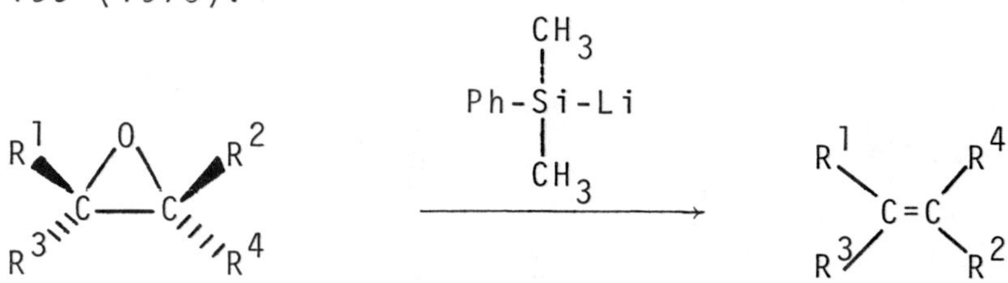

Epoxidized Substrate	Yield
trans-stilbene	75% (97% *cis*)
cis-stilbene	83% (99% *trans*)
Styrene	polymer
1-Pentene	60%
1-Octene	64%

III.G-3 T. Kempe, T. Norin, and R. Caputo, *Acta Chem. Scand. B*, **30**, 366 (1976).

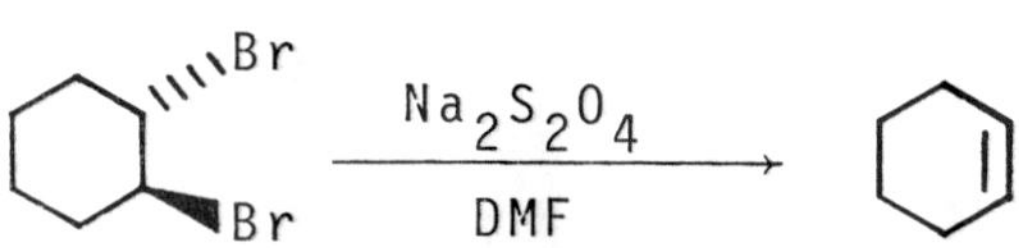

60%

III.G-4 J. E. McMurray and M. P. Fleming, J. Org. Chem., 41, 896 (1976).

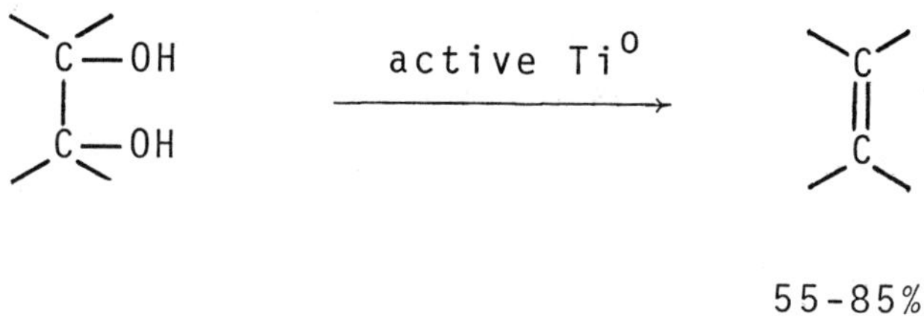

Reaction fails with glycols which cannot form a five-membered ring:

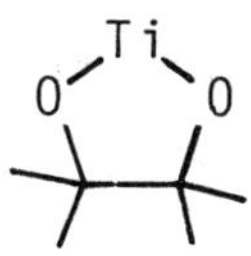

III.G-5 V. Calo *et al.*, Synthesis, 200 (1976).

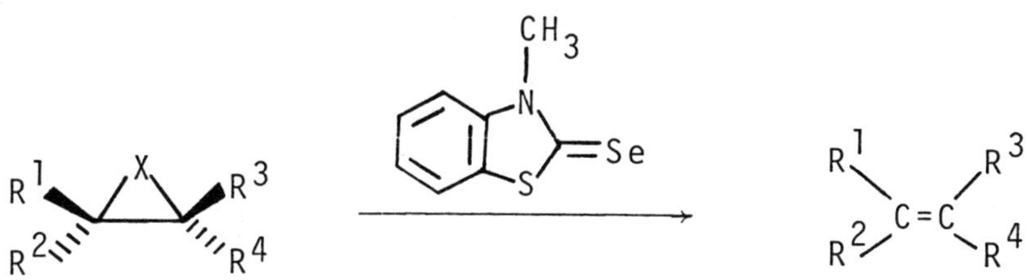

R's = alkyl, aryl, cyclic, Cl

X = O, S

87-100% (VPC)

III.G-6 P. Dowd and L. K. Marwaha, J. Org. Chem., 41, 4035 (1976).

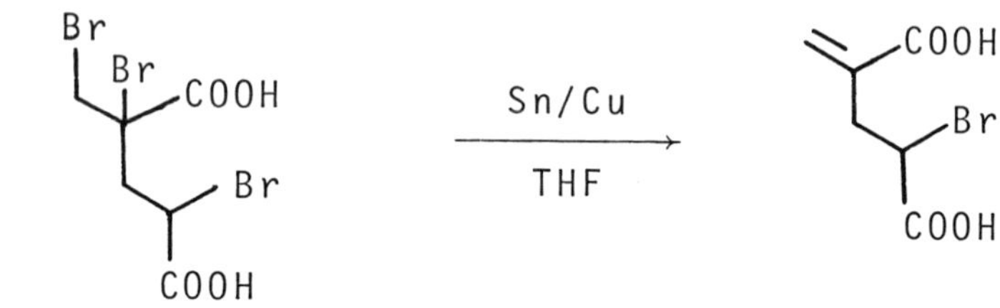

81%

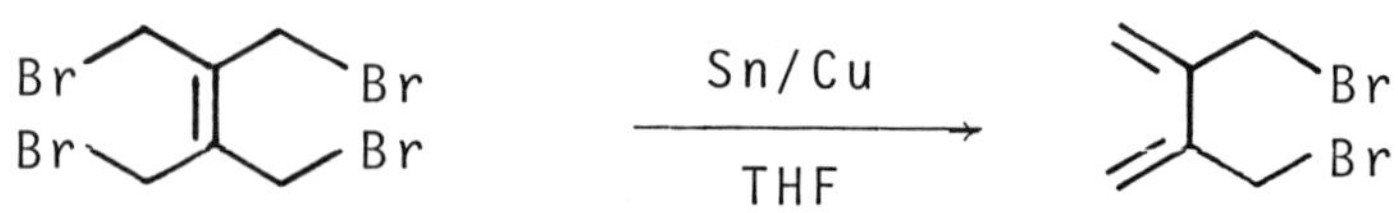

53%

III.G-7 S. Mageswaran and M.U.S. Sultanbawa, J. Chem. Soc. Perkin I, 884 (1976).

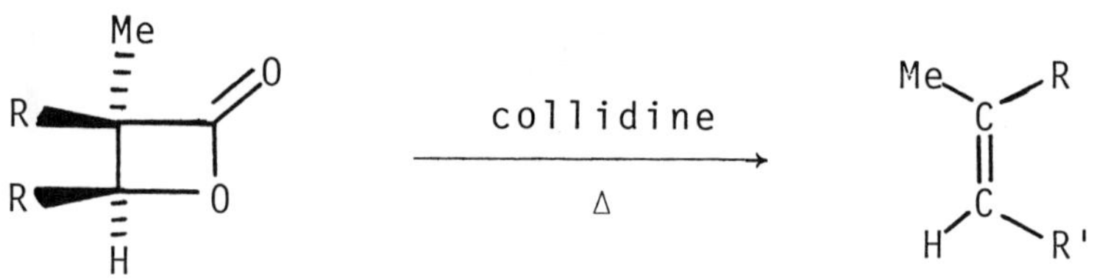

R,R' = alkyl

~60%

III.H-1 A. E. Greene, A. Cruz, and P. Crabbé, <u>Tetrahedron Lett</u>., 2707 (1976).

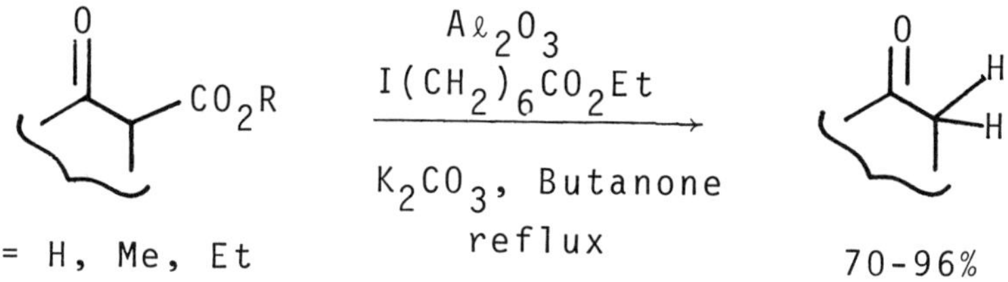

III.H-2 M. Mori and Y. Ban, <u>Chem. Pharm. Bull</u>., **24**, 1992 (1976).

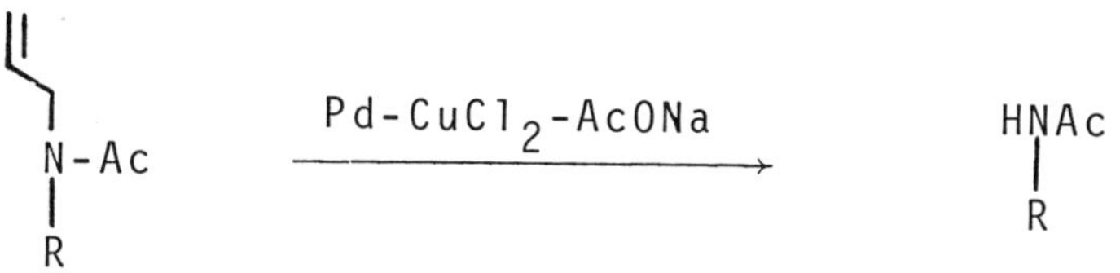

59%, R = Bz

61%, R = $PhCH_2CH_2$

III.H-3 M. T. Thomas, E. G. Breitholle, and A. G. Fallis, <u>Synth. Commun</u>., **6**, 113 (1976).

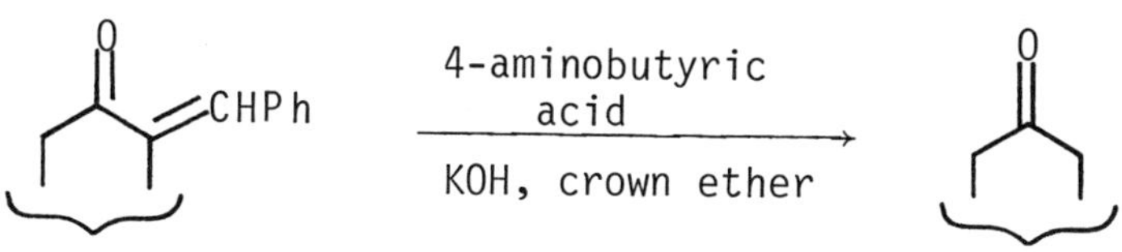

40-56%

III.I-1 H. C. Brown and N. Ravindran, J. Am. Chem. Soc., 98, 1785, 1798 (1976).

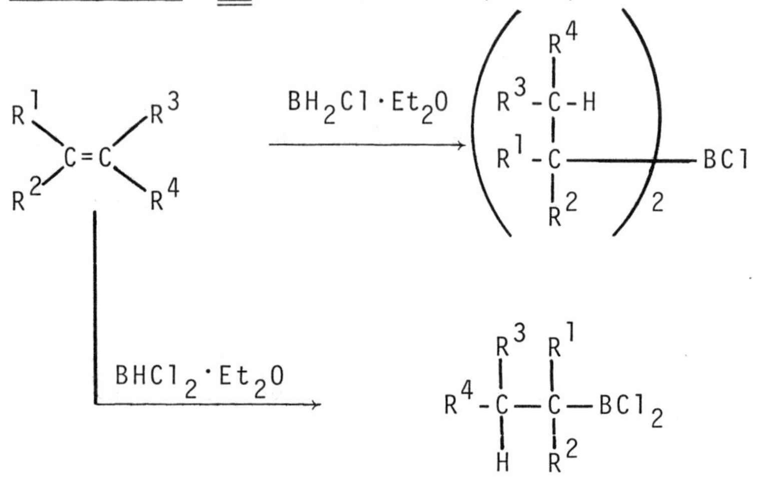

Usual sense of addition, yields generally >80%.

III.I-2 G. W. Kabalka, D.T.C. Yang, and J. D. Baker, Jr., J. Org. Chem., 41, 574 (1976).

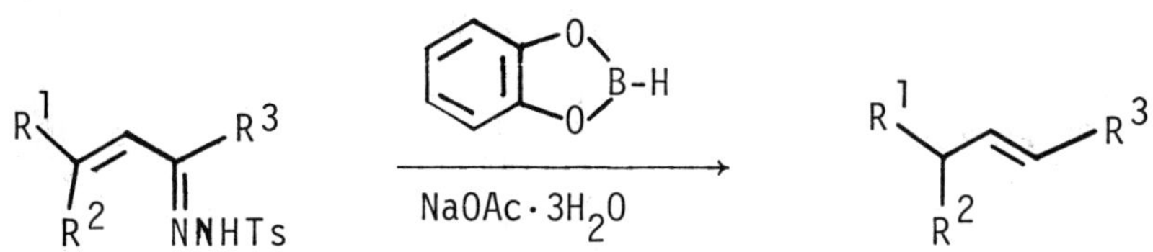

R^1, R^2 = H, alkyl, aryl 53-77%

R^3 = H, alkyl

III.I-3 A. Pelter, Chem. and Ind., 888 (1976).

Review: "Reductions Involving Diborane and Its Derivatives"

III.I-4 C. F. Lane and G. W. Kabalka, *Tetrahedron,* **32**, 981 (1976).

Review: "Catecholborane. A new hydroboration reagent."

III.I-5 H. C. Brown, S. Krishnamurthy, and N. M. Yoon, *J. Org. Chem.*, **41**, 1778 (1976).

An exhaustive review of the uses of 9-BBN as a reducing agent in organic chemistry.

III.I-6 C. F. Lane, *Chem. Rev.*, **76**, 773 (1976).

Review: "Reduction of Organic Compounds with Diborane"

III.J-1 H. Ishikawa and T. Mukaiyama, *Chem. Lett.*, 737 (1976).

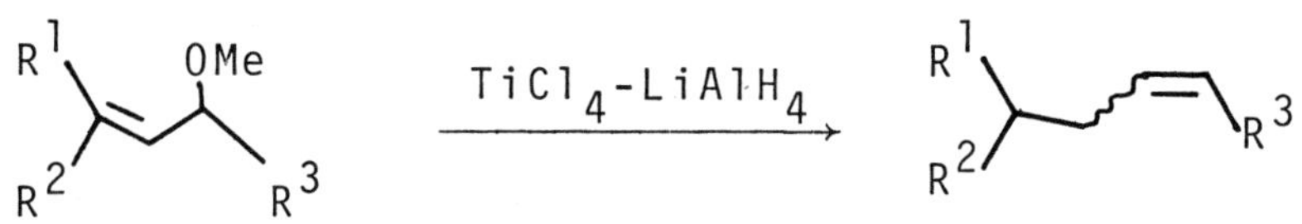

R^1, R^2 = H, Me, Et ~80%

R^3 = alkyl, Bz, cyclohexyl

III.J-2 D. J. Raber and W. C. Guida, J. Org. Chem., 41, 690 (1976).

Use of $Bu_4\overset{\oplus}{N}\overset{\ominus}{B}H_4$ as a reducing agent. Very similar to $NaBH_4$, but may be used in CH_2Cl_2 solution.

III.J-3 E.R.H. Walker, Chem. Soc. Rev., 5, 23 (1976).

Review: "The Functional Group Selectivity of Complex Hydride Reducing Agents"

III.J-4 O. N. Chupakhin and I. Ya. Postovskii, Russ. Chem. Rev., 45, 454 (1976).

Review: "Nucleophilic Substitution of Hydrogen in Aromatic Systems"

III.J-5 H. B. Kagan, Pure and Appl. Chem., 43, 401 (1975).

Review: "Asymmetric catalysis by chiral rhodium complexes in hydrogenation and hydrosilylation reactions"

IV.A-1 Y. Girault, M. Decouzon, and M. Azzaro, Tetrahedron Lett., 1175 (1976).

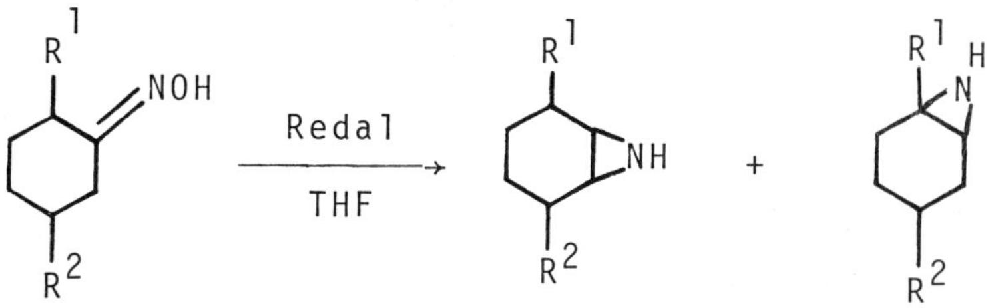

R^1, R^2 = alkyl

IV.A-2 K. Imai. Y, Kawazoe, and T. Taguchi, Chem. Pharm. Bull., 24, 1083 (1976).

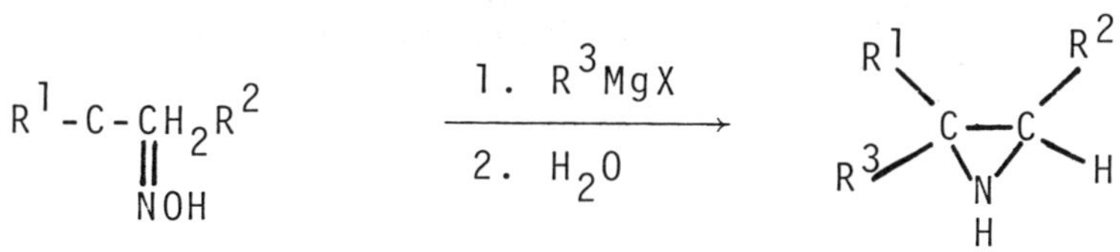

R's = Ph, Bz, alkyl, etc. Widely varying yields.

IV.A-3 Y. Tamura et al., Synthesis, 35 (1976).

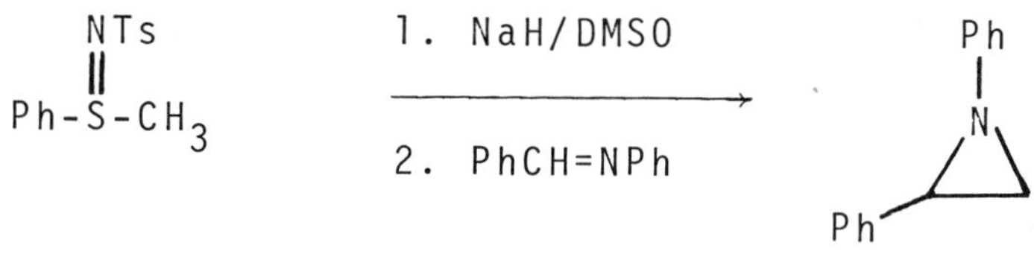

47%

IV.A-4 I. Shahak, Y. Ittah, and J. Blum, Tetrahedron Lett., 4003 (1976).

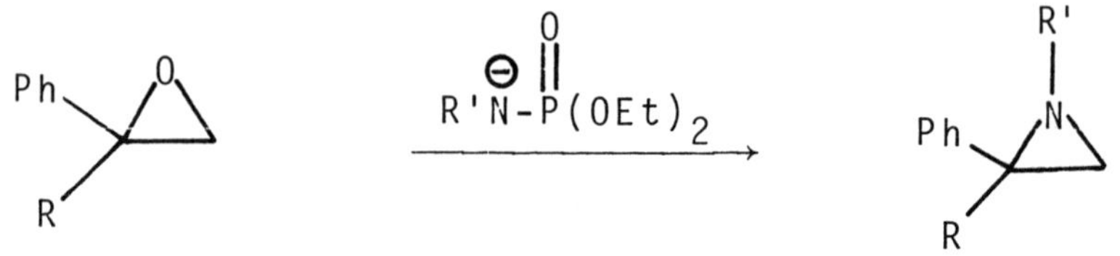

R = H, CH_3
R' = Ph, Bz, t-Bu

generally low yields (10-58%)

IV.A-5 A. Zwierzak and K. Osowska, Angew. Chem. Int. Ed., 15, 302 (1976).

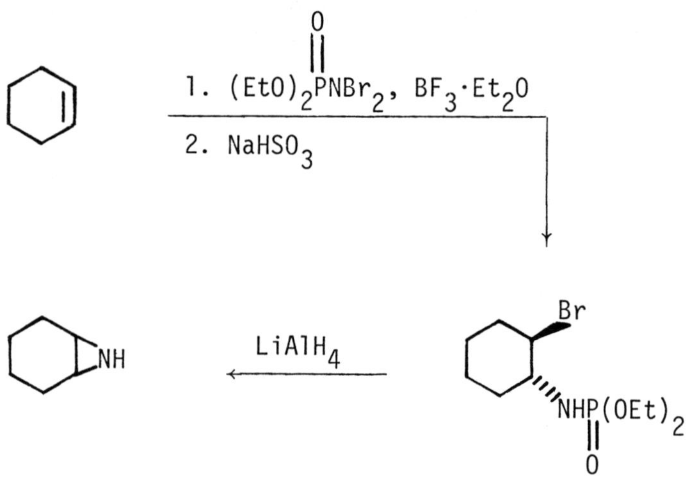

90%

no yield

IV.A-6 Y. Nomura, N. Hatanaka, and Y. Takeuchi, Chem. Lett., 901 (1976).

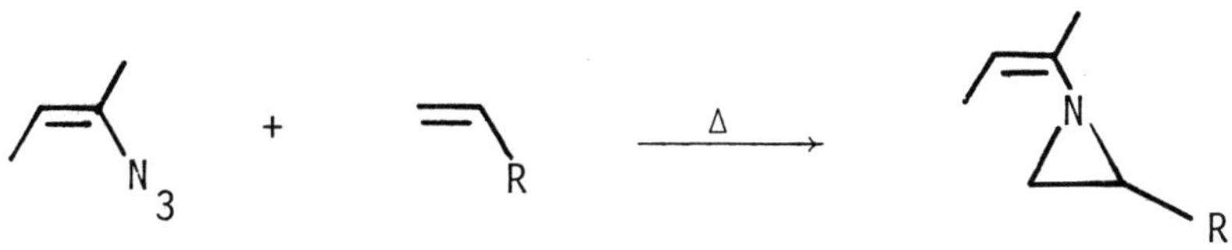

73% (R=CN)

83% (R=CO_2Me)

IV.B-1 H.-D. Scharf and E. Wolters, Angew. Chem. Int. Ed., 15, 682 (1976).

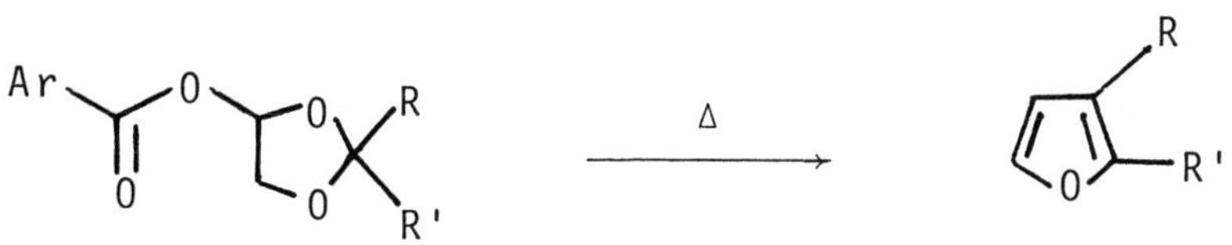

R,R' = H, Me, Et, $-(CH_2)_3-$

30-63%

IV.B-2 G. M. El-Naggar and M. A. Eldawy, Indian J. Chem., 14B, 47 (1976).

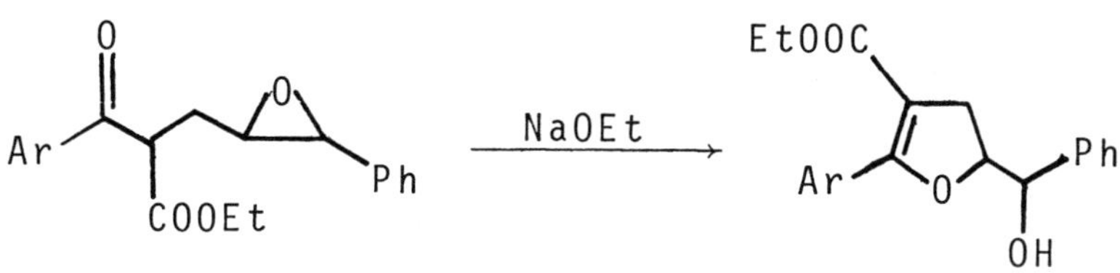

60-80%

IV.B-3 P. Cagniant and G. Kirsch, <u>Comptes Rendus (C)</u>, <u>282</u>, 465 (1976).

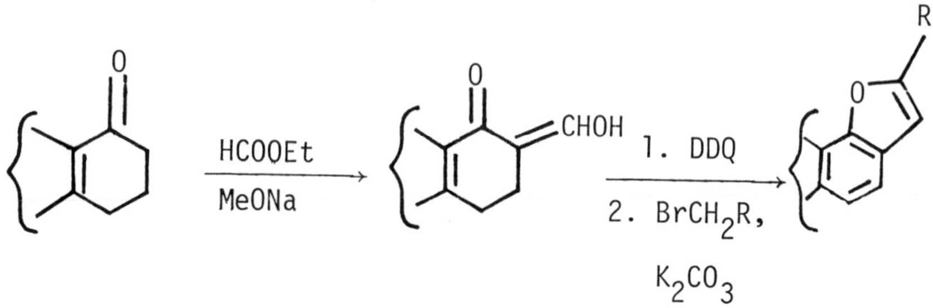

R = COR, COOR, CN, NO_2, etc.

~35-55% overall

IV.B-4 M. Yamamoto, <u>J. Chem. Soc. Perkin I</u>, 1688 (1976).

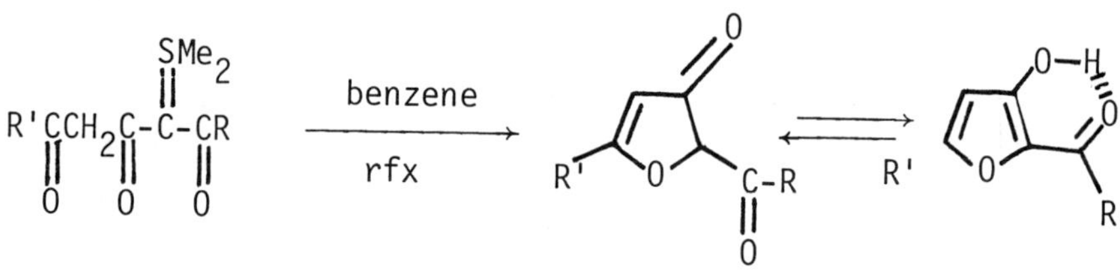

R = Me, Ph, OEt

80-95%

R' = Me, Ph, OEt, NHPh

IV.B-5 W. K. Anderson, E. J. LaVoie, and J. C. Bottaro, <u>J. Chem. Soc. Perkin I</u>, 1 (1976).

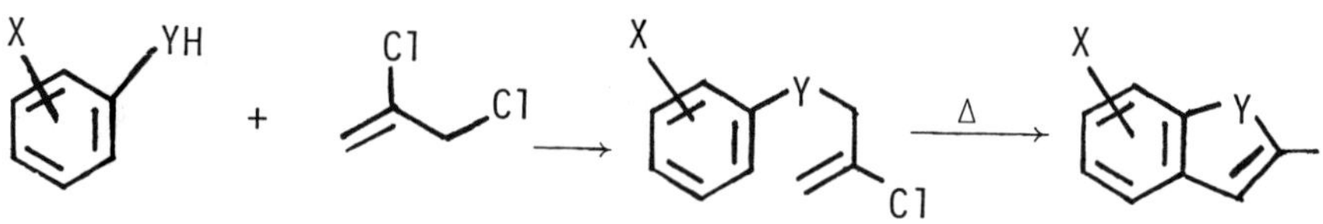

X = H, Cl, Me, OMe

Y = O, S

Widely varying yields.

IV.B-6 L. Cadona and P. D. Croce, Synthesis, 800 (1976).

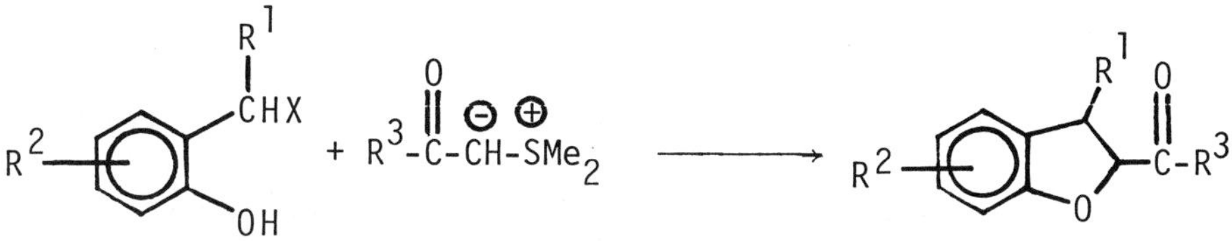

R^3 = OEt, Ar

50-85%

R^1 = H, Me, Ph

R^2 = H, Cl, NO_2, Me

X = $^+NMe_3I^-$, OH

IV.C-1 L. S. Hegedus, G. F. Allen, and E. L. Waterman, J. Am. Chem. Soc., 98, 2674 (1976).

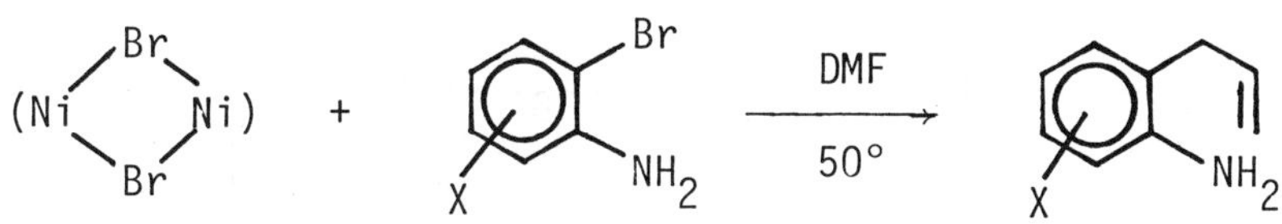

X = H, CH_3, OCH_3, COOR

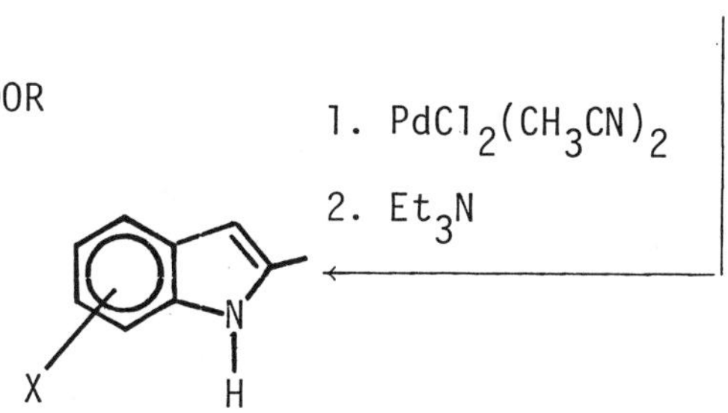

74-84%

IV.C-2 C. Germain and J. Bourdais, J. Het. Chem., 13, 1209 (1976).

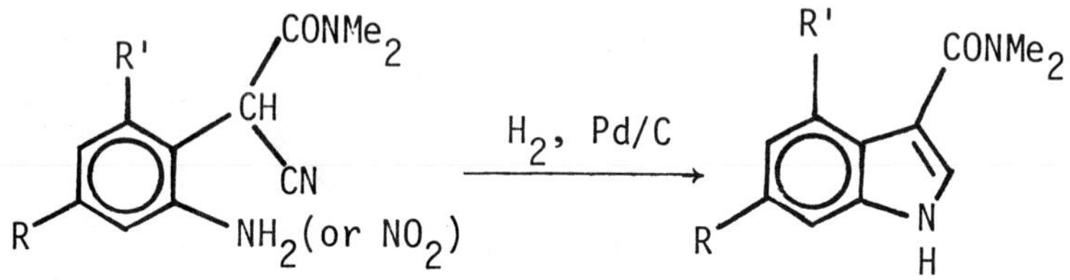

$R = H, Me, OMe, SO_2NMe_2$

$R' = H, OMe, NH_2$

~40-67%

IV.C-3 H. Bartsch, Monat. für Chem., 107, 663 (1976).

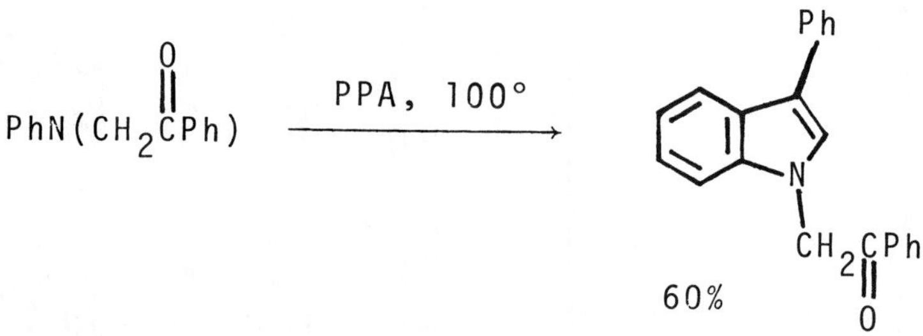

60%

Further reaction to the pyrroloindole can be effected.

IV.C-4 Y. Makisumi and S. Takada, <u>Chem. Pharm. Bull.</u>, <u>24</u>, 770 (1976).

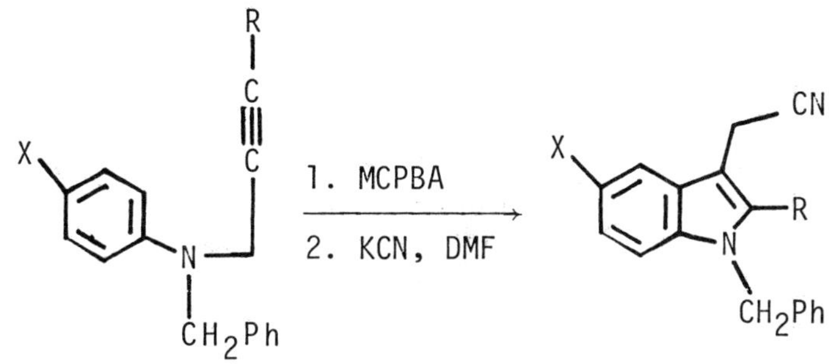

R = H, CH_3

X = H, Cl, OCH_3

~30-50%

IV.C-5 D. E. Ames and O. Ribeiro, <u>J. Chem. Soc. Perkin I</u>, 1073 (1976).

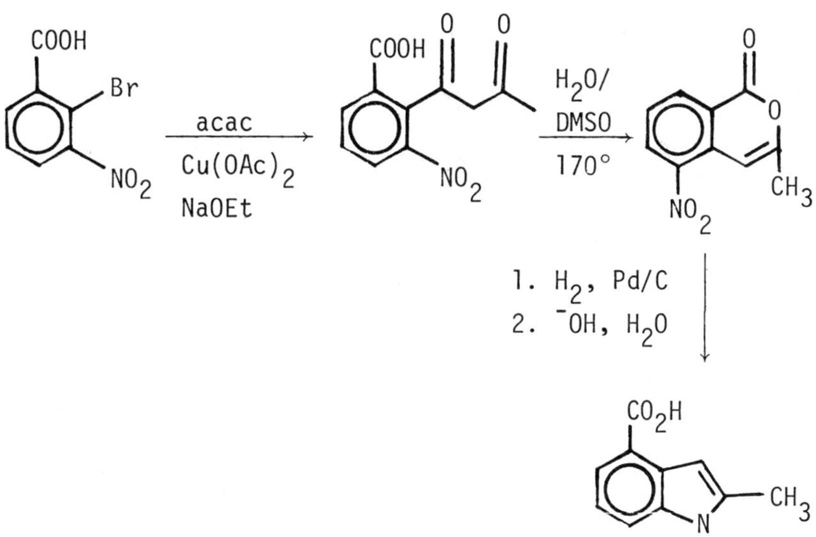

individual yields ~68-70%

IV.C-6 K. Freter, F. Hess, and K. Grozinger, Liebigs Ann. Chem., 241 (1976).

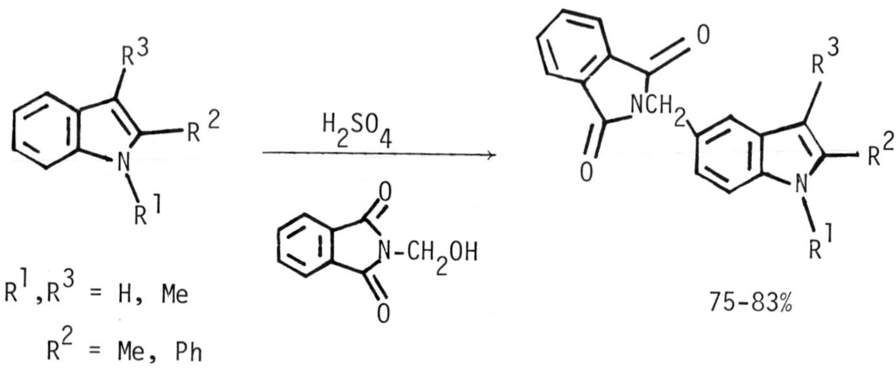

IV.C-7 M. K. Eberle and L. Brzechffa, J. Org. Chem., 41, 3775 (1976).

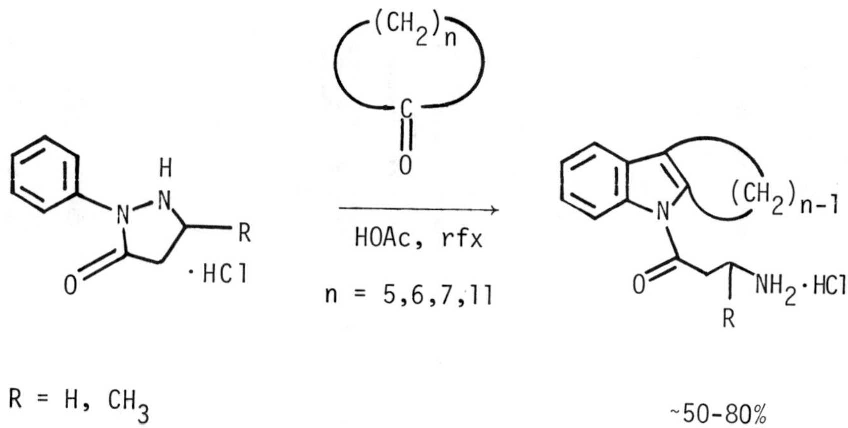

IV.C-8/IV.D-1 H. Aoyama *et al.*, *J. Chem. Soc. Perkin I*, 1556 (1976).

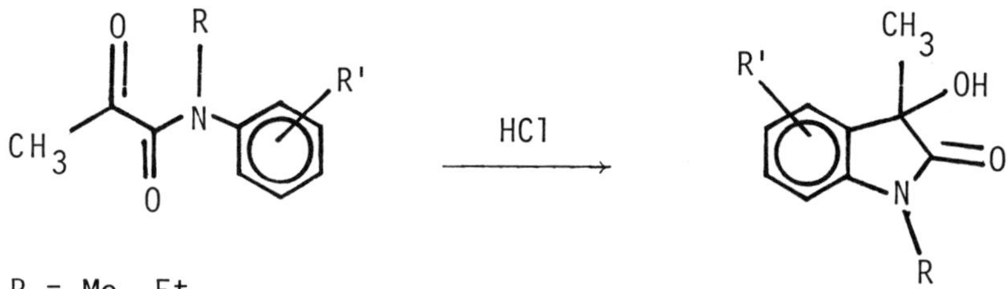

R = Me, Et

R' = Me, OMe, Cl

73-96%

IV.D-2 T. Hasegawa, H. Aoyama, and Y. Omote, *J. Chem. Soc. Perkin I*, 2054 (1976).

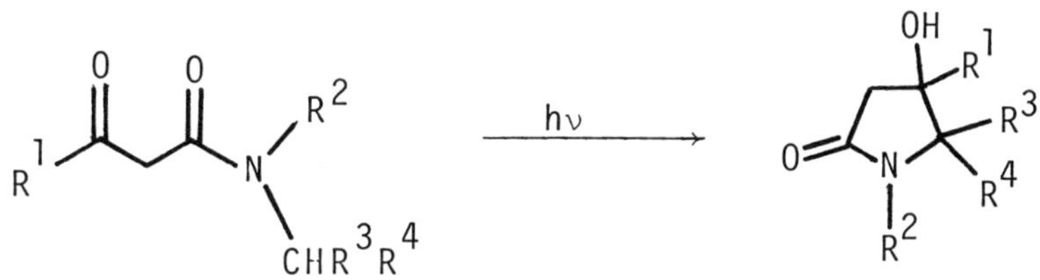

R^1 = Me, Ph

R^2,R^3,R^4 = Me_2, Et_2, *i*-Pr_2, Bz_2, $-(CH_2)_5-$, etc.

~70-90%

IV.D-3 A. Kasahara, *Chem. and Ind.*, 1032 (1976).

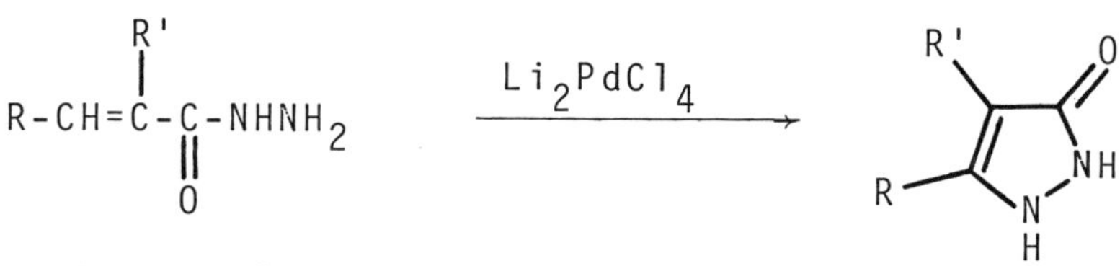

R's = H, Me, Ph

33-42%

IV.D-4 A. Baba, Y. Oshiro, and T. Agawa, Chem. Lett., 11 (1976).

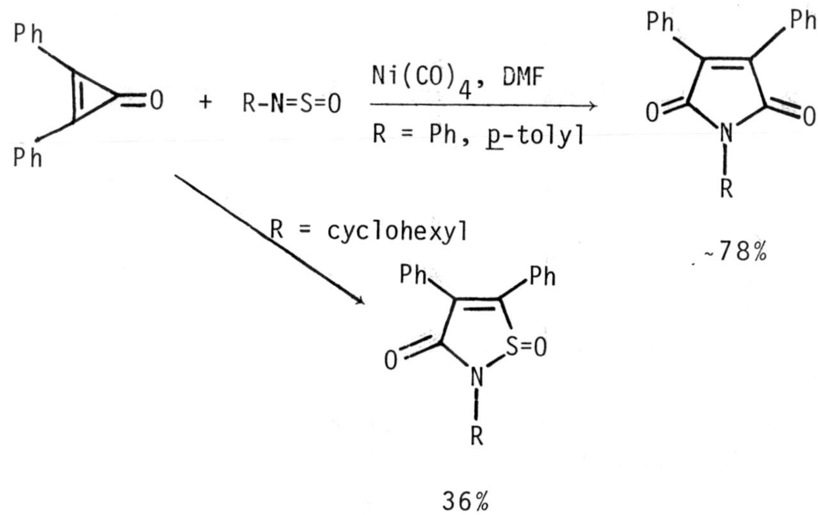

IV.D-5 M. A. Ponomareva-Stepnaya et al., J. Gen. Chem. (USSR), 45, 2451 (1975).

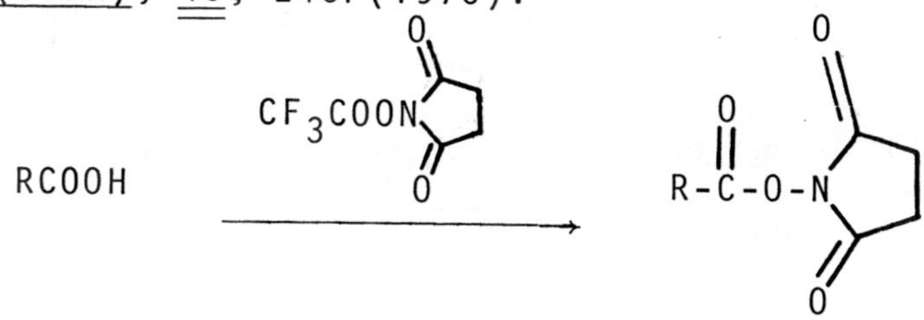

RCOOH = amino acid

52-99%

racemization not observed

IV.D-6 G. Bettoni *et al.*, *J. Org. Chem.*, **41**, 2780 (1976).

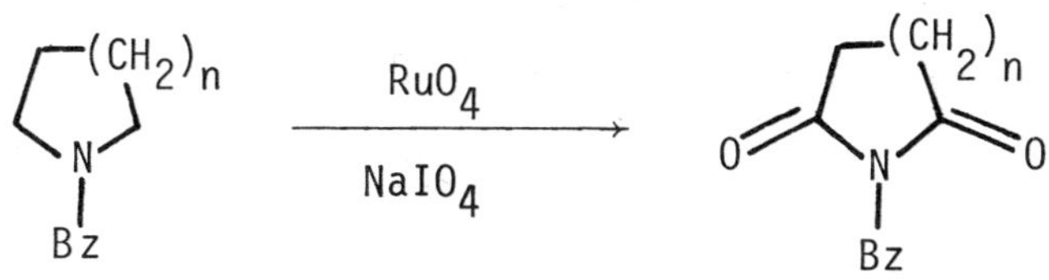

n = 1, 2

Several examples

yields ~30-40% (isolated)

IV.D-7 J.-C. Gramain, R. Remuson, and Y. Troin, *J. Chem. Soc., Chem. Commun.*, 194 (1976).

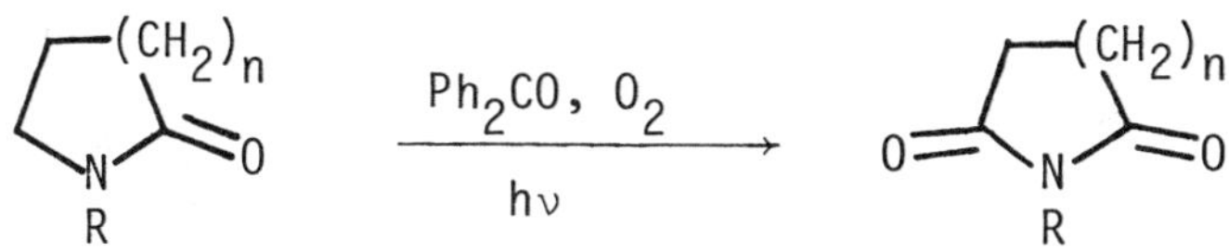

n = 1, 2

~60% (isolated)

IV.D-8 M. Ogata and H. Matsumoto, *Chem. and Ind.*, 1067 (1976).

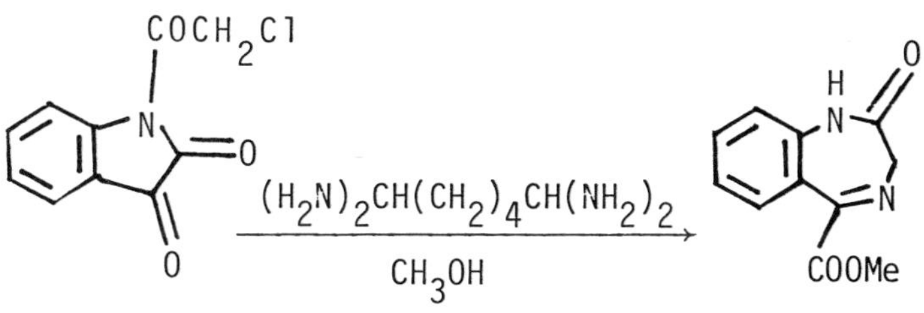

51%

IV.D-9 M. S. Manhas *et al.*, *Synthesis*, 689 (1976).

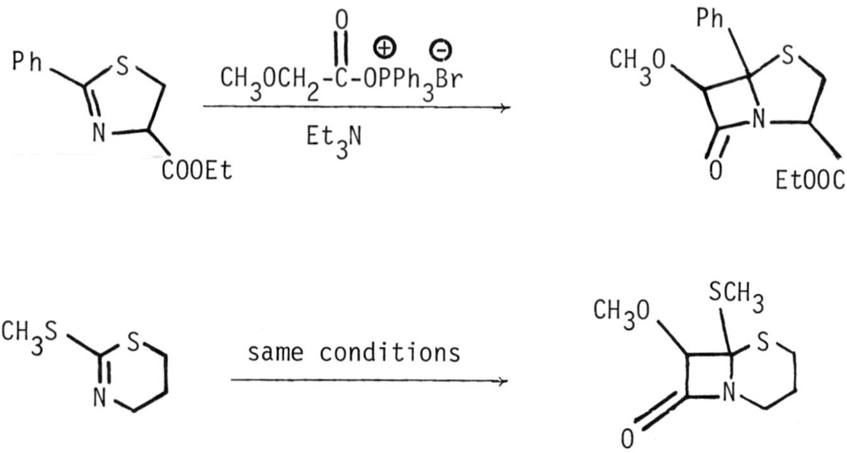

IV.D-10 D. F. Sullivan *et al.*, *J. Org. Chem.*, **41**, 1112 (1976).

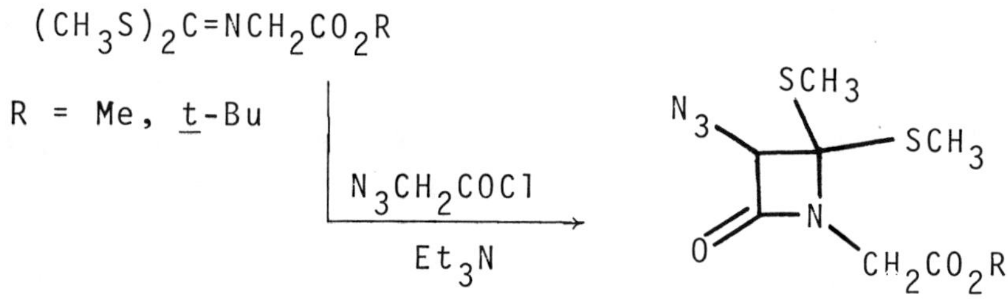

~80%

IV.D-11 H. H. Wasserman and B. H. Lipshutz, Tetrahedron Lett., 4613 (1976).

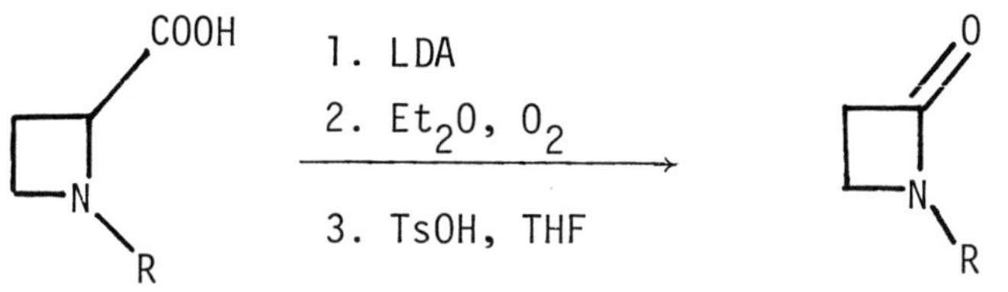

R = 1° alkyl, t-butyl, acetal, cyclic 50-61%

IV.D-12 M. S. Manhas et al., Synth. Commun., 6, 435 (1976).

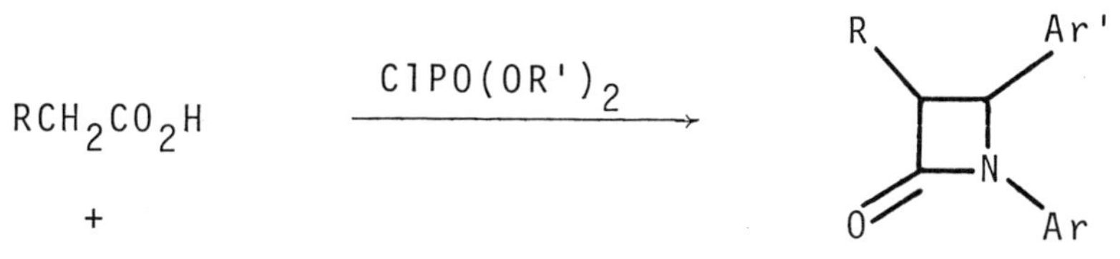

50-80%

IV.D-13 N. S. Isaacs, Chem. Soc. Rev., 5, 181 (1976).

Review: "Synthetic Routes to β-Lactams"

IV.D-14 P. G. Sammes, Chem. Rev., 76, 113 (1976).

Review: "Recent Chemistry of the β-Lactam Antibiotics"

IV.D-15 Y. Kishi, <u>Pure and Appl. Chem.</u>, **43**, 423 (1975).

Review: "Synthetic Studies in the Fields of Natural Products"

(penicillins and cephalosporins)

IV.D-16 H. Bhagwatheeswaran, S. P. Gaur, and P. C. Jain, <u>Synthesis</u>, 615 (1976).

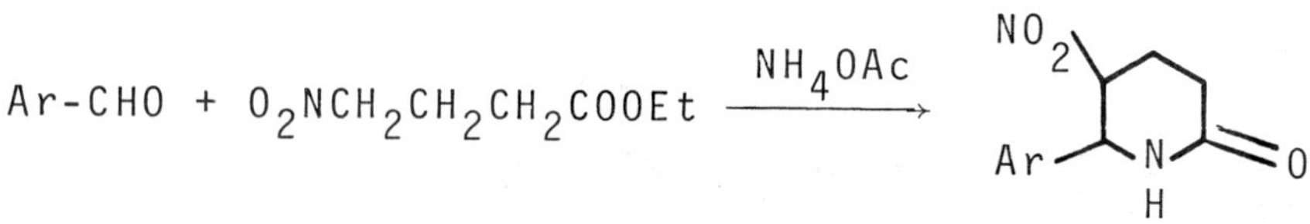

widely varying yields

IV.D-17 W. F. Keir and H.C.S. Wood, <u>J. Chem. Soc., Perkin I</u>, 1847 (1976).

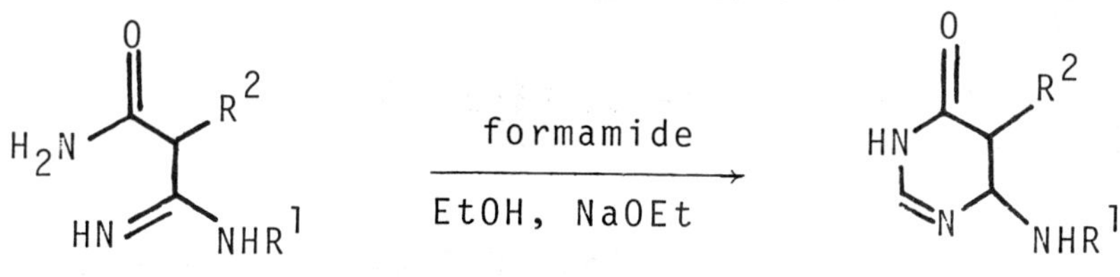

R^1 = H, alkyl, aryl

R^2 = H, PhN_2, NH_2, NHCHO

yields ~70-90%

IV.D-18 J. Becher and E. G. Frandsen, Tetrahedron Lett., 3347 (1976).

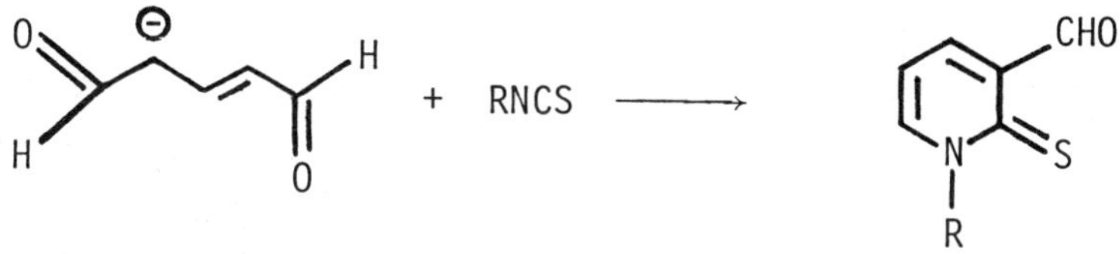

R = alkyl, aryl

yields generally 60-90%

IV.D-19 C. Bischoff and H. Herma, J. Prakt. Chem., 318, 891 (1976).

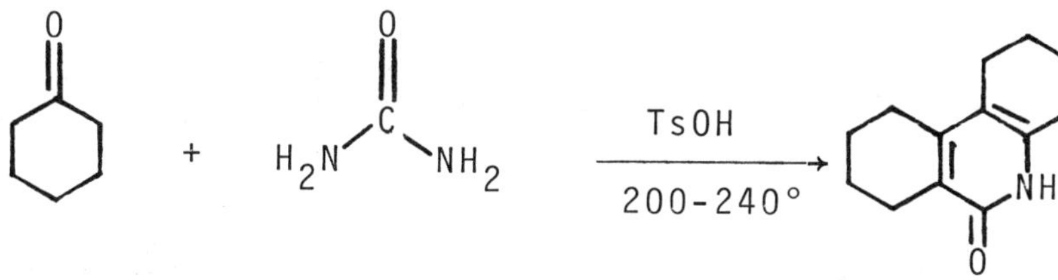

64%

IV.D-20 H. L. Wehrmeister, J. Het. Chem., 13, 61 (1976).

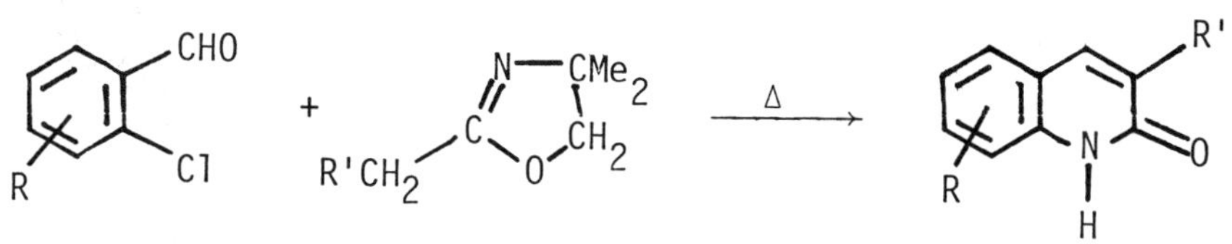

R = Cl, NO_2 R' = Me, Ph

Yields vary widely

IV.D-21 M. Y. Shandala and N. H. Al-Jobour, Aust. J. Chem., 29, 1583 (1976).

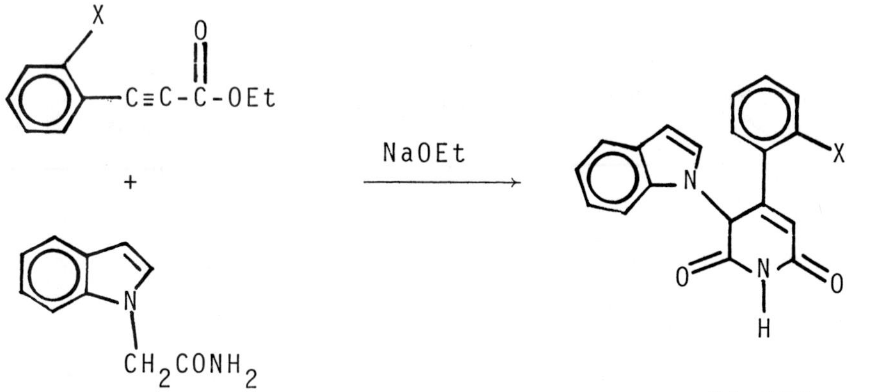

~25-30%

IV.D-22/IV.E-1 H. J. Bestmann, G. Schmid, and D. Sandmeier, Angew. Chem. Int. Ed., 15, 115 (1976).

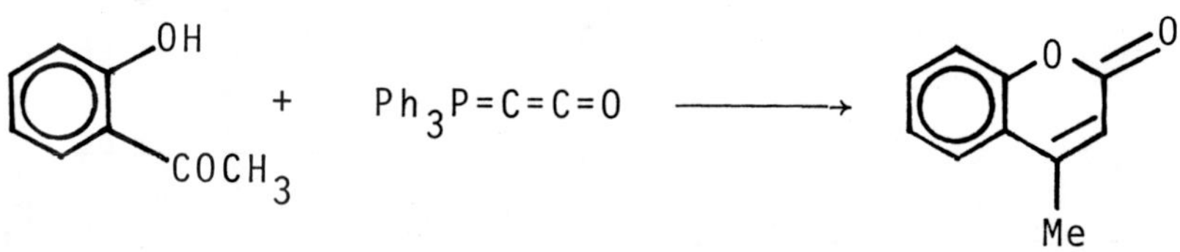

78%

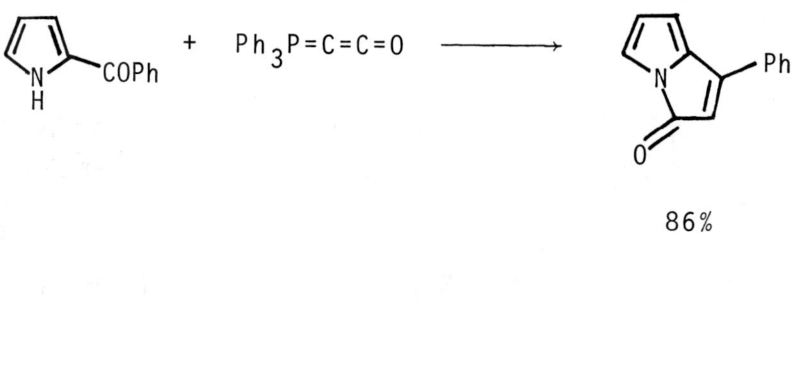

86%

IV.D-23/IV.E-2 W. Ott, G. Kollenz, and E. Ziegler, Synthesis, 546 (1976).

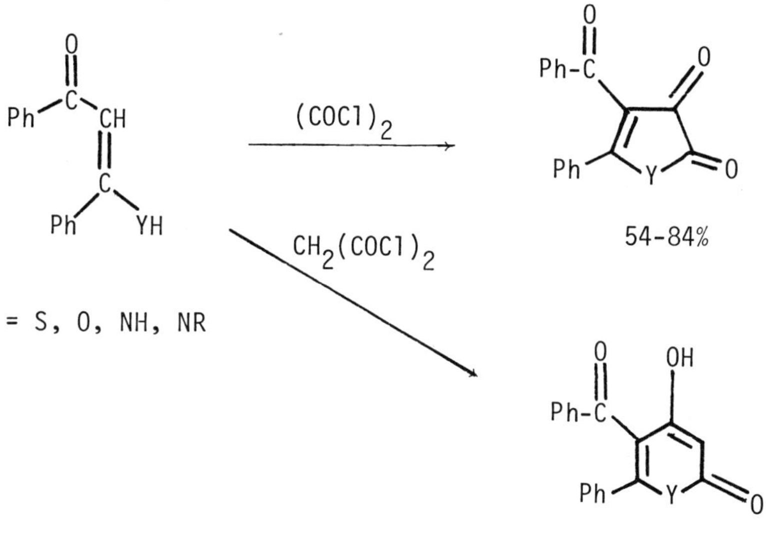

Y = S, O, NH, NR

IV.D-24/IV.E-3 F. A. L'Eplattenier et al., Synthesis, 543 (1976).

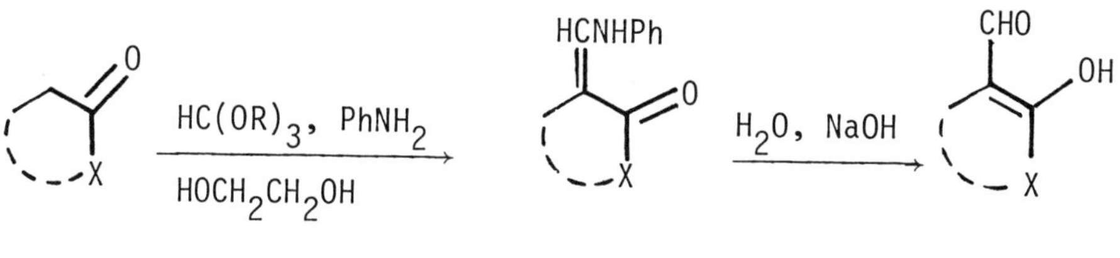

X = O, NH

~70-90%

IV.E-4 S. Danishefsky, T. Kitahara, M. Tsai and J. Dynak, J. Org. Chem., 41, 1669 (1976).

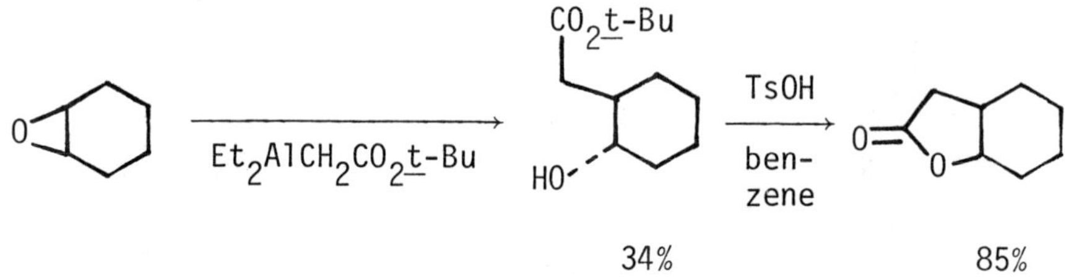

IV.E-5 T. F. Murray, V. Varma, and J. R. Norton, J. Chem. Soc., Chem. Commun., 907 (1976).

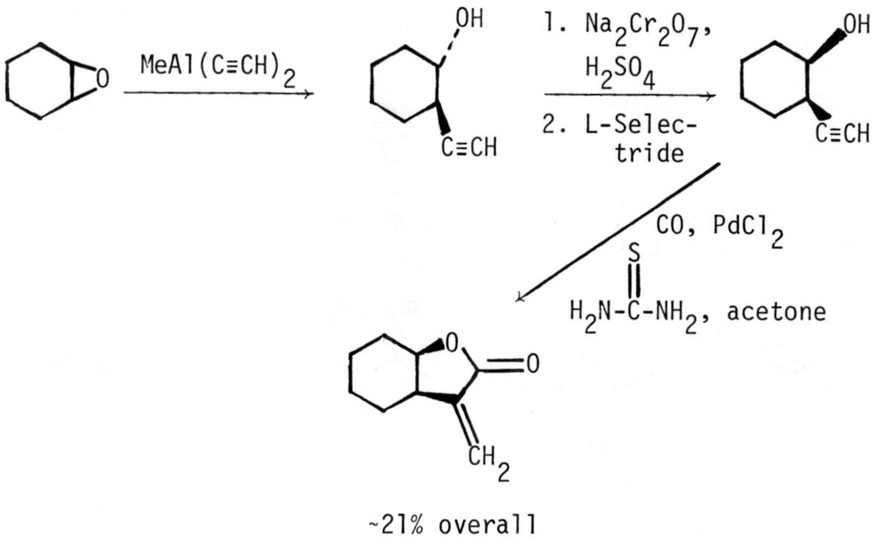

IV.E-6 M. Bertrand *et al.*, *Tetrahedron Lett.*, 3305 (1976).

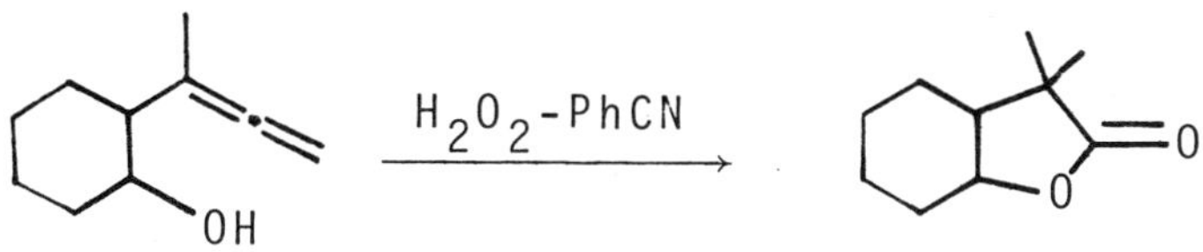

est. 80% yield

IV.E-7 J. F. LeBorgne *et al.*, *Synthesis*, 238 (1976).

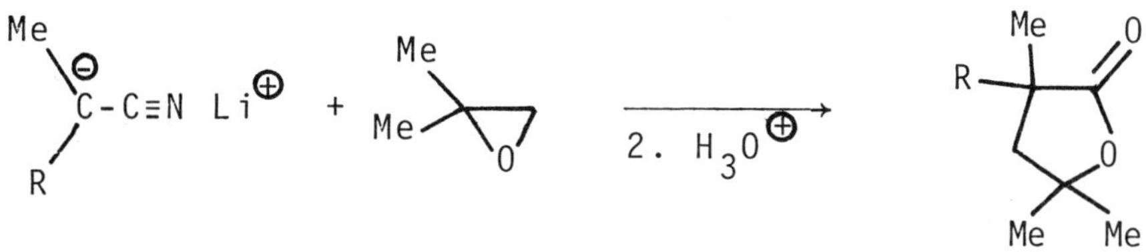

72%, R=H

89%, R=Me

IV.E-8 G. Domschke, *Z. Chem.*, **16**, 13 (1976).

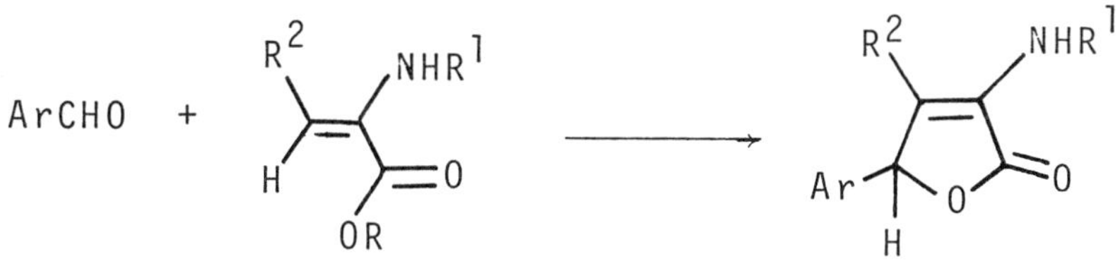

R^1 = H, alkyl, aryl

R^2 = COR, COOR

~60-90%

IV.E-9 R. C. Larock and B. Riefling, Tetrahedron Lett., 4661 (1976).

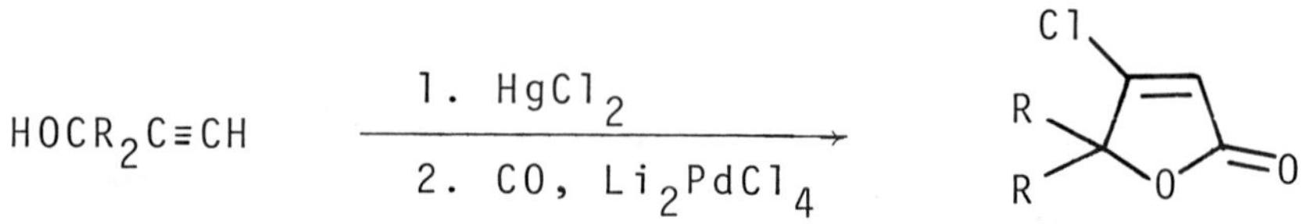

$R_2 = H_2$, Me_2, $-(CH_2)_{4,5}$ ~90%

IV.E-10 S. F. Martin and D. R. Moore, Tetrahedron Lett., 4459 (1976).

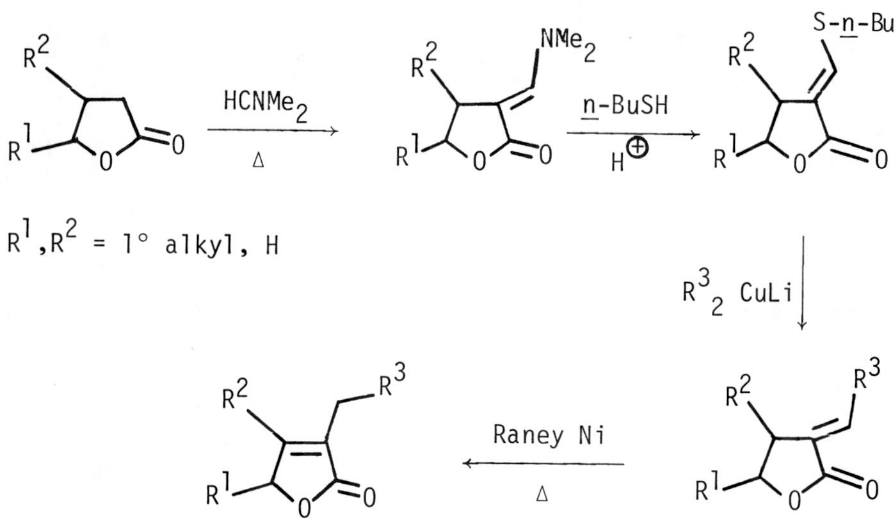

R^1,R^2 = 1° alkyl, H

Overall yields ~50-60% R^3 = Me, n-Bu

IV.E-11 R. M. Carlson and A. R. Oyler, J. Org. Chem., 41, 4065 (1976).

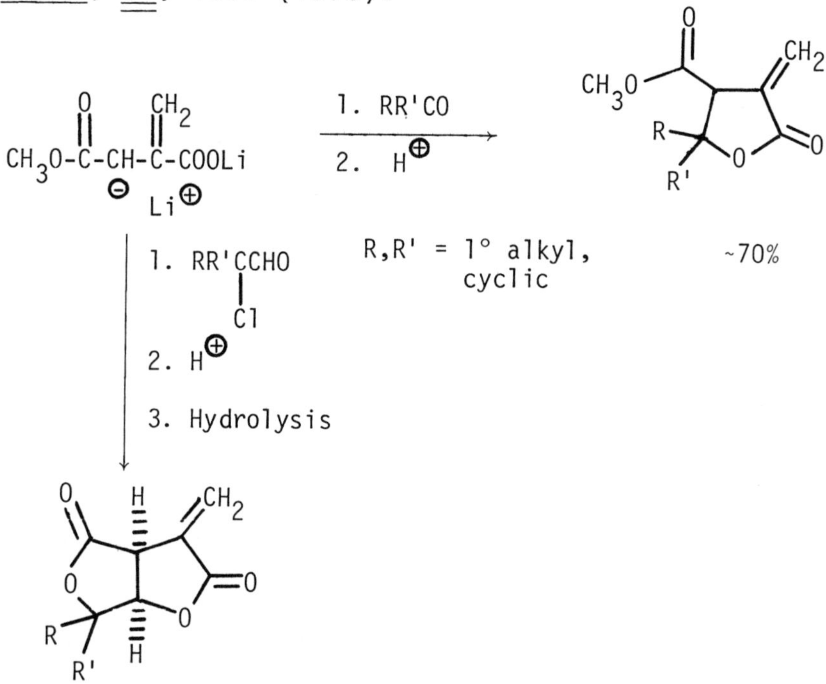

R,R' = 1° alkyl, cyclic

~70%

~70%

IV.E-12 K. Kurata, S. Tanaka, and K. Takahashi, Chem. Pharm. Bull., 24, 538 (1976).

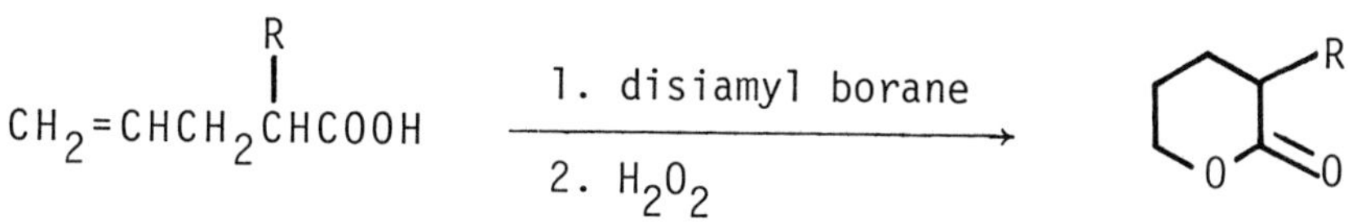

(obtained from malonic ester)

54-84%

R = H, 1°,2°,3° alkyl

IV.E-13 R. H. Carter *et al.*, *J. Chem. Soc. Perkin I*, 1438 (1976).

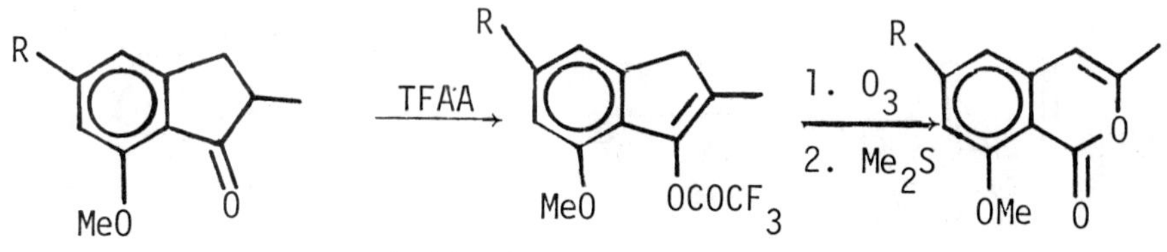

R = H, OMe

40% overall

IV.E-14 J. Ficini *et al.*, *Tetrahedron Lett.*, 683 (1976).

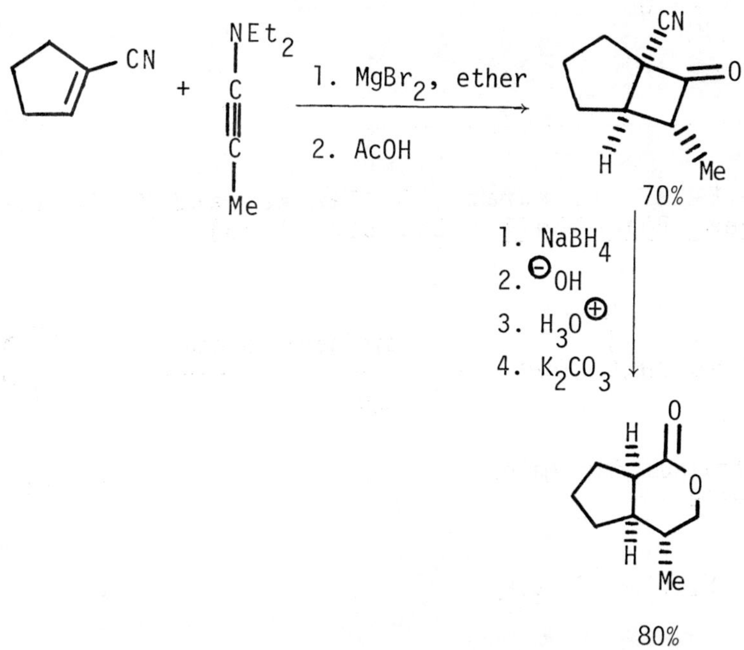

IV.E-15 A. I. Meyers and C. E. Whitten, <u>Tetrahedron Lett.</u>, 1947 (1976).

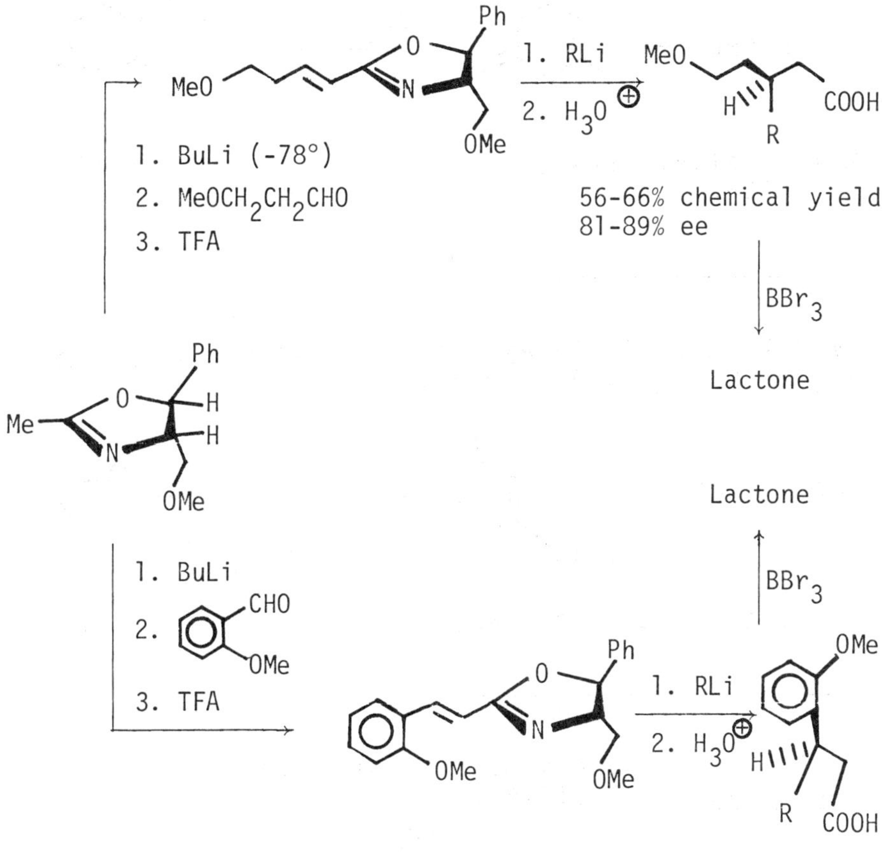

R = Et, <u>n</u>-Bu, Ph

70-87% chemical yield
75-85% ee

IV.E-16 T. Sato *et al.*, *Bull. Chem. Soc. Japan*, *49*, 1055 (1976).

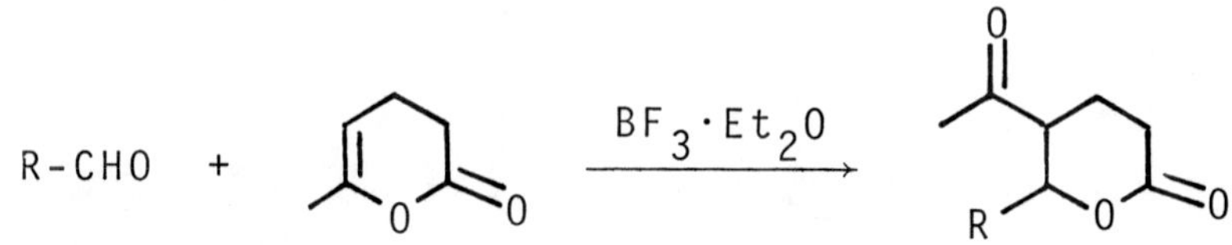

R = H, alkyl, aryl, $ClCH_2$, etc. 60-90%

IV.E-17 E. J. Corey and D. J. Brunelle, *Tetrahedron Lett.*, 3409 (1976).

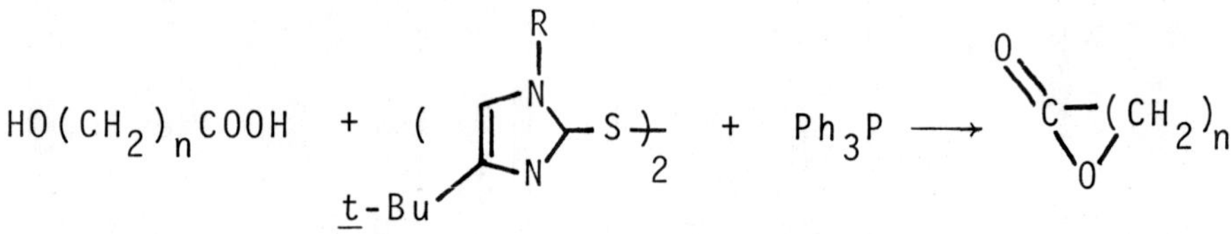

n≃15 R = Me, *i*-Pr ~80%

IV.E-18 T. Mukaiyama, M. Usui, and K. Saigo, *Chem. Lett.*, 49 (1976).

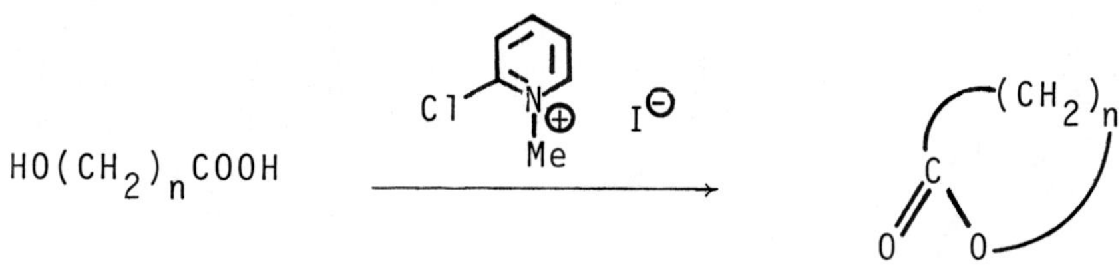

n = 5,7,10,11,14 Reaction fails for n = 6.

IV.E-19 T. Kurihara, Y. Nakajima, and O. Mitsunobu, Tetrahedron Lett., 2455 (1976).

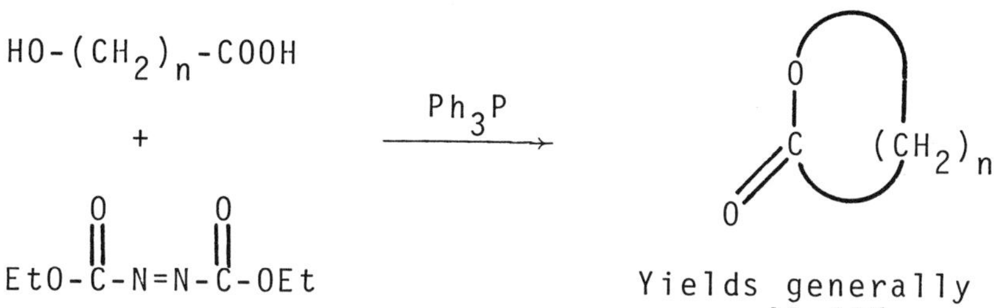

Yields generally 40-60%

n = 2-11

IV.E-20 L. T. Scott and J. O. Naples, Synthesis, 738 (1976).

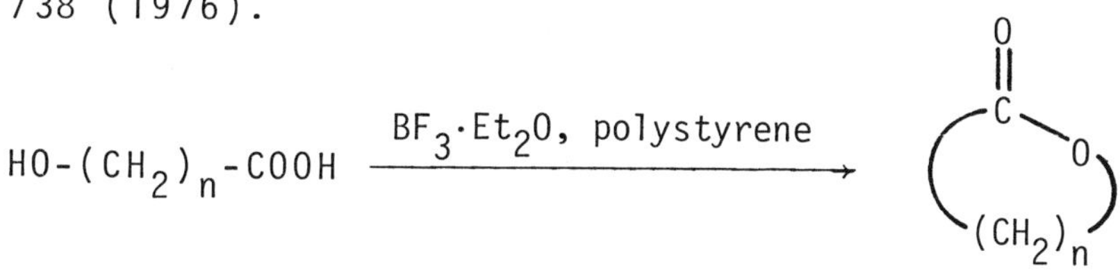

41% (n=11) - 78% (n=17)

IV.E-21 Y. S. Rao, Chem. Rev., 76, 625 (1976).

Review: "Recent Advances in the Chemistry of Unsaturated Lactones"

IV.E-22 J. R. Mahajan, <u>Synthesis</u>, 110 (1976).

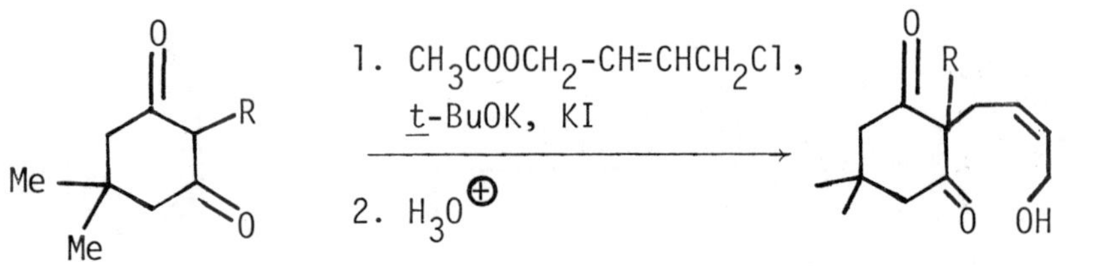

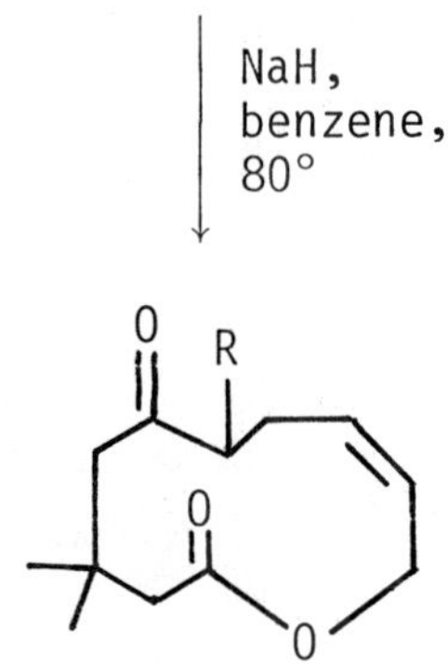

R	Yield (pure)
methyl	34%
allyl	48%
benzyl	50%

IV.F-1 Y. Wakatsuki and H. Yamazaki, <u>Synthesis</u>, 26 (1976).

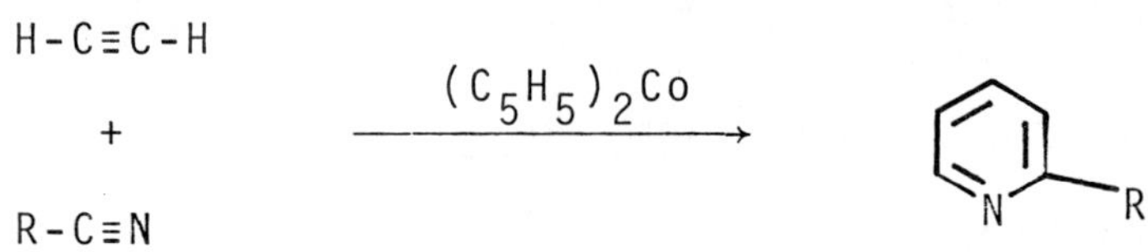

R = alkyl, aryl, vinyl 20-73%

IV.F-2 A. L. Cossey *et al.*, Aust. J. Chem., **29**, 1039 (1976).

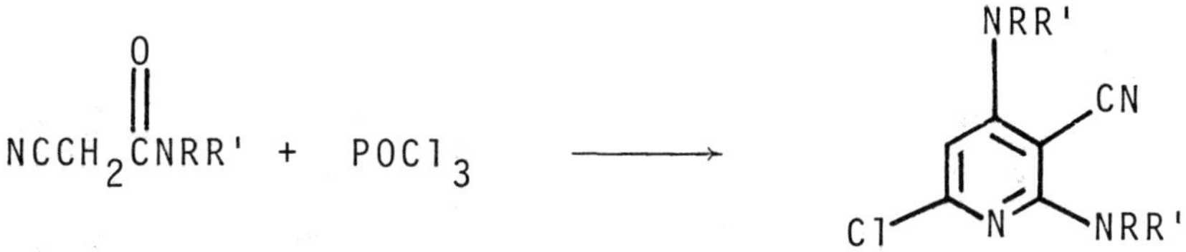

R,R' = alkyl

76-86%

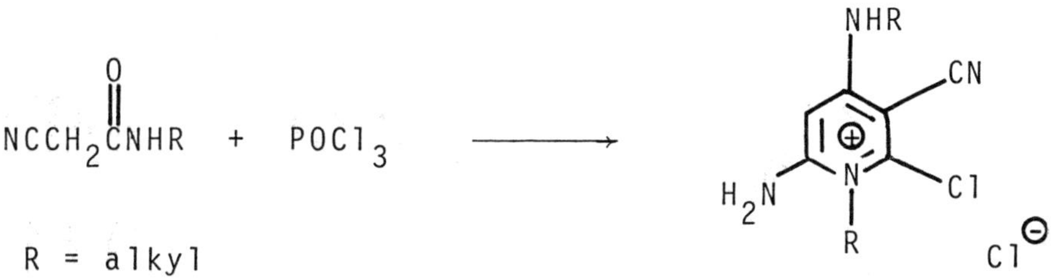

R = alkyl

39-86%

IV.F-3 J. N. Bonfiglio *et al.*, J. Am. Chem. Soc., **98**, 2344 (1976).

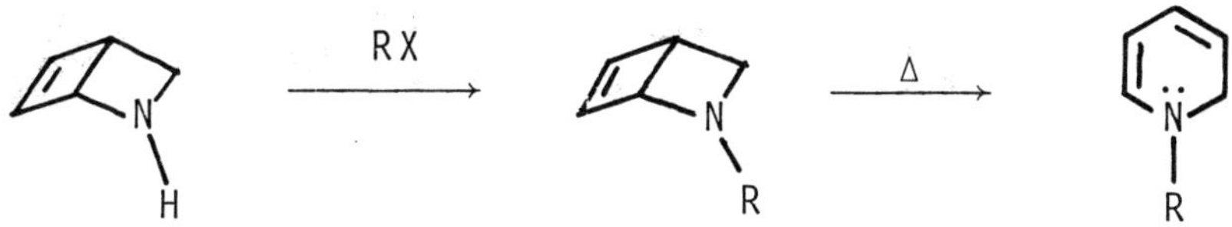

R = alkyl, benzyl, unsaturated, etc.

18-54% overall

IV.F-4 Y. Akita *et al.*, *Chem. Pharm. Bull.*, **24**, 1839 (1976).

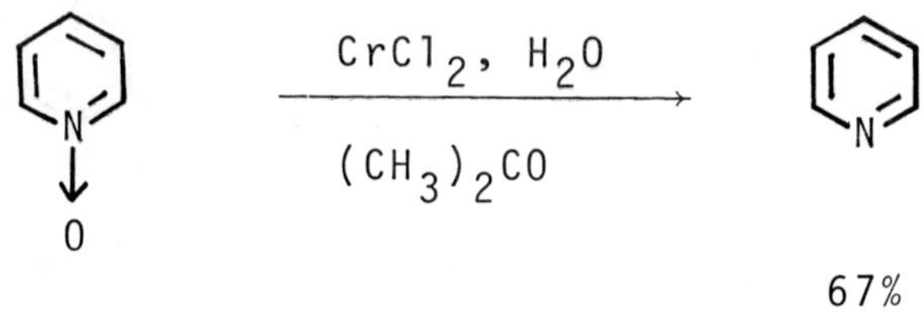

(Also works for chloropyridine, quinoline, and pyrazine derivative N-oxides, yields 41-95%.)

IV.F-5 G. Dauphin and D. Planat, *Tetrahedron Lett.*, 4065 (1976).

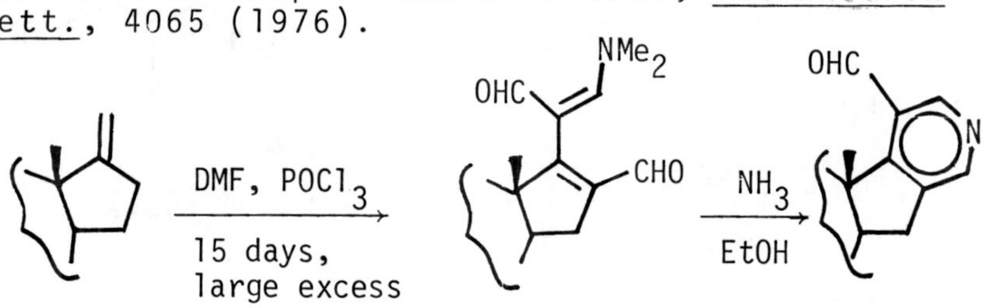

IV.F-6 E. B. Pedersen, *Acta Chem. Scand. B*, **30**, 136 (1976).

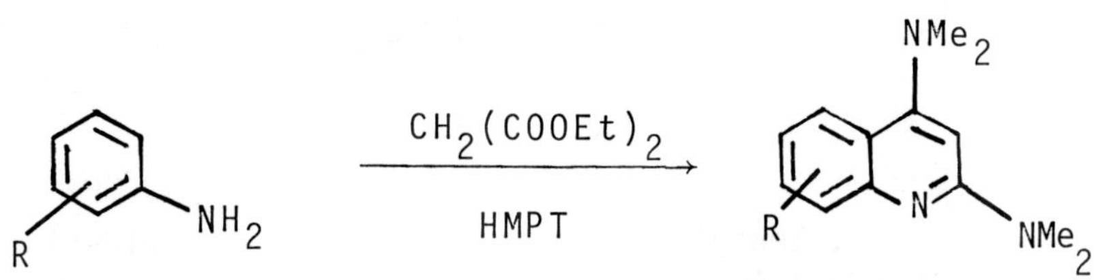

R = H, Me, OMe, Cl 25-30%

IV.F-7 A. Walser, G. Zenchoff, and R. I. Fryer, J. Het. Chem., 13, 131 (1976).

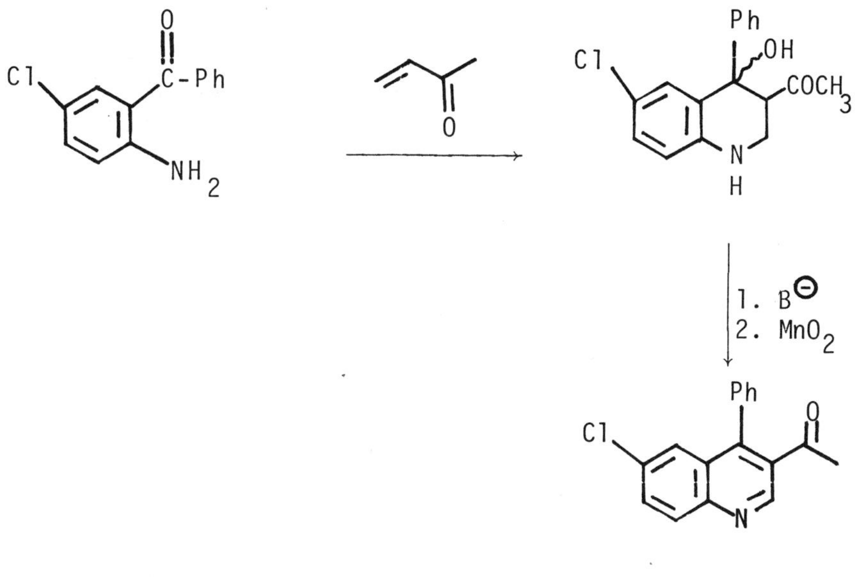

overall ~40%

IV.F-8 F. X. Smith and G. G. Evans, J. Het. Chem., 13, 1025 (1976).

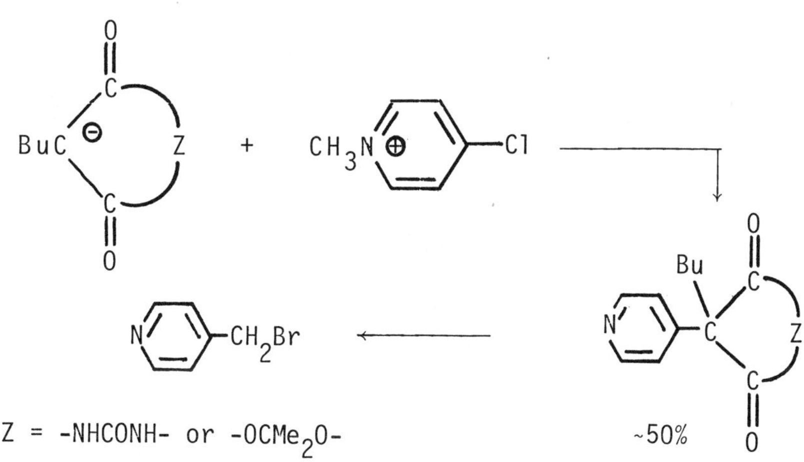

Z = -NHCONH- or $-OCMe_2O-$

IV.F-9 S. Kwon *et al.*, *Synthesis*, 249 (1976).

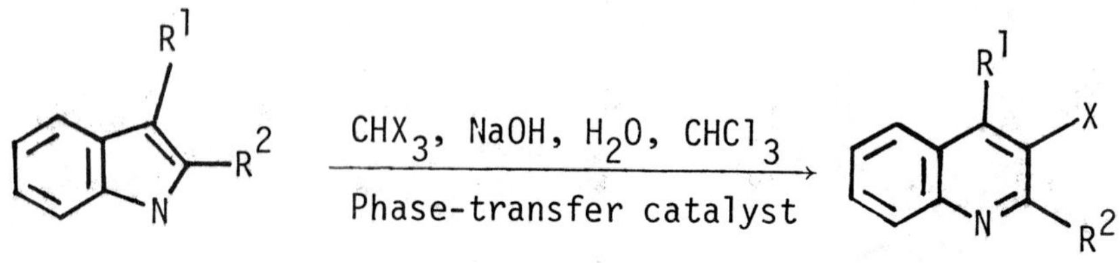

X = Cl, Br

R^1 = H, CH_3
R^2 = H, CH_3, Ph

24-68%

IV.F-10 G. Coudert *et al.*, *Synthesis*, 764 (1976).

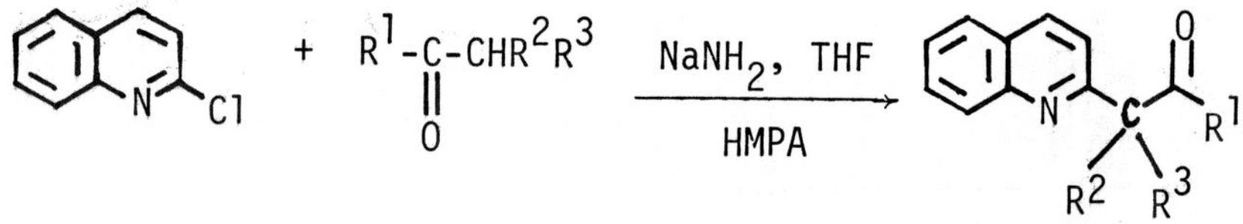

60-78%

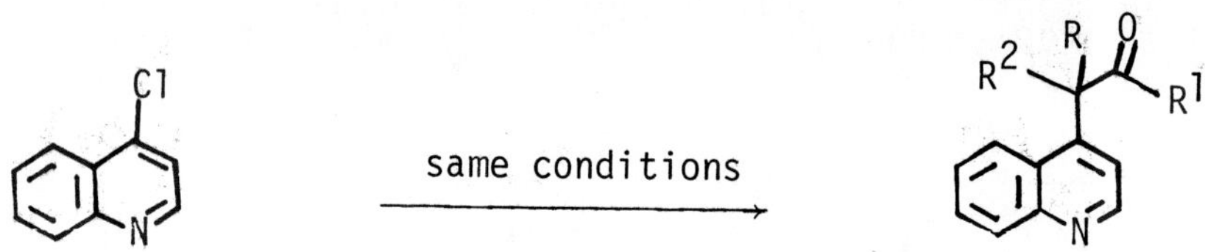

42-90%

R's = H, aryl, 1°,2° alkyl, cyclic

IV.F-11 F. Kröhnke, *Synthesis*, 1 (1976).

Review: "The Specific Synthesis of Pyridines and Oligopyridines"

IV.G-1 H. Wamhoff and B. Wehling, Synthesis, 51 (1976).

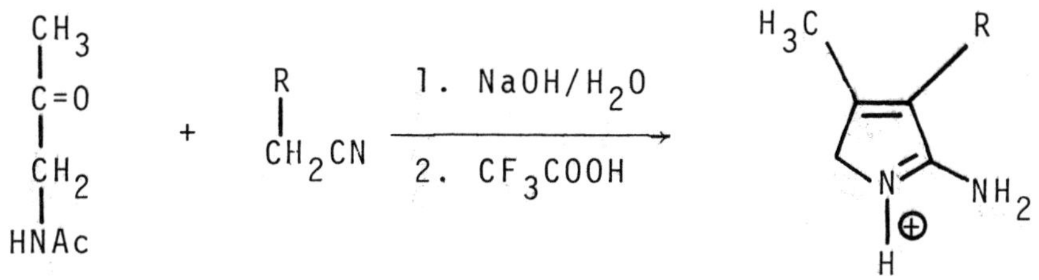

R = CN 73%

R = COO*t*-Bu 41%

IV.G-2 Z. Yoshida, T. Harada, and Y. Tamaru, Tetrahedron Lett., 3823 (1976).

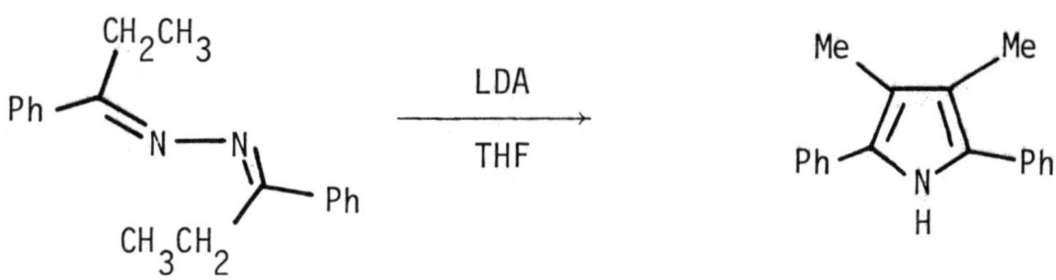

52%

IV.G-3 H. Sliwa and D. Blondeau, Tetrahedron Lett., 933 (1976).

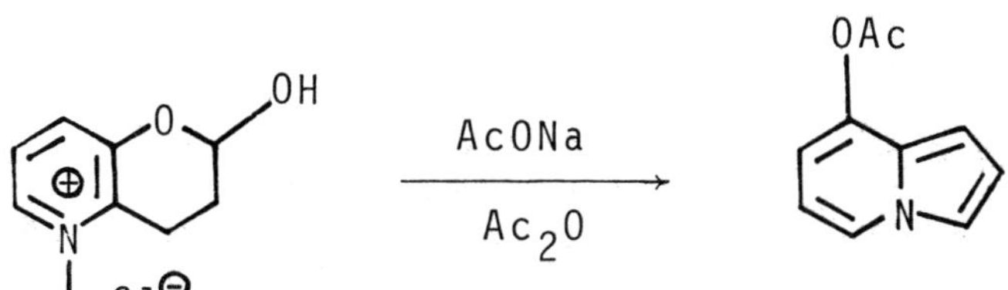

70% isolated

IV.G-4 R. R. Fraser and S. Passannanti, Synthesis, 540 (1976).

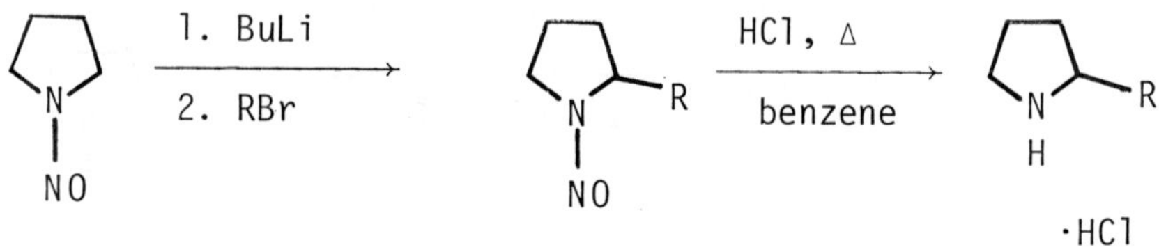

R = benzyl, 1° alkyl

(may be repeated)

IV.G-5 J. Leroy and C. Wakselman, Can. J. Chem., 54, 218 (1976).

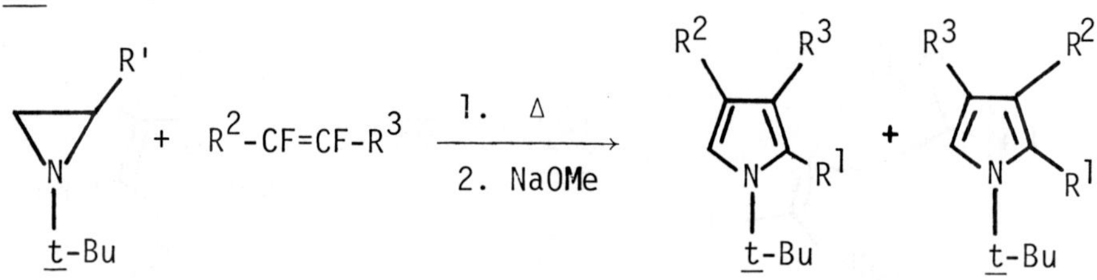

R^1 = CN, COOR

R^2,R^3 = H, F, CF_3

Yields ~20-60%

IV.G-6 T. Uchida and K. Matsumoto, Synthesis, 209 (1976).

Review: "Methods for the Construction of the Indolizine Nucleus"

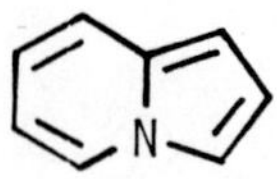

IV.G-7 J. M. Patterson, Synthesis, 281 (1976).

Review: "Recent Synthetic Methods for Pyrroles and Pyrrolenines (2H- or 3H-Pyrroles)"

IV.H-1 A. A. Akhrem, F. A. Lakhvich, and V. A. Khripach, J. Gen. Chem. (USSR), 45, 2534 (1975).

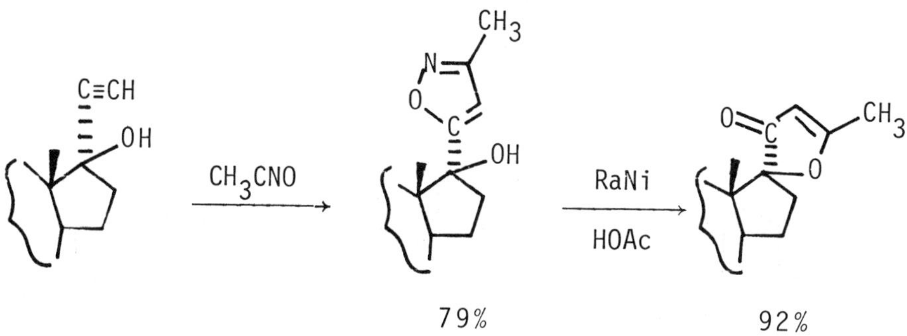

IV.H-2 R. J. Bass, J. Chem. Soc., Chem. Commun., 78 (1976).

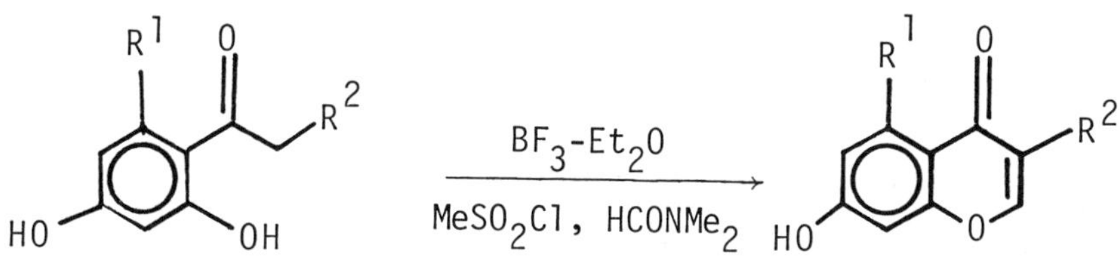

R^1 = H, OH

R^2 = aryl, Bz, OMe, cyclopentyl

60-98%

IV.H-3 J. Martens, K. Praefcke, and U. Schulze, Synthesis, 532 (1976).

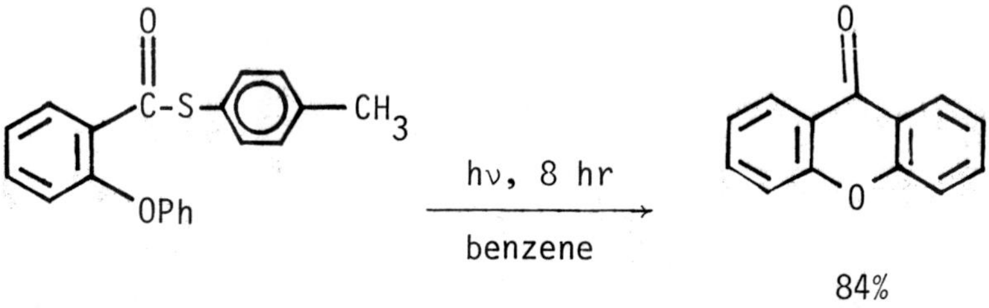

IV.H-4 G.J.P. Becket and G. P. Ellis, Tetrahedron Lett., 719 (1976).

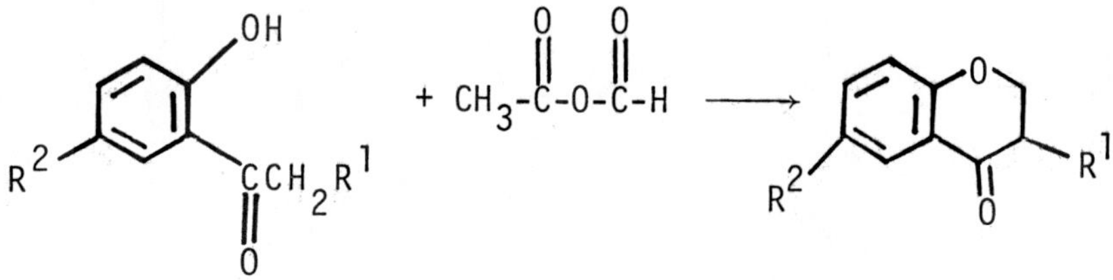

R^1 = Ac, NO_2

R^2 = H, NO_2

"excellent" yields

IV.H-5 G. Pfeiffer and H. Bauer, Liebigs Ann. Chem., 383 (1976).

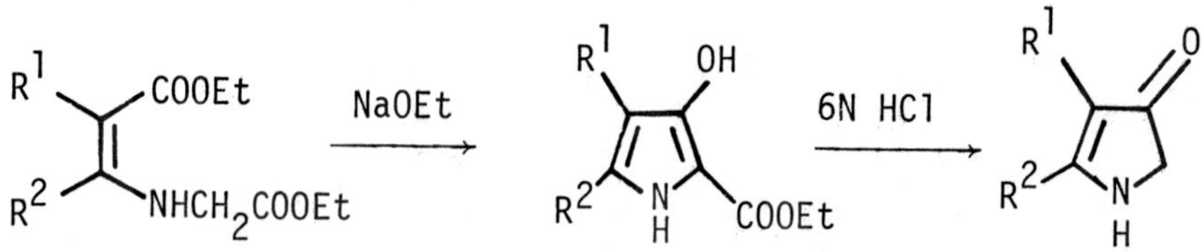

R^1,R^2 = Me, n-Bu, cyclic

>90%

IV.H-6   D. S. Black and A. B. Boscacci, <u>Aust. J. Chem.</u>, <u>29</u>, 2561 (1976).

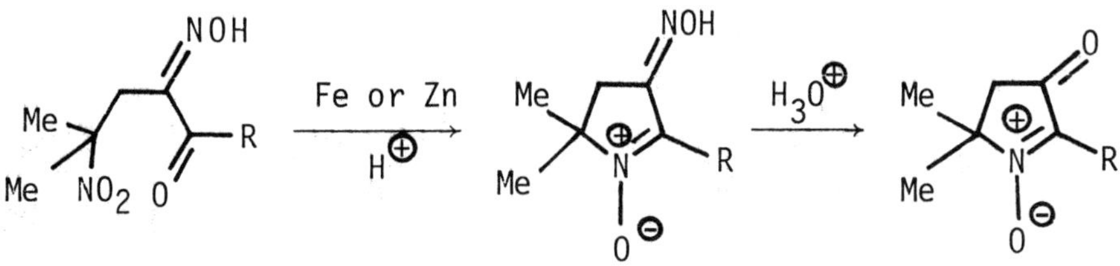

~50%

IV.H-7   H. O. House and L. F. Lee, <u>J. Org. Chem.</u>, <u>41</u>, 863 (1976).

$CH_2=CH(CH_2)_2CH(CH_3)NHOH$

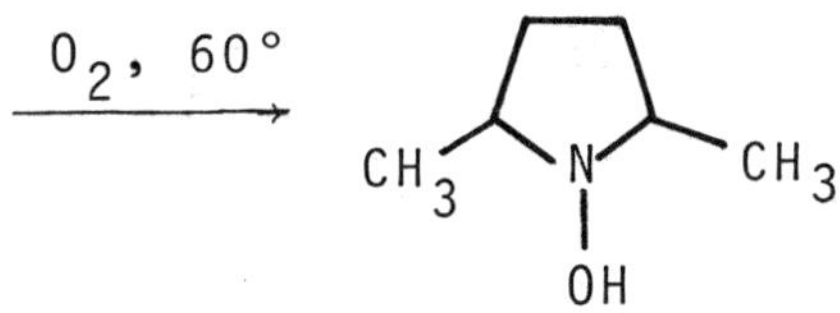

~80%

$CH_2=CH(CH_2)_3CH_2NHOH$

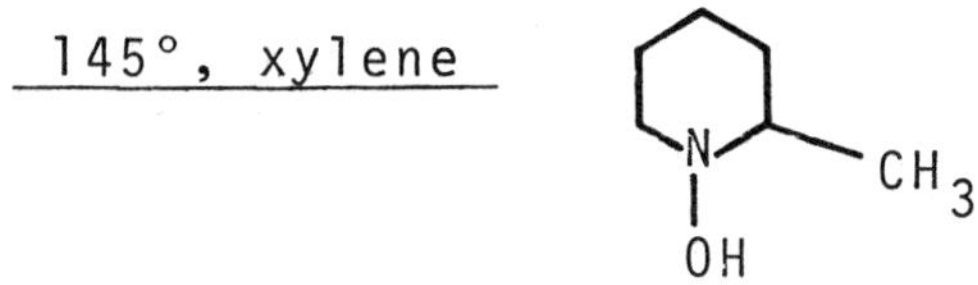

~70-80%

IV.H-8 P. K. Larsen and H. Hjeds, Acta Chem. Scand. (B), 30, 884 (1976).

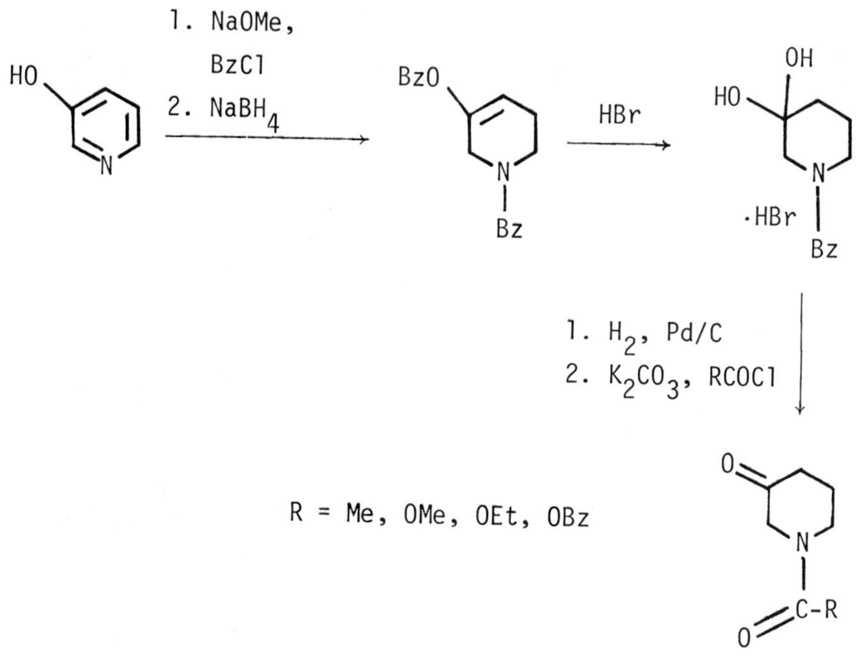

R = Me, OMe, OEt, OBz

~30% overall

IV.H-9 C. R. Johnson and K. Tanaka, Synthesis, 413 (1976).

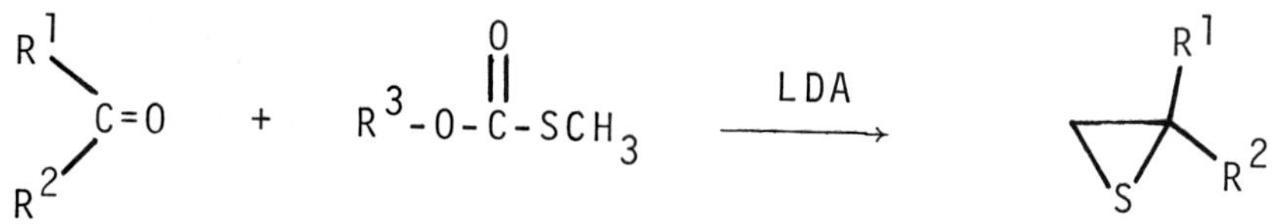

R^1, R^2 = H, alkyl, aryl

R^3 = (-)-menthyl

55-77% yield
13-34% ee

VI.H-10 A. I. Meyers and M. E. Ford, J. Org. Chem., 41, 1735 (1976).

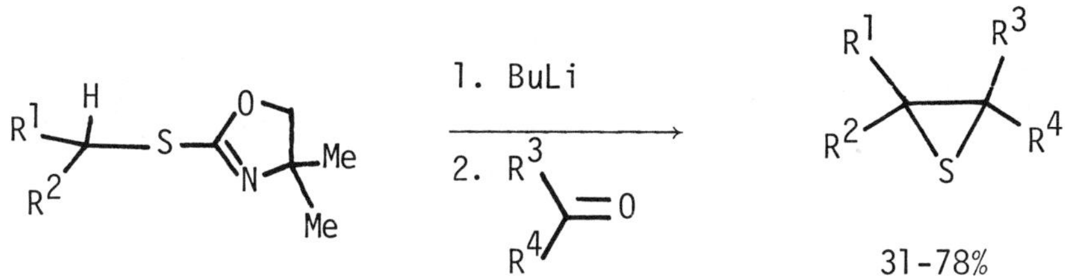

31-78%

Full paper, many examples cited.

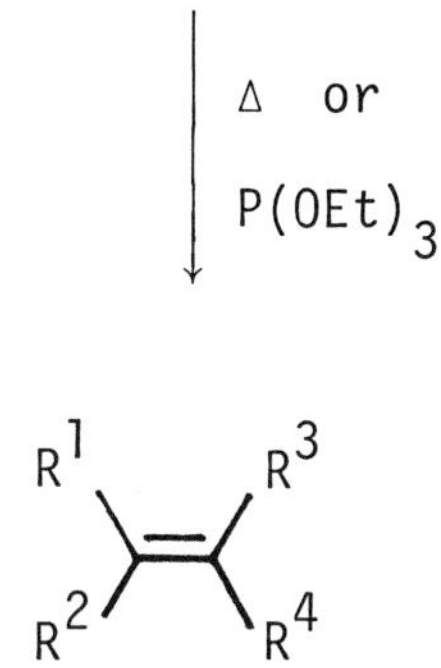

40-81%

IV.H-11 I. L. Knunyants, Doklady Chem., 224, 521 (1975).

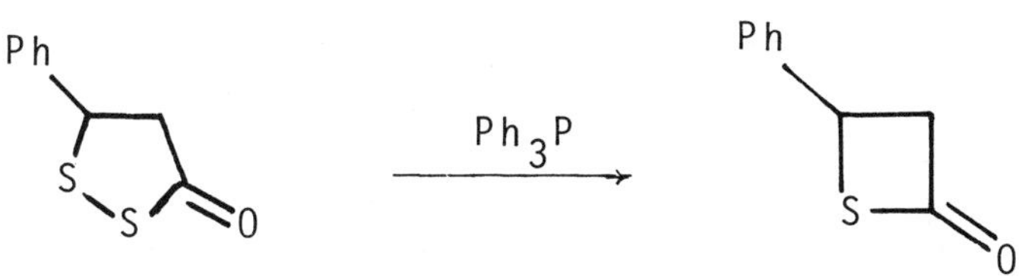

98%

IV.H-12 F. Duus, Tetrahedron, 32, 2817 (1976).

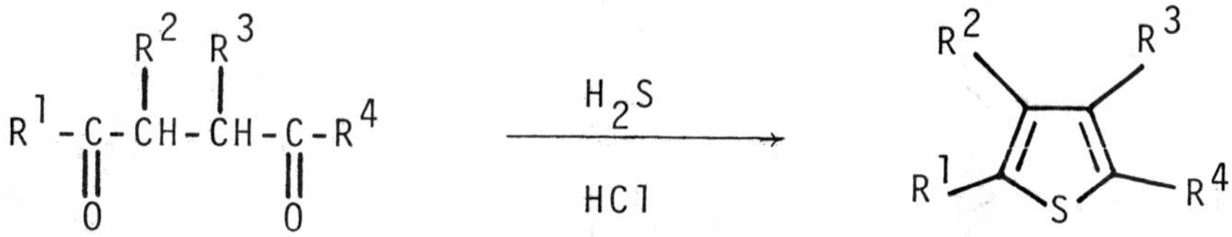

R's = Me, Ph, t-Bu, COOR

Widely varying yields.

IV.H-13 W. Ried and L. Kaiser, Synthesis, 120 (1976).

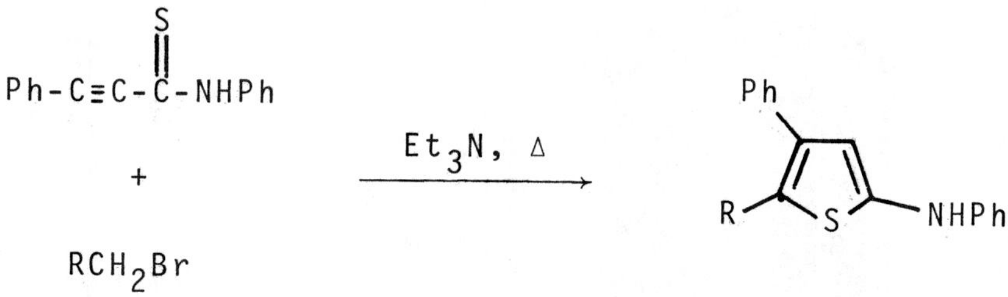

45-83%

R = NO_2, CN, p-NO_2Ph

IV.H-14 H. Gotthardt, M. C. Weisshuhn, and B. Christl, Chem. Ber., 109, 753 (1976).

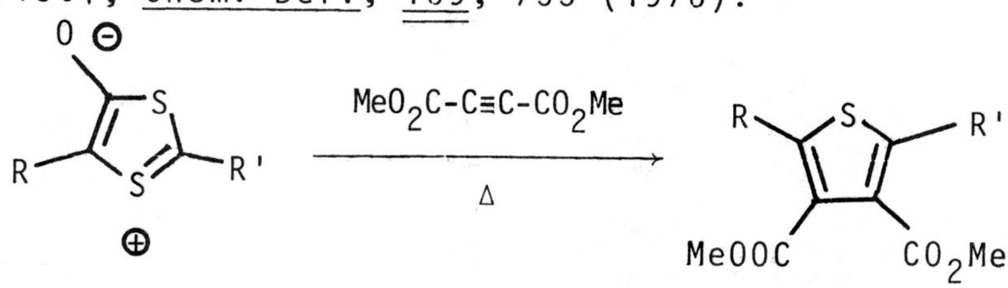

R = Me, aryl

R' = aryl

~70-90%

IV.H-15 H. Blatt et al., Aust. J. Chem., 29, 883 (1976).

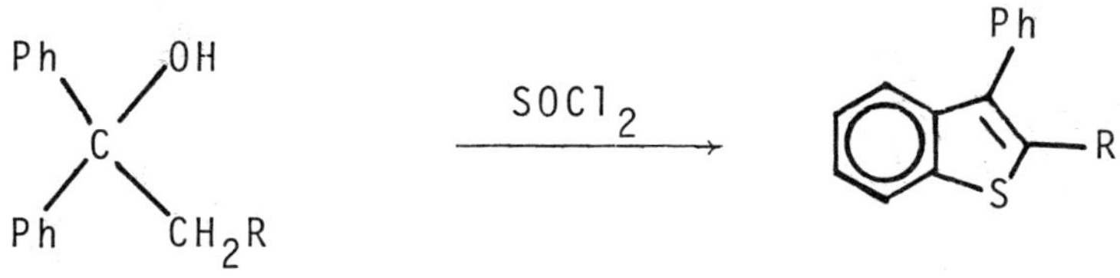

~50%

R = H, Me, Et, Ph

IV.H-16 T. Higa and A. J. Krubsack, J. Org. Chem., 41, 3399 (1976).

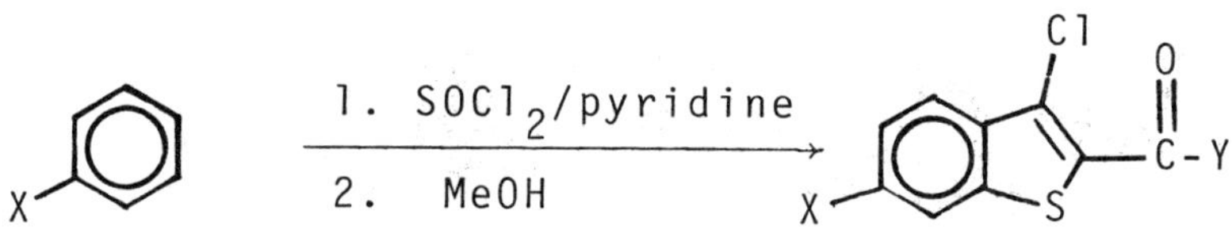

X = H, CH_3, OCH_3, NO_2

Y = Cl, OCH_3

46-69%

IV.H-17 J. M. McIntosh and H. Khalil, Can. J. Chem., 54, 1923 (1976).

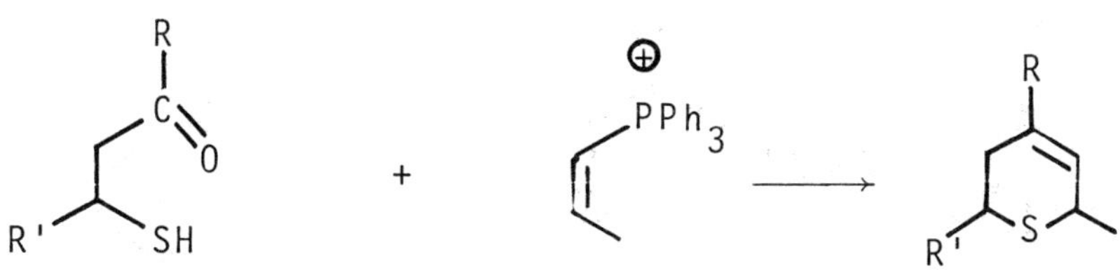

generally ~50-80%

R = H, Me

R' = Me, Ph

IV.I.1.a-1 J. P. Ferris and R. W. Trimmer, J. Org. Chem., 41, 19 (1976).

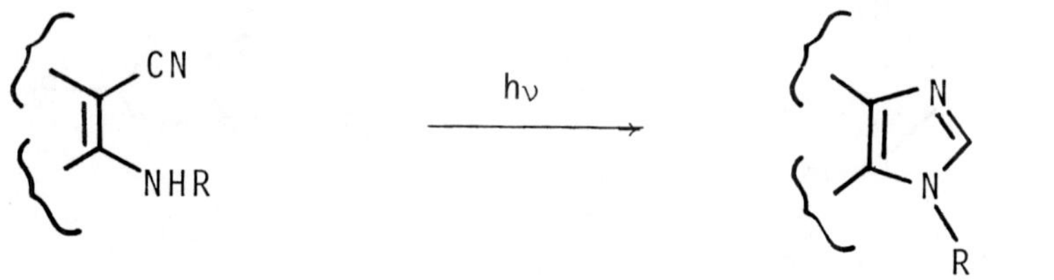

R = H, CH_3, t-Bu

Yields ~50-80%

IV.I.1.a-2 D. Pocar and R. Stradi, Tetrahedron Lett., 1839 (1976).

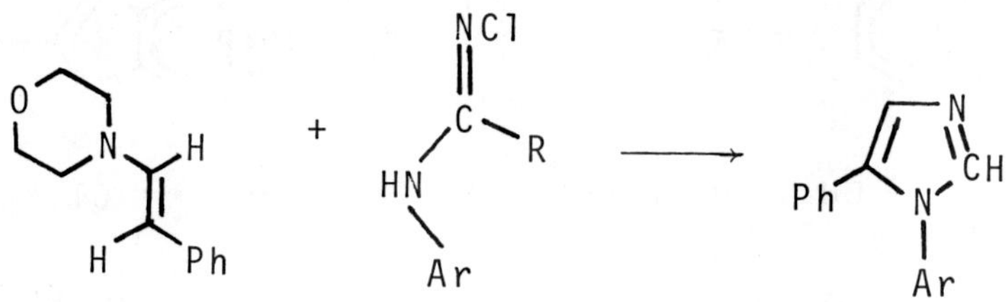

"good yields"

IV.I.1.a-3 G. M. Devasia, Tetrahedron Lett., 571 (1976).

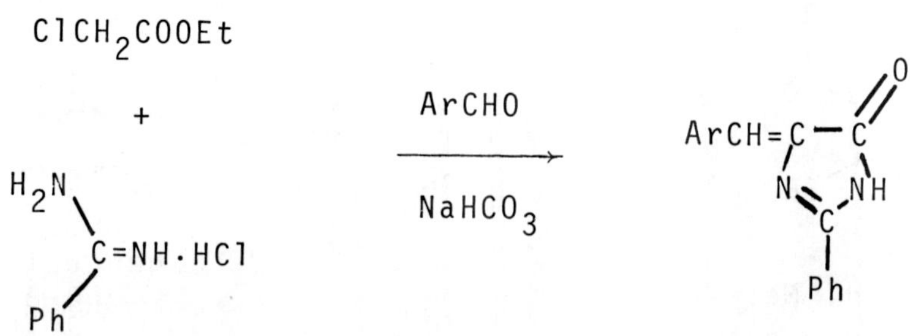

Yields ~40-55%

IV.I.1.a-4 Y. Kikugawa and L. A. Cohen, Chem. Pharm. Bull., 24, 3205 (1976).

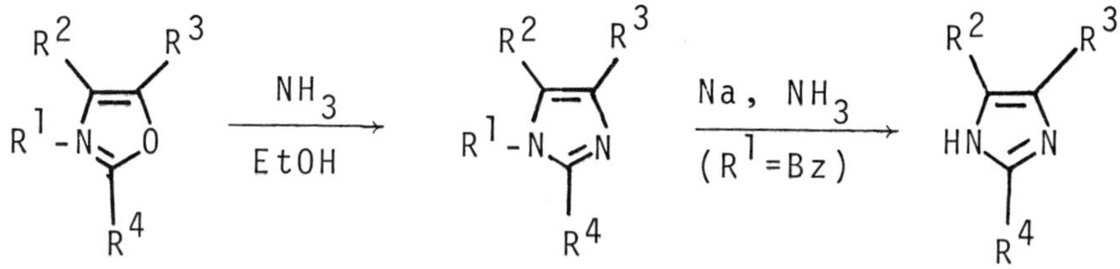

R^1 = Me, Bz 61-97%

R^{2-4} = H, Me, Et, Ph

IV.I.1.a-5 N. Latif, N. Mishriky, and N. S. Girgis, Chem. and Ind., 28 (1976).

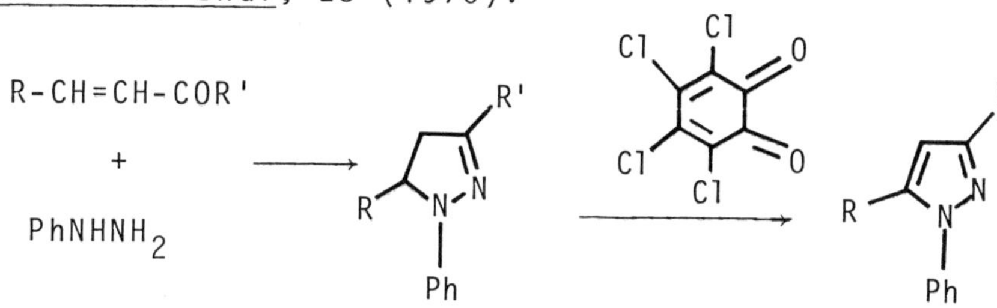

No yields given.

IV.I.1.a-6 R. Baumes, R. Jacquier, and G. Tarrago, Bull. Soc. Chim. France, 260 (1976).

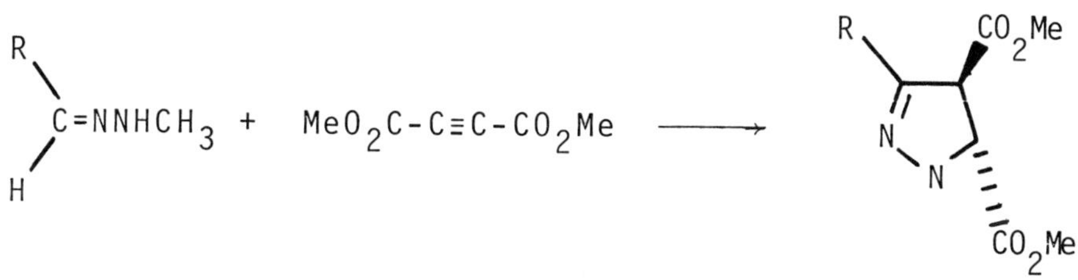

R = alkyl, Ph 45-60%

IV.I.1.a-7 J. F. Hansen and D. E. Vietti, J. Org. Chem., 41, 2871 (1976).

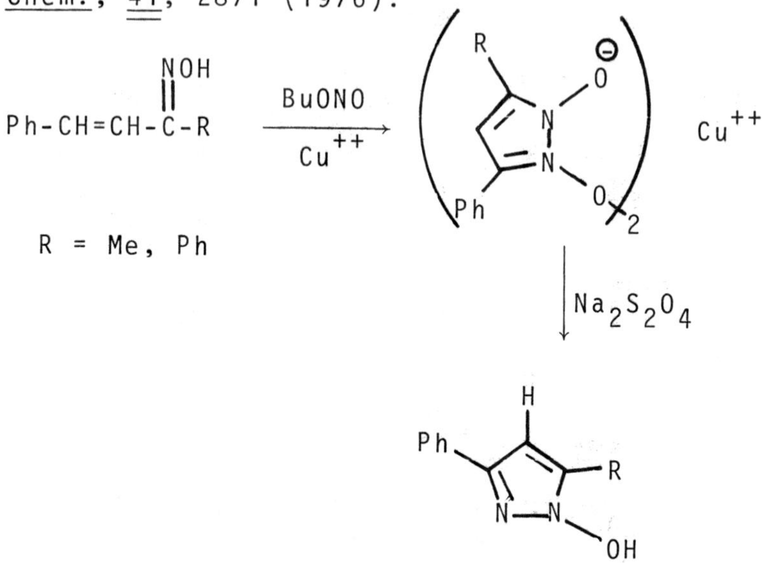

R = Me, Ph

IV.I.1.a-8 P. Frêche, A. Gorgues, and E. Levas, Tetrahedron Lett., 1495 (1976).

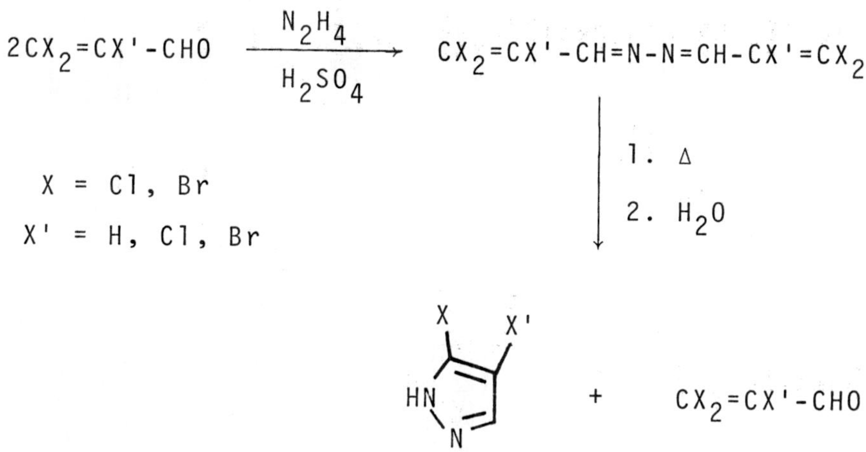

X = Cl, Br

X' = H, Cl, Br

Overall yields 30-70%

IV.I.1.a-9 C. F. Beam et al., Synth. Commun., 6, 5 (1976).

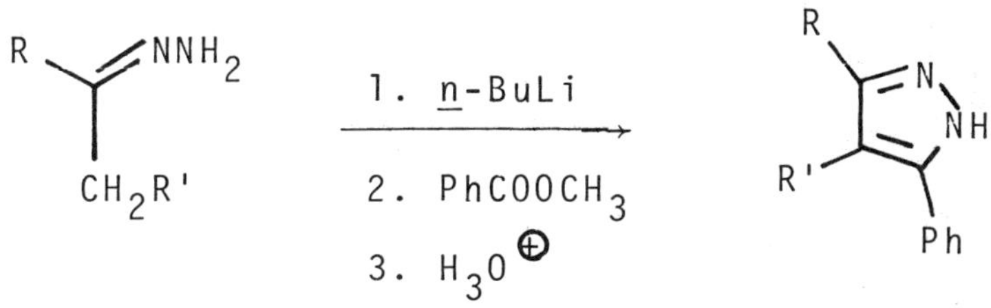

R,R' = 1° alkyl, aryl, cyclic

27-73%

IV.I.1.a-10 A. Kasahara, Chem. and Ind., 1032 (1976).

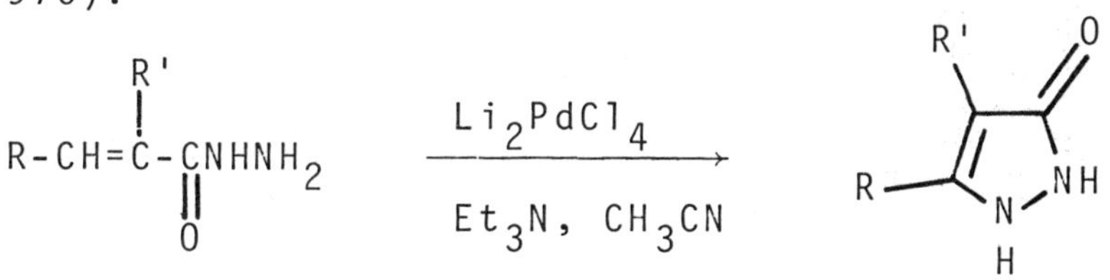

R = H, Me, Ph

R' = H, Me

33-42%

IV.I.1.a-11 N. Suzuki, T. Yamabayashi, and Y. Izawa, Bull. Chem. Soc. Japan, 49, 353 (1976).

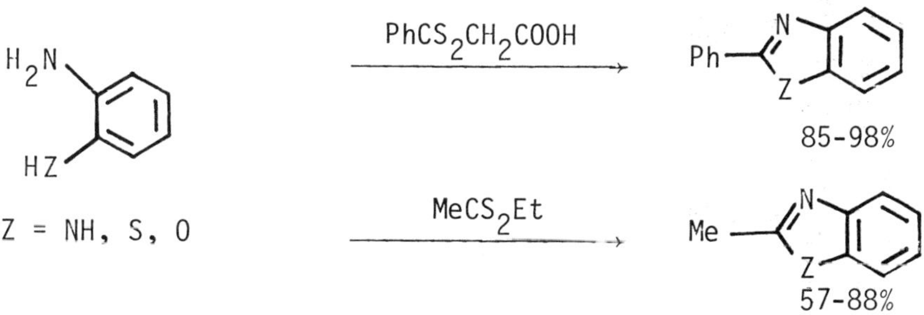

Z = NH, S, O

85-98%

57-88%

IV.I.1.a-12 A. Botta, Liebigs Ann. Chem., 336 (1976).

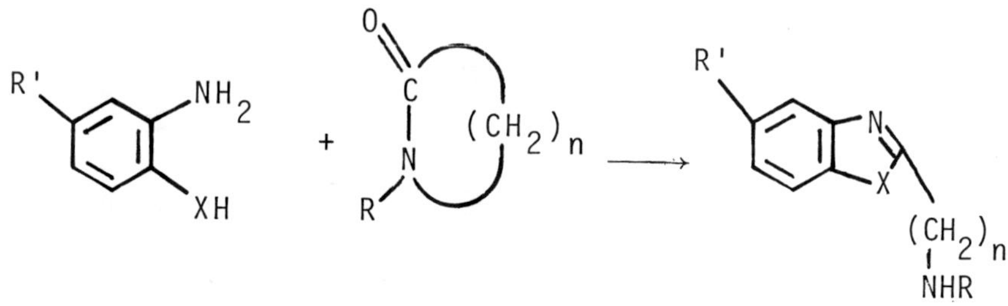

X = NH, O, S   R = H, Me, Ph   ~50-80%

R' = H, Me   n = 3,4,5,11

IV.I.1.b-1 A. Kreutzberger and U.-H. Tesch, Chem. Ber., 109, 3255 (1976).

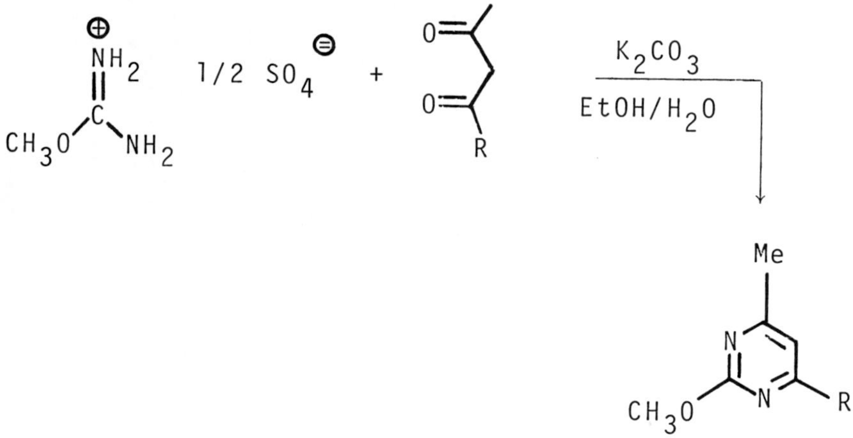

70%, R = Me

56%, R = Et

IV.I.1.b-2 J.-P. Gallemaers. D. Christophe, and R. Promel, <u>Tetrahedron Lett.</u>, 693 (1976).

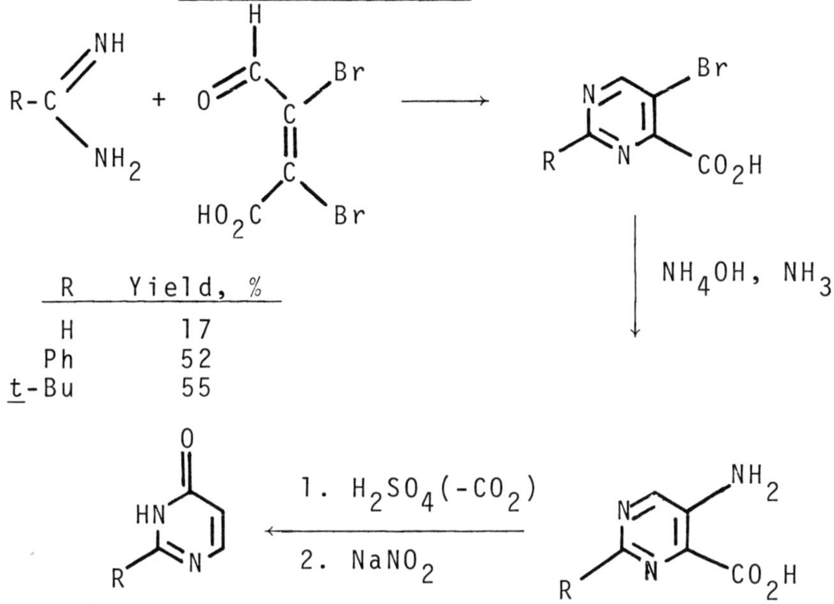

R	Yield, %
H	17
Ph	52
t-Bu	55

IV.I.1.b-3 T. Hirayama *et al.*, <u>Chem. Pharm. Bull.</u>, **24**, 26, 507 (1976).

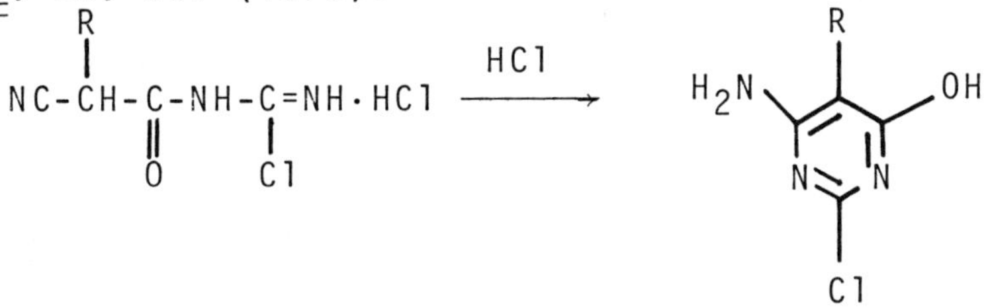

R = H, Me, Ph 56-97%

Also several similar pyrimidine syntheses.

IV.I.1.b-4 S.M.S. Chauhan and H. Junjappa, Tetrahedron, 32, 1779, 1911 (1976).

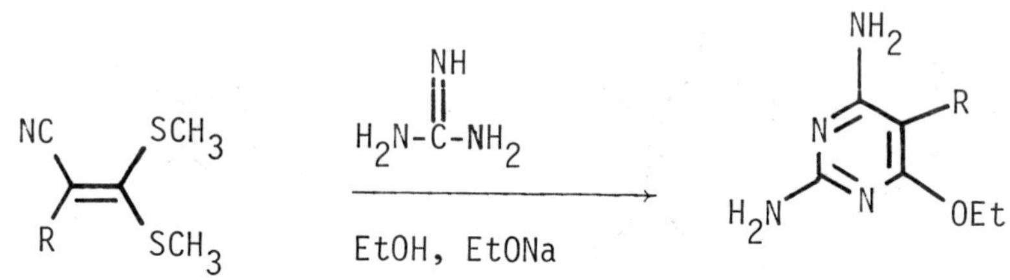

R = CN, Ar

50-60%

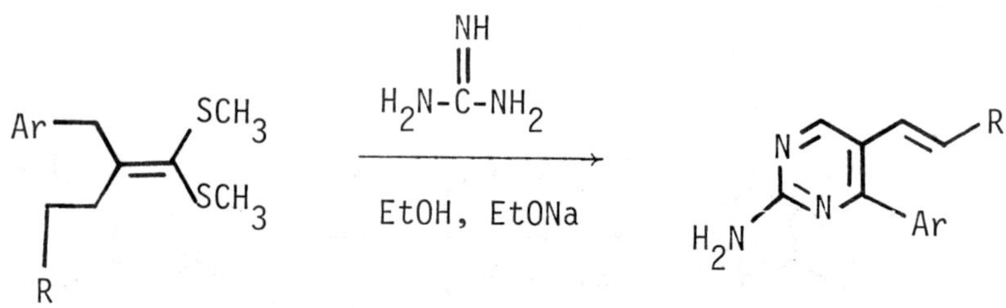

"fair" yields

R = H, CH_3

IV.I.1.b-5 C. H. Foster and E.U. Elam, J. Org. Chem., 41, 2646 (1976).

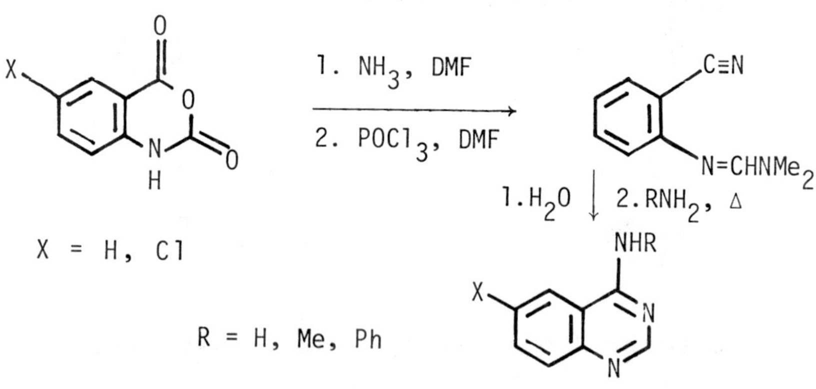

X = H, Cl

R = H, Me, Ph

44-79%

IV.I.1.b-6 D. G. McMinn, Synthesis, 824 (1976).

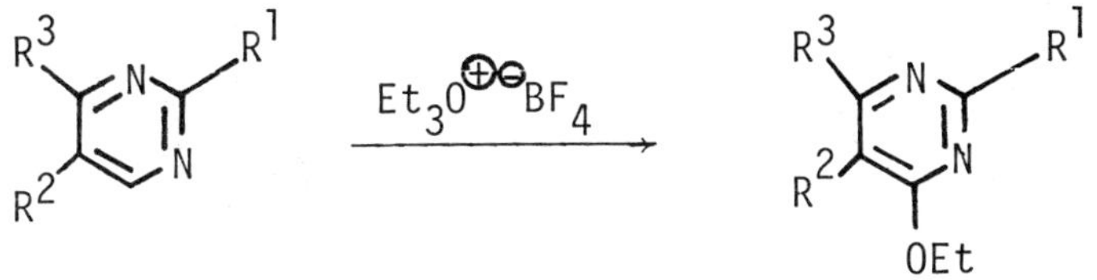

R^1 = Me, OEt, OH, NH_2

R^2 = H, COOR

R^3 = H, Me, Ph

~60-75%

IV.I.1.b-7 R. H. Fischer and H. M. Weitz, Synthesis, 53 (1976).

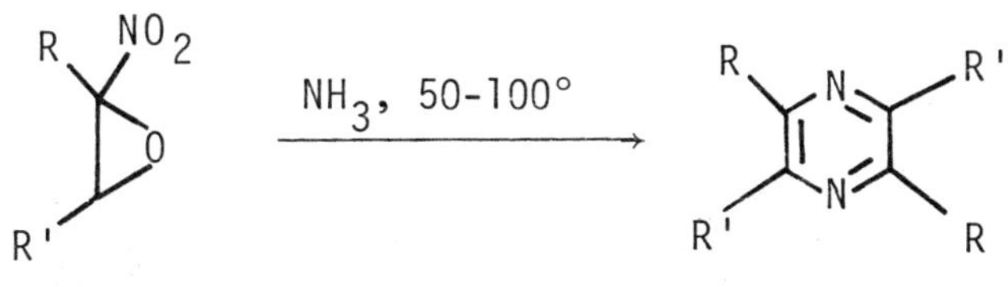

R,R' = alkyl, aryl, cyclic

44-94%

IV.I.1.b-8 T. Kato, A. Takada, and T. Ueda, Chem. Pharm. Bull., 24, 431 (1976).

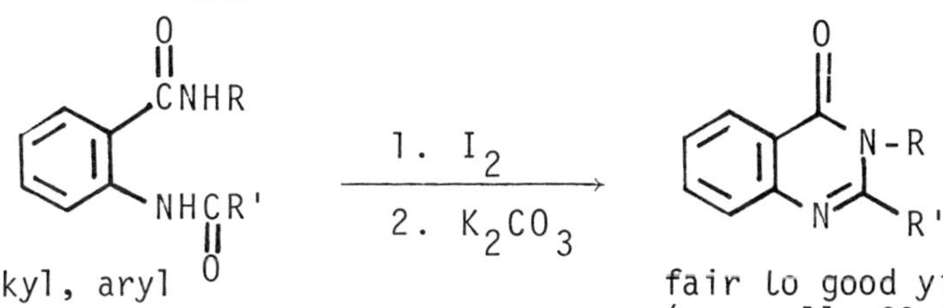

R = alkyl, aryl

R' = alkyl

fair to good yields (generally 30-60%)

IV.I.1.b-9 Y. F. Shealy and C. A. O'Dell, J. Het. Chem., 13, 1041 (1976).

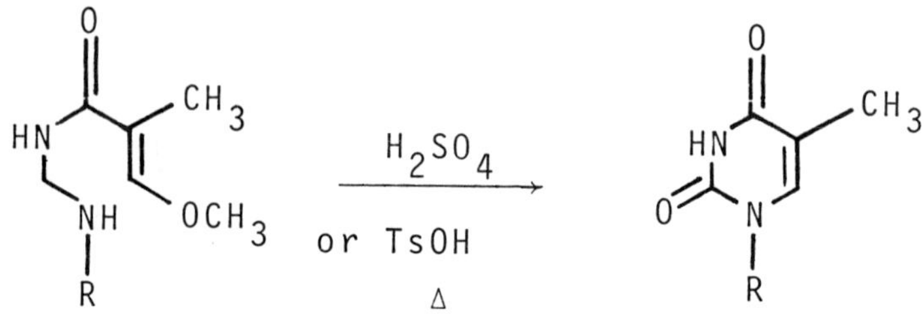

~85%

R = Ph,

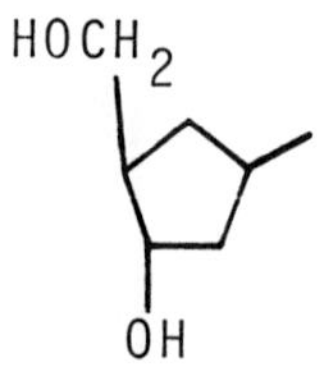

IV.I.1.b-10 J. R. Beck and R. P. Gajewski, J. Het. Chem., 13, 605 (1976).

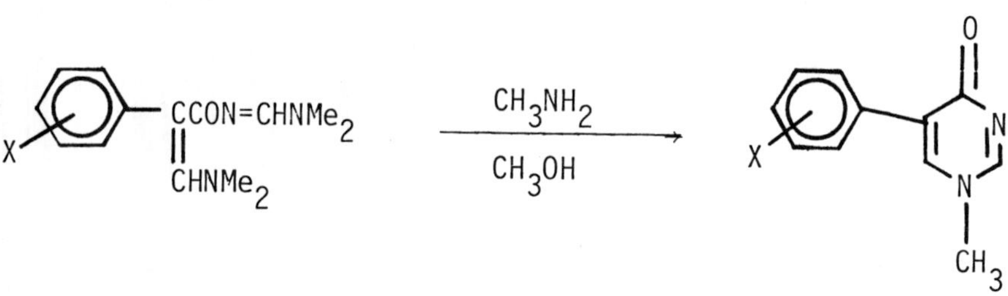

X = CF_3, Br, Cl, H

20-71%

IV.I.1.b-11 J.F.W. Keana *et al.*, *J. Org. Chem.*, **41**, 2124 (1976).

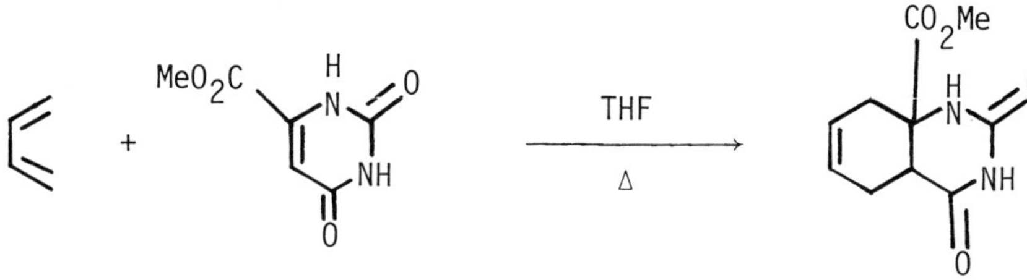

several dienes used.

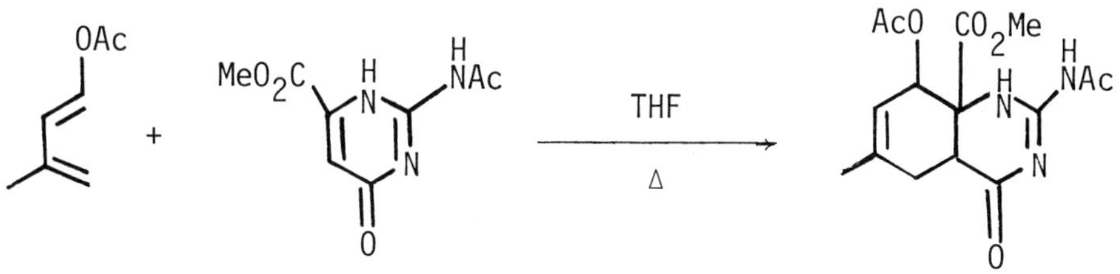

IV.I.1.b-12 K. Nagahara, K. Takagi, and T. Ueda, *Chem. Pharm. Bull.*, **24**, 1197 (1976).

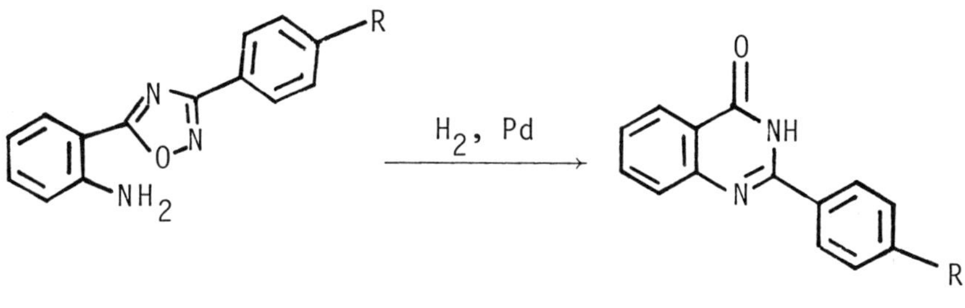

R = H, OCH_3 ~70%

IV.I.1.b-13 H. Wamhoff and B. Wehling, Chem. Ber., 109, 2983 (1976).

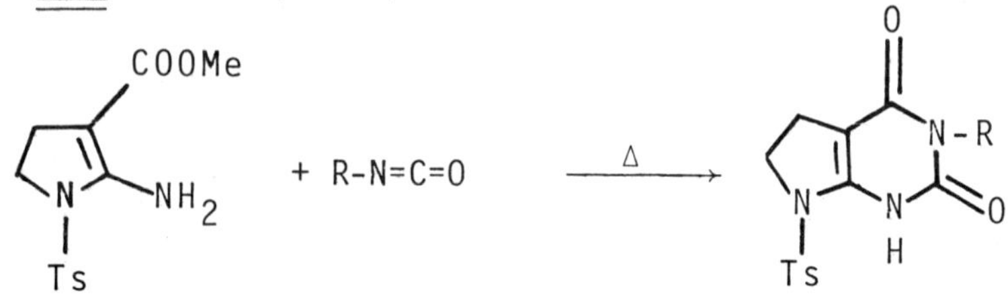

R = Ph, Bz, CH_2OCH_3 ~50%

IV.I.1.b-14 R. F. Abdulla, Tetrahedron Lett., 521 (1976).

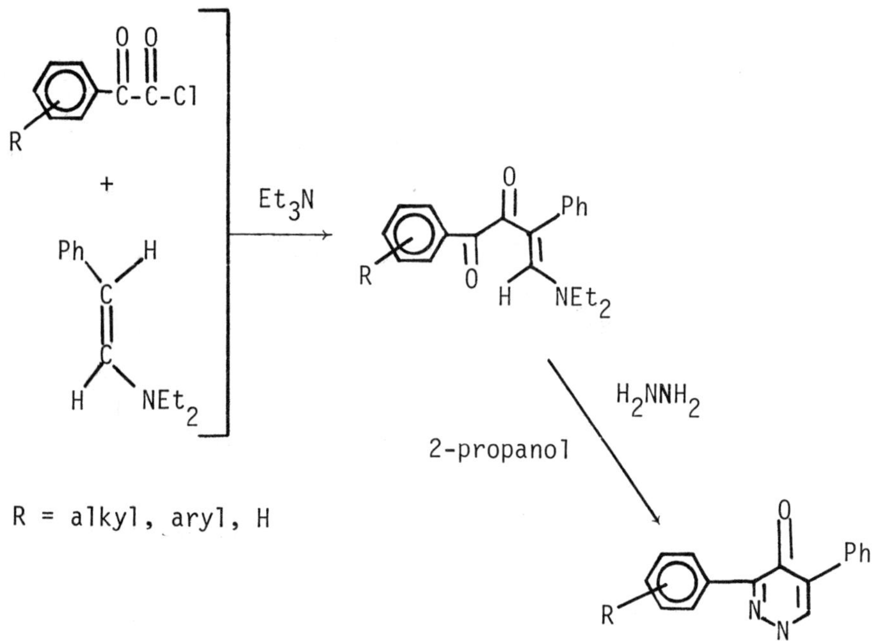

R = alkyl, aryl, H

~60%

IV.I.1.c-1 M. Ogata and H. Matsumoto, <u>Chem. and Ind.</u>, 1067 (1976).

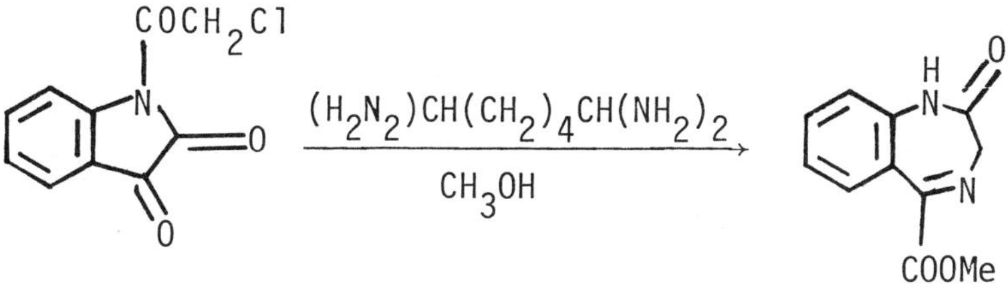

51%

IV.I.1.c-2 Y. Ohtsuka, <u>J. Org. Chem.</u>, <u>41</u>, 629 (1976).

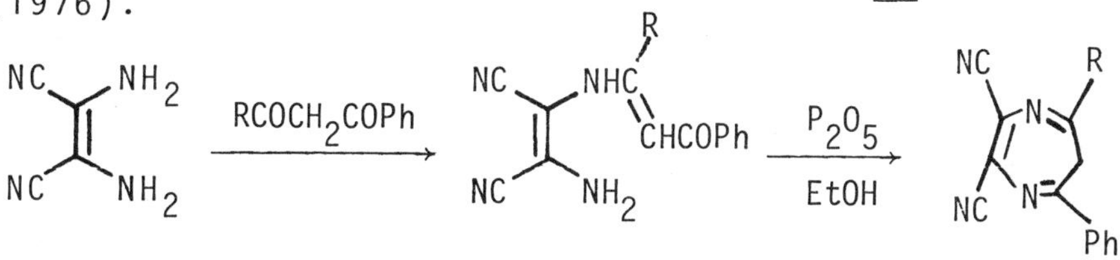

18% (R=Me)

73% (R=H)

IV.I.2-1 M. Miocque <u>et al.</u>, <u>Comptes Rendus (C)</u>, <u>282</u>, 469 (1976).

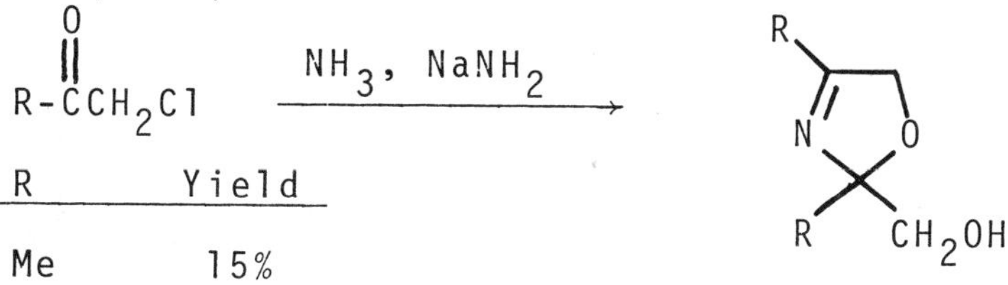

R	Yield
Me	15%
Ph	62%
Bz	66%

IV.I.2-2 U. Schollkopf et al., Liebigs Ann. Chem., 183 (1976).

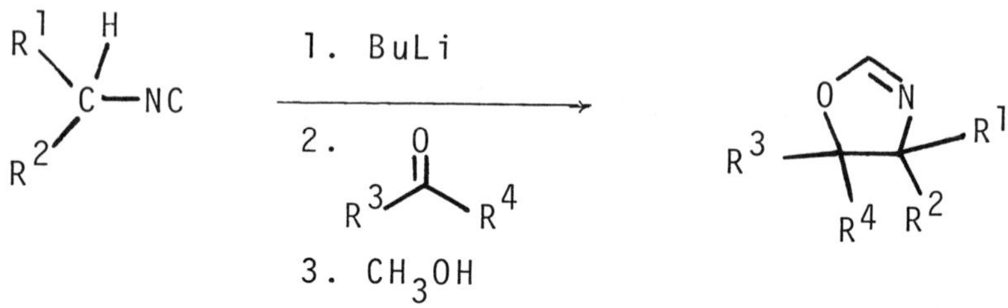

~60-80%

R's = H, 1° alkyl, aryl, cyclic

IV.I.2-3 A. Wolloch and E. Zbiral, Tetrahedron, 32, 1289 (1976).

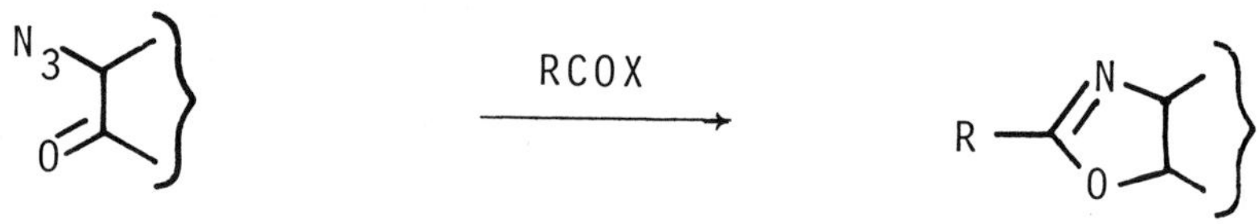

X = Cl, Br

29%, R=Me

45%, R=Ph

IV.I.2-4 C. A. Park et al., J. Het. Chem., 13, 449 (1976).

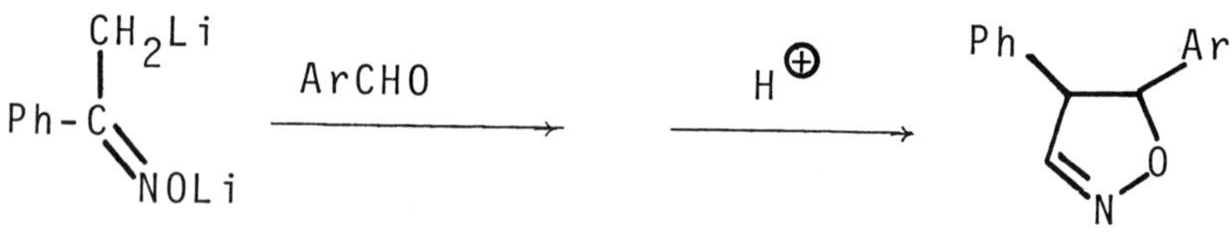

yields ~30-60%

IV.I.2-5 A. F. Cockerill *et al.*, *Synthesis*, 591 (1976).

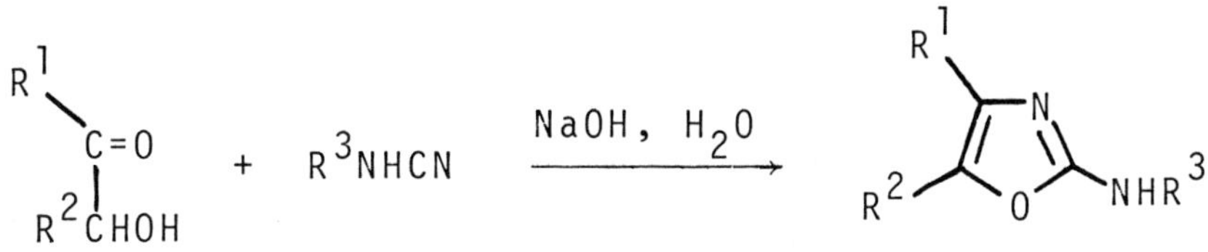

R^1,R^3 = 1°,2° alkyl, H, aryl ~40-90%

R^2 = H, Me

IV.I.2-6 S. P. McManus *et al.*, *J. Org. Chem.*, **41**, 1642 (1976).

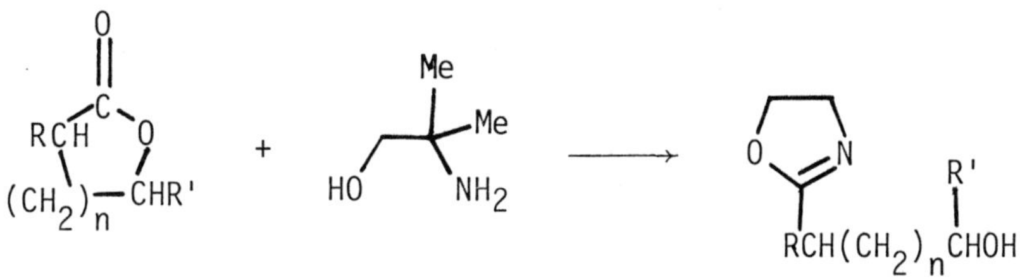

n = 1,2,3

R,R' = H, CH_3

45-100% (VPC)

IV.I.2-7 A. I. Meyers and E. D. Mihelich, *Angew. Chem. Int. Ed.*, **15**, 270 (1976).

Review: "The Synthetic Utility of 2-Oxazolines"

IV.I.2-8 R. A. Kretchmer and P. J. Daly, J. Org. Chem., 41, 192 (1976).

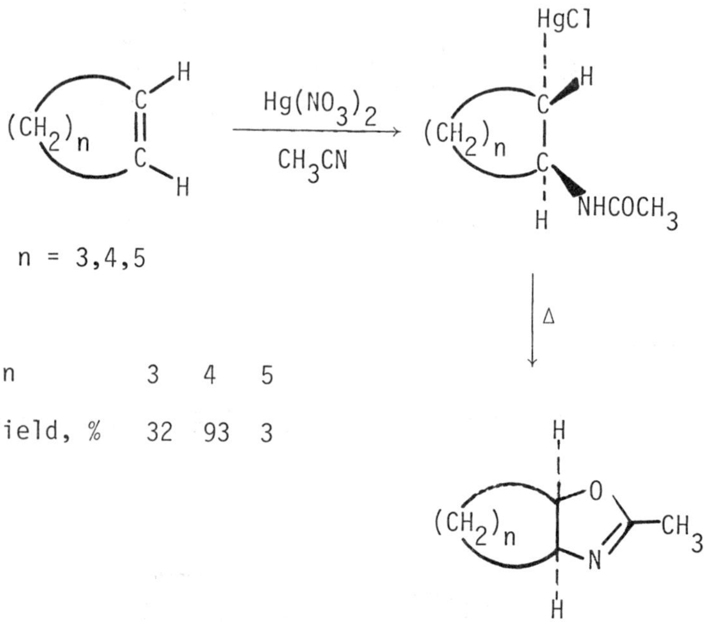

n = 3,4,5

n	3	4	5
Yield, %	32	93	3

IV.I.2-9 M. I. Shevchuk, S. T. Shpak, and A. V. Dombrovskii, J. Gen. Chem. (USSR), 45, 2571 (1975).

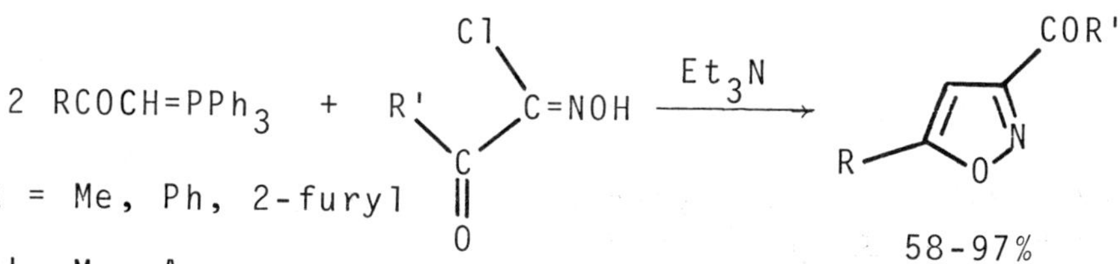

R = Me, Ph, 2-furyl

R' = Me, Ar

IV.I.2-10 R. M. Sandifer _et al_., _J. Het. Chem._, 13, 607 (1976).

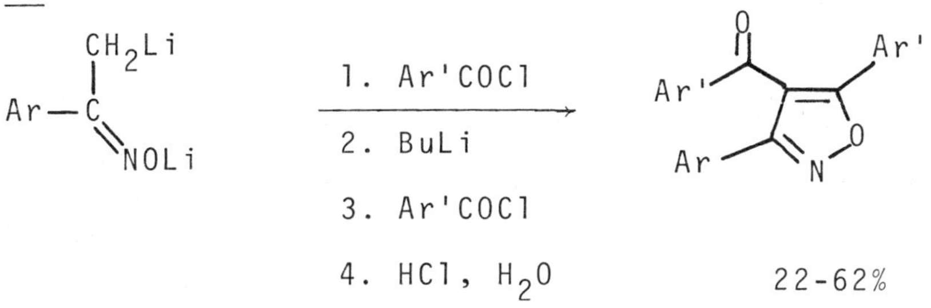

IV.I.2-11/IV.I.3-1 N. Suzuki, T. Yamabayashi, and Y. Izawa, _Bull. Chem. Soc. Japan_, 49, 353 (1976).

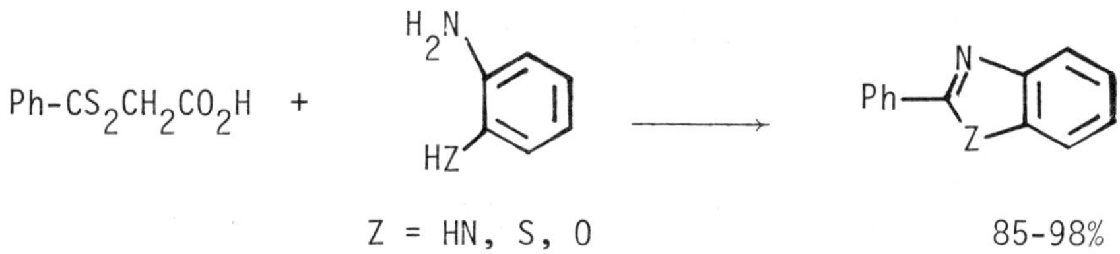

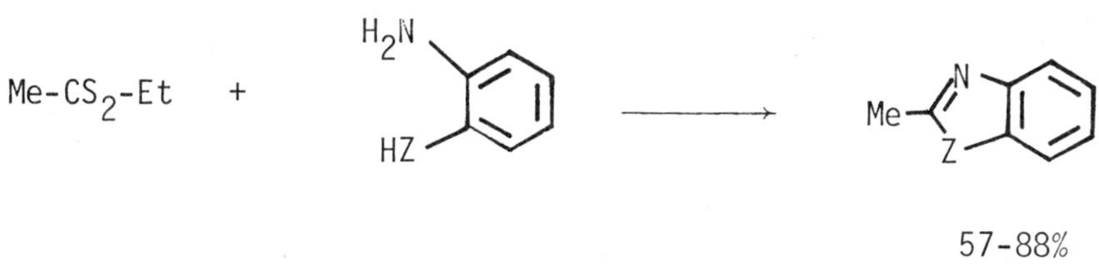

IV.I.2-12/IV.I.3-2 A. Botta, Liebigs Ann. Chem., 336 (1976).

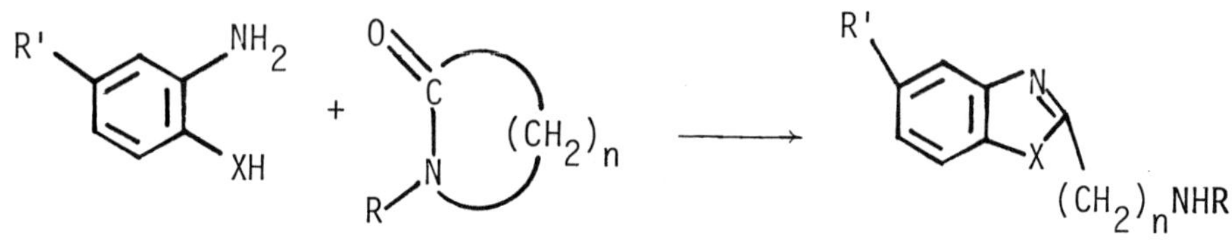

X = NH, O, S R = H, Me, Ph ~50-80%

R' = H, Me n = 3,4,5,11

IV.I.3-3 P. Dubs and R. Stuessi, Synthesis, 696 (1976).

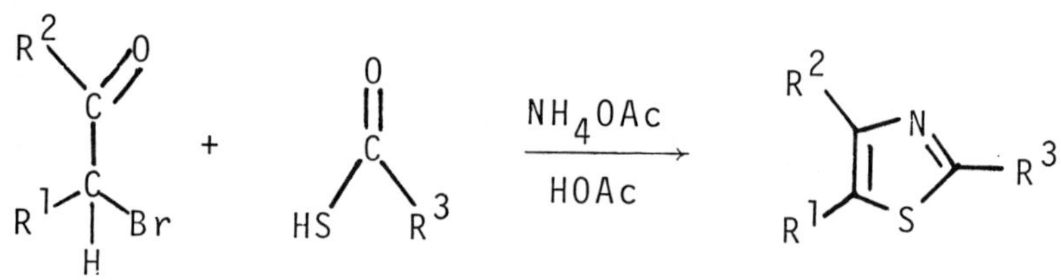

R^1, R^2 = H, CH_3 37-72%

R^3 = CH_3, i-Bu

IV.I.3-4 N. Suzuki and Y. Izawa, Bull. Chem. Soc. Japan, 49, 3155 (1976).

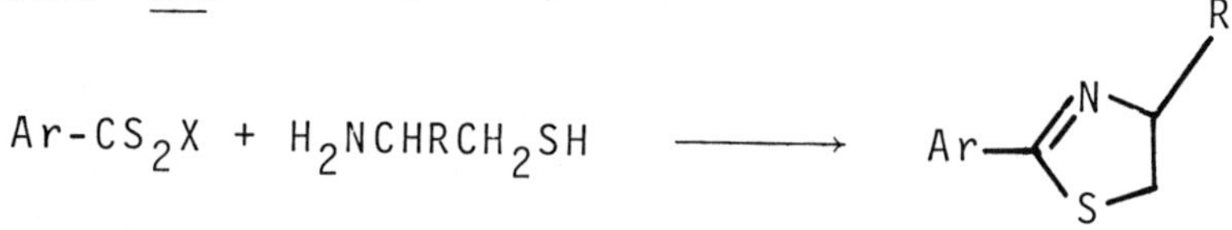

Ar = Ph, p-Cl-Ph

R = H, Et, COOR, CH_2COOR generally ~70-80%

IV.I.3-5 W. Ried and L. Kaiser, *Liebigs Ann. Chem.*, 395 (1976).

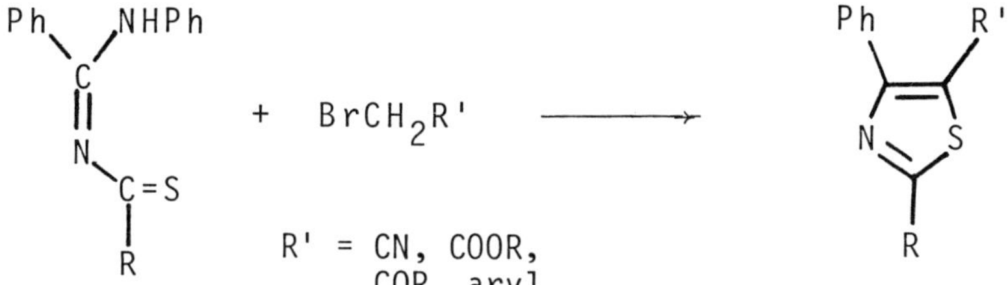

R' = CN, COOR, COR, aryl

R = OEt, morpholino

generally ~60-90%

IV.I.3-6 M. Ferrey, A. Robert, and A. Foucaud, *Synthesis*, 261 (1976).

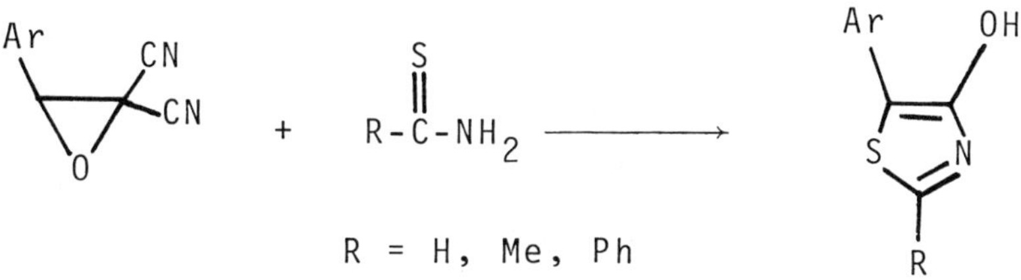

R = H, Me, Ph

36-55%

IV.I.3-7 R. D. Haugwitz *et al.*, *Synthesis*, 336 (1976).

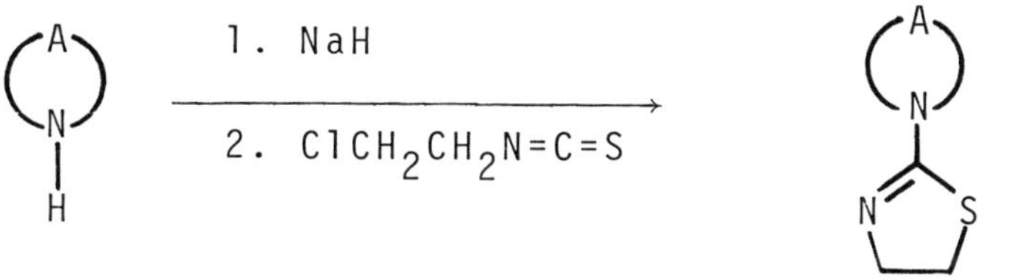

A = various heterocycles

46-93%

IV.I.3-8 G. D. Hartman and L. M. Weinstock, Synthesis, 681 (1976).

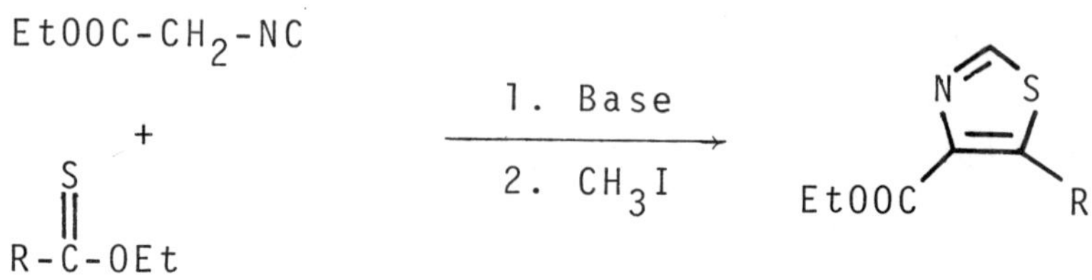

R = H, alkyl, Bz 68-92%

IV.I.3-9 M. J. Spitulnik, Synthesis, 730 (1976).

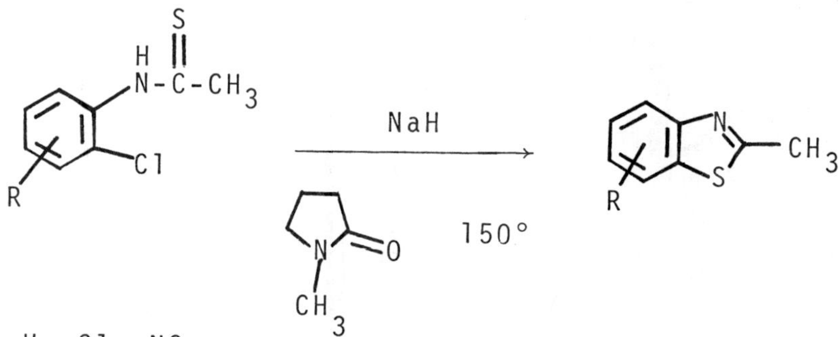

R = H, Cl, NO_2 ~50-70%

IV.I.3-10 G. Westphal et al., J. Prakt. Chem., 318, 875 (1976).

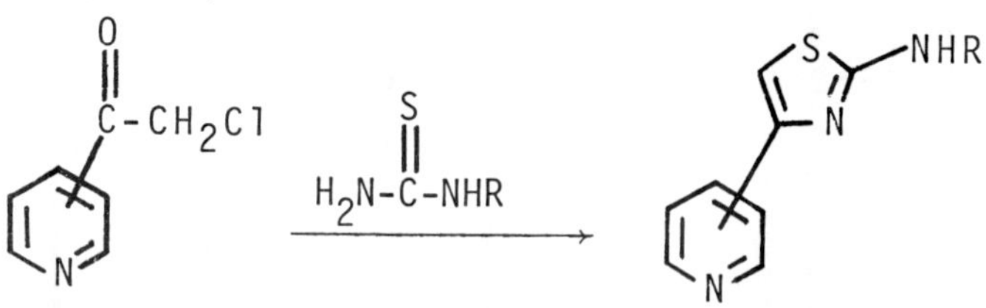

R = 1° alkyl, aryl ~30%

IV.I.3-11 M. Baudy and A. Robert, J. Chem. Soc., Chem. Commun., 23 (1976).

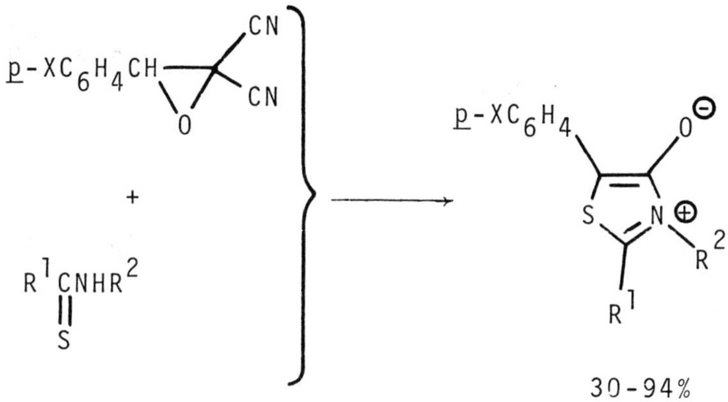

X = H, Cl, NO_2

R^1 = Me, Ph

R^2 = Ph, Bz

IV.I.3-12 K. Peseke, Synthesis, 386 (1976).

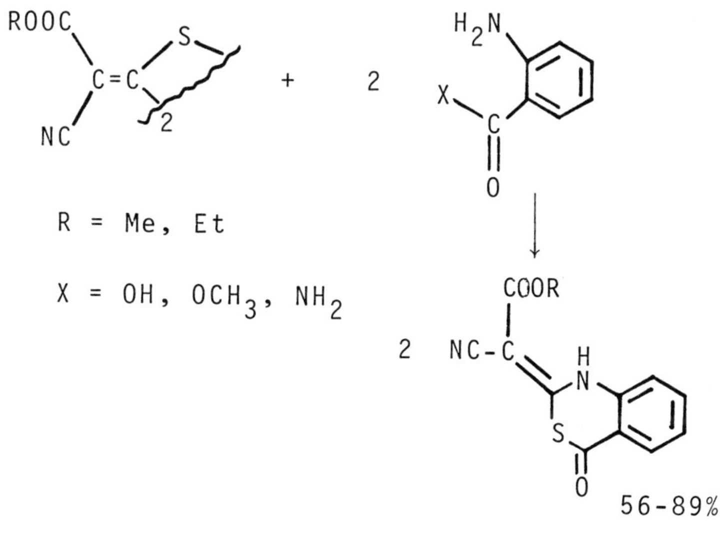

R = Me, Et

X = OH, OCH_3, NH_2

IV.I.4-1 S. Cabiddu, A. Maccioni, and M. Secci, Synthesis, 797 (1976).

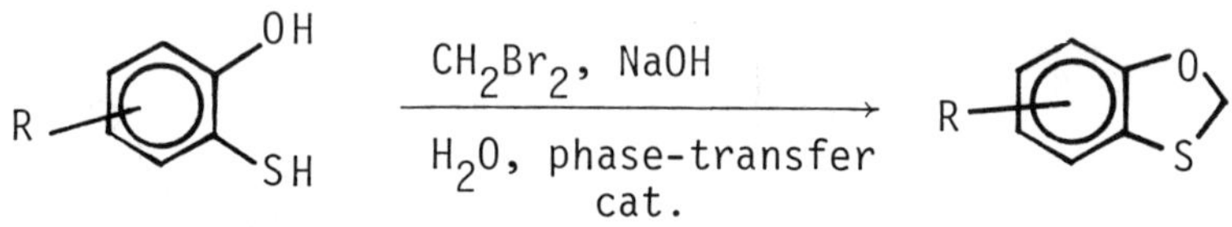

70-81%

R = Me, Cl

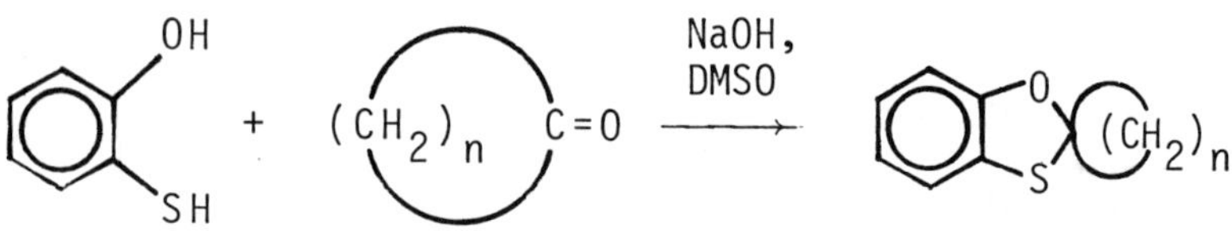

n = 4,5,6 65-71%

IV.I.5-1 N. E. Alexandrou and S. Adamopoulos, Synthesis, 482 (1976).

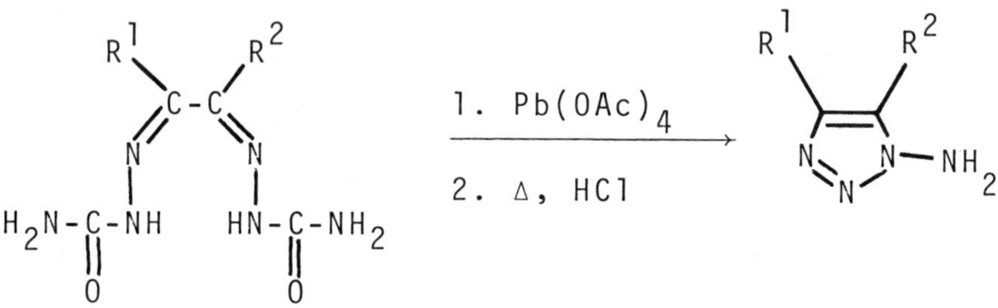

~20-30%

R^1, R^2 = H, Me, aryl

IV.I.5-2 J. Stein, J. Prakt. Chem., 318, 693

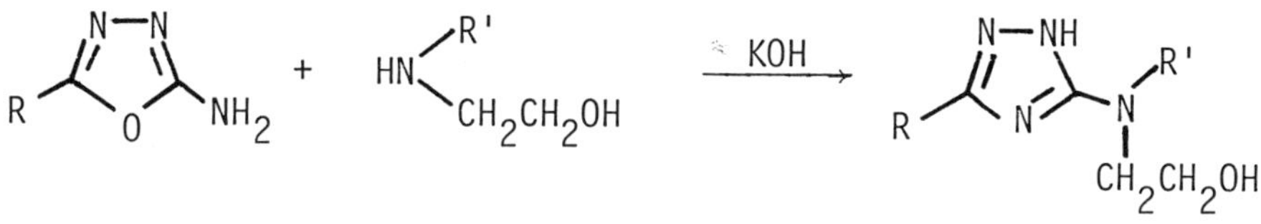

R = n-Bu, aryl

R' = CH_3, CH_2CH_2OH

40-95%

IV.I.5-3 S. Gorjan et al., Monat. fur Chem., 107, 1199 (1976).

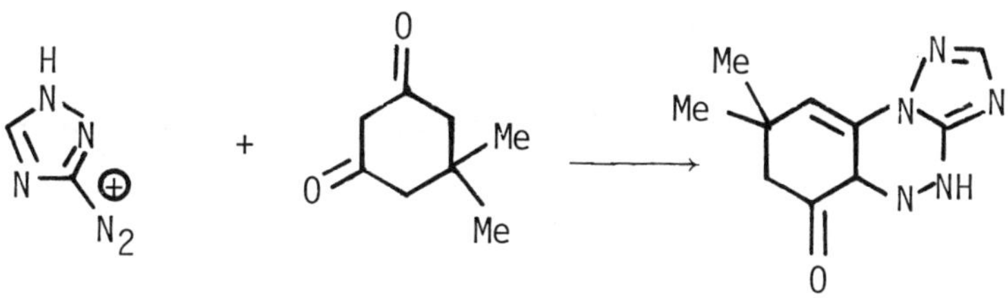

53%

IV.I.5-4 L. Garanti, A. Scandroglio and G. Zecchi, J. Het. Chem., 13, 1339 (1976).

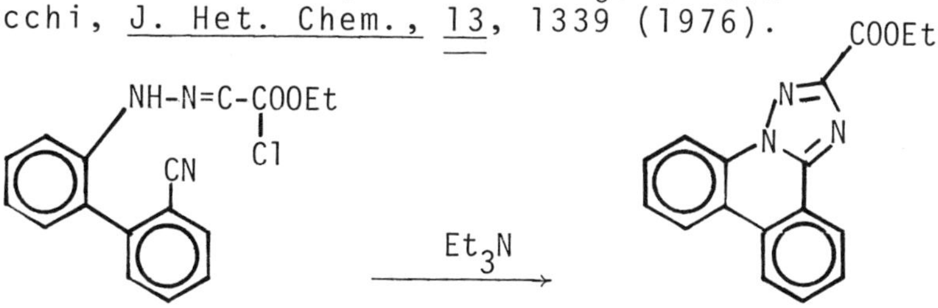

76%

IV.I.5-5 H. Neunhoeffer and V. Boehnisch, Liebigs Ann. Chem., 153 (1976).

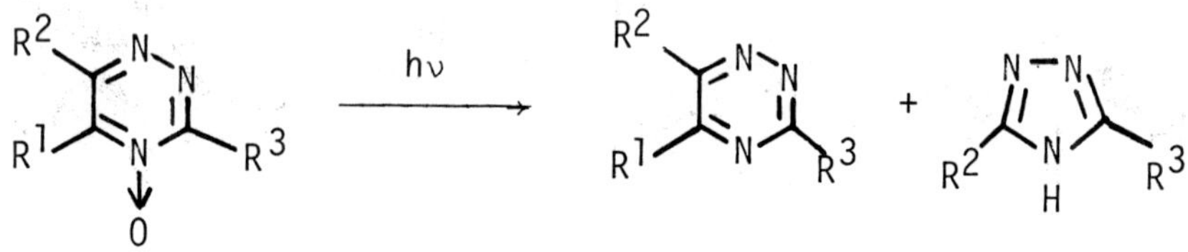

48-82% total yield

IV.I.5-6 D. Alsofrom, H. Grossberg, and H. Sheffer, J. Het. Chem., 13, 917 (1976).

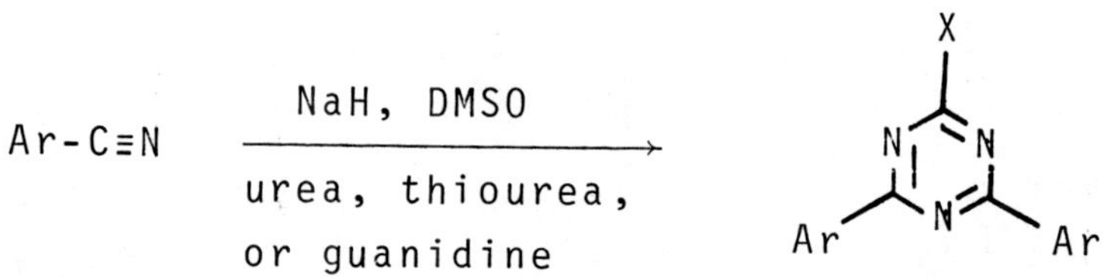

X = OH, SH, NH_2

35-93%

IV.I.6-1 T. Fuchigami and K. Odo, Bull. Chem. Soc. Japan, 49, 3607 (1976).

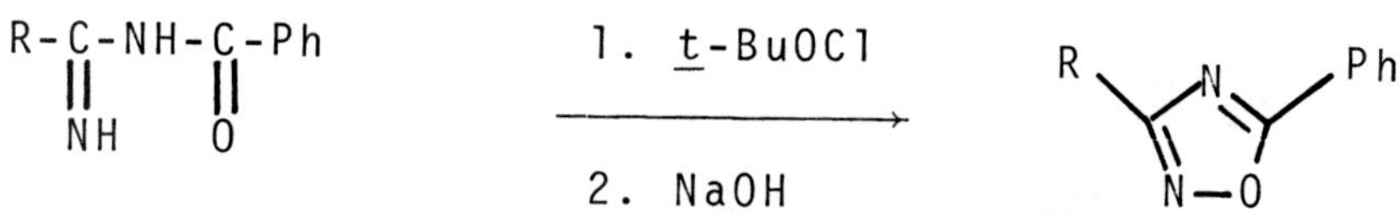

R = Ph, OR, N-piperidino

58-91%

IV.I.6-2 G. Rembarz, E. Fischer, and R. Behm, J. Prakt. Chem., 318, 479 (1976).

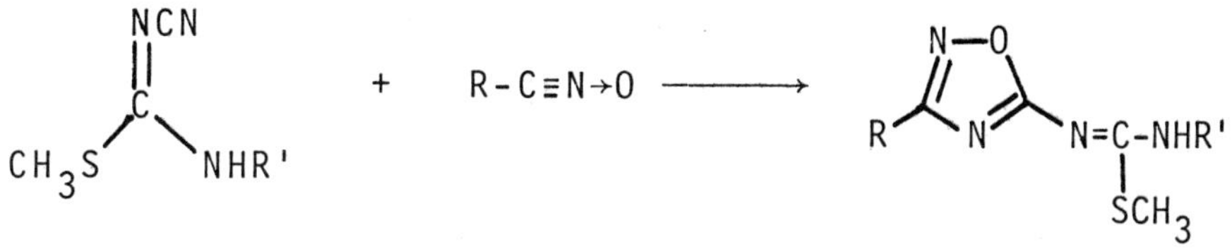

R = Me, Ar ~40-70%

R' = H, 1°,2° alkyl, aryl

IV.I.6-3 I. T. Kay and I. T. Streeting, Synthesis, 38 (1976).

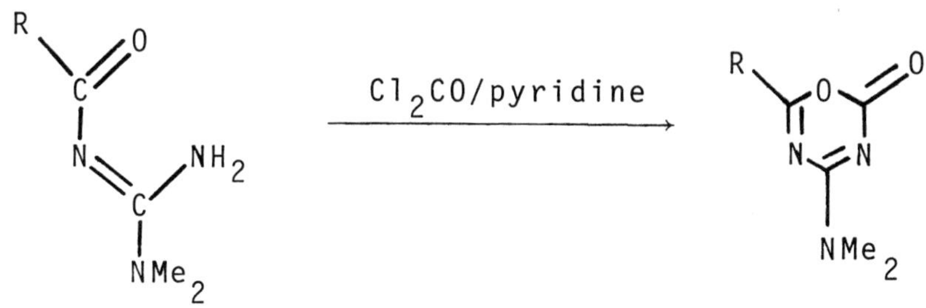

R = aryl, t-Bu 43-69%

IV.I.6-4 E. Lippmann and A. Könneche, Z. Chem., 16, 90 (1976).

Several methods of forming 2-aryl tetrazoles, e.g.:

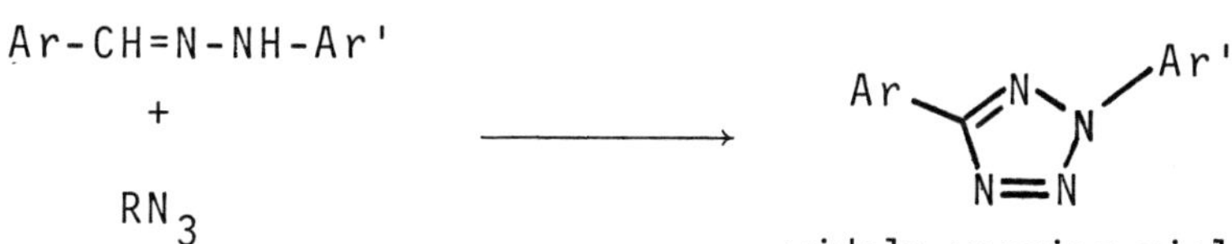

widely varying yields

IV.J-1 V. Y. Pochinok *et al.*, *Russ. Chem. Rev.*, *45*, 183 (1976).

Review: "Condensed Tetrazoles"

IV.J-2 M. Narita and C. U. Pittman, Jr., *Synthesis*, 489 (1976).

Review: "Preparation of Tetrathiafulvalenes (TTF) and their Selenium Analogs - Tetraselenafulvalenes (TSeF)"

IV.J-3 W. D. Munslow and T. J. Delia, *J. Het. Chem.*, *13*, 675 (1976).

Review: "Syntheses of the Isomeric Benzoquinazolines"

V.A-1 C. G. Kruse, N.L.J.M. Broekhof, and A. van der Gen, <u>Tetrahedron</u> <u>Lett</u>., 1725 (1976).

Protection of hydroxyl groups as 2'-O-tetrahydrofuranyl ether. Similar to THP ethers, but removed faster and more easily.

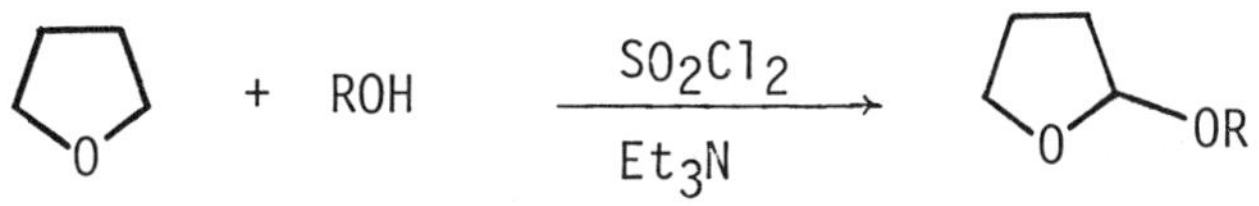

Hydrolyzed under mildly acidic conditions.

V.A-2 E. J. Corey, J.-L. Gras, and P. Ulrich, <u>Tetrahedron Lett</u>., 809 (1976).

Use of the β-methoxyethoxymethyl (MEM) group for the protection of alcohols.

$$\text{MEM} = -CH_2OCH_2CH_2OCH_3$$

Stable to strong bases, reducing agents, many oxidizing agents, and mild acids.

$$RO^{\ominus} + \text{MEM-Cl} \longrightarrow \text{RO-MEM}$$

$$ROH + \text{MEM-}\overset{\oplus}{N}Et_3\ {}^{\ominus}Cl \xrightarrow[\text{reflux}]{CH_3CN} \text{RO-MEM}$$

$$ROH + \text{MEM-Cl} + (\underline{i}\text{-Pr})_2NEt \longrightarrow \text{RO-MEM}$$

Removed by $ZnBr_2$ or $TiCl_4$ in CH_2Cl_2.

V.A-3 E. Seymour and J.M.J. Fréchet, *Tetrahedron Lett.*, 1149 (1976).

Use of a polystyrylboronic acid resin as a selective protecting group for hydroxyl functions in glycosides.

V.A-4 J.M.F. Fréchet and L. J. Nuyens, *Can. J. Chem.*, **54**, 926 (1976). Also T. M. Fyles and C. C. Leznoff, *Can. J. Chem.*, **54**, 935 (1976).

Use of polymers containing trityl chloride residues to block one primary alcohol functional group of a polyhydroxy alcohol. Cleaved by acid.

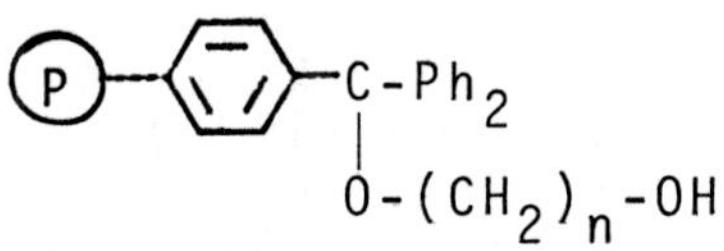

V.A-5 P. M. Pojer and S. J. Angyal, *Tetrahedron Lett.*, 3067 (1976).

$$\text{R-OH} \xrightarrow[\text{Ac}_2\text{O}]{\text{DMSO, HOAc}} \text{R-OCH}_2\text{SCH}_3 \xrightarrow[\text{acetone/H}_2\text{O}]{\text{CH}_3\text{I, NaHCO}_3} \text{ROH}$$

R = 1°, 2°, 3°

V.A-6 K. Yamada, *et al.*, *Tetrahedron Lett.*, 65 (1976).

Use of the methylthiomethyl ether function for protection of tertiary alcohols. Protected under basic conditions, stable to base. Removed under mildly acidic or neutral (Ag^+ or Hg^{++}) conditions.

V.A-7 J. P. Yardley and H. Fletcher, Synthesis, 244 (1976).

$$\text{Ar-OH} + H_2C(OCH_3)_2 \xrightarrow[\text{molecular sieves}]{TsOH,\ CH_2Cl_2} \text{Ar-}OCH_2OCH_3$$

60-80%
if no *ortho*-
substituents present

V.A-8 E. Nakamura *et al*., J. Am. Chem. Soc., **98**, 2346 (1976).

Use of ethyl trimethylsilylacetate-tetra-*n*-butylammonium fluoride (ETSA-TBAF) for silylation of alcohols, e.g.:

92%

V.A-9 J. H. Clark, H. L. Holland, and J. M. Miller, Tetrahedron Lett., 3361 (1976).

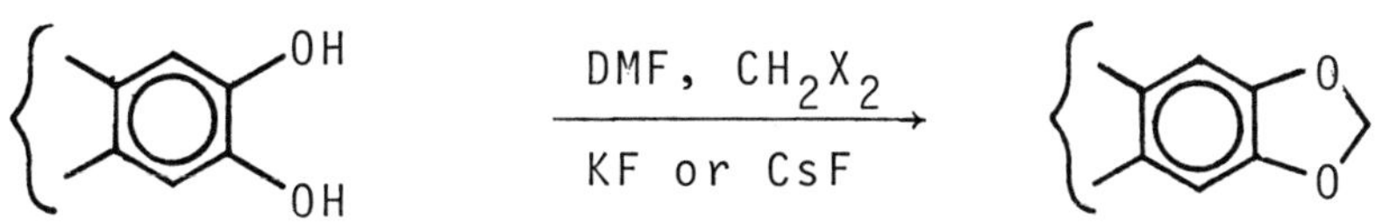

X = Cl, Br 70-90%

V.A-10 C. Hansson and B. Wickberg, *Synthesis*, 191 (1976).

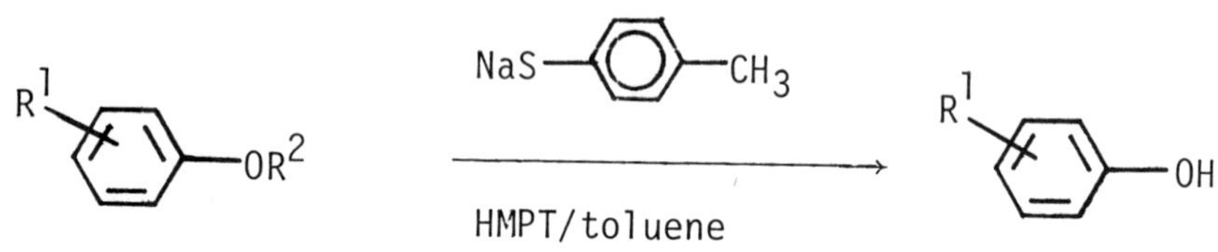

R^1 = Br, CHO

R^2 = Me, Bz

generally 80-90%

V.A-11 G. A. Olah, J. Welch, and T.-L. Ho, *J. Am. Chem. Soc.*, **98**, 6717 (1976).

$$\text{R-O-Bz} \xrightarrow{UF_6} \text{R-OH}$$

R=alkyl 44-69%

V.A-12 C. A. Smith and J. B. Grutzner, *J. Org. Chem.*, **41**, 367 (1976).

$$\text{ROCH}_3 \xrightarrow[\substack{CH_3CN \\ CH_3OCHI_2}]{HI} \text{ROH}$$

Widely varying yields

V.A-13 T. Kametani *et al.*, *J. Org. Chem.*, **41**, 2545 (1976).

$$\text{ArOR} \xrightarrow{NaAlH_2(OC_2H_4OCH_3)_2} \text{Ar-OH}$$

R = Ph, $CH{=}CMe_2$

V.A-14 M. Node, H. Hori, and E. Fujita, J. Chem. Soc. Perkin I, 2237 (1976).

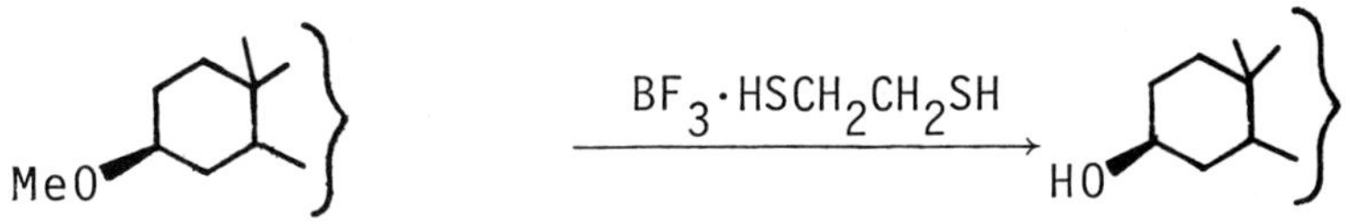

~60-90%
(retention of config.)

V.A-15 R. Boss and R. Scheffold, Angew. Chem. Int. Ed., 15, 558 (1976).

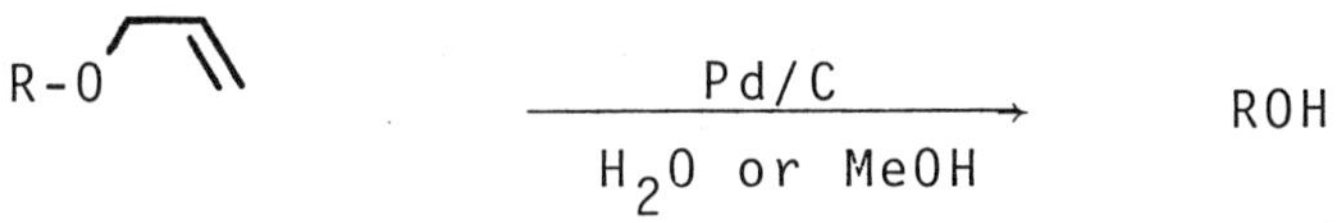

R = alkyl, aryl 78-95%

R may contain a wide variety of functional groups.

V.A-16/V.B-1 F. D. King and D.R.M. Walton, Synthesis, 40 (1976).

Use of trimethylsilyl derivatives of phenols, anilines, and carboxylic acids in organocopper coupling reactions.

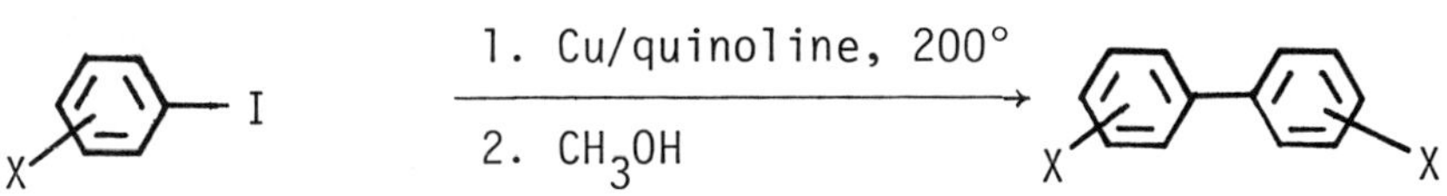

X = $-OSiMe_3$ 55-70%

$-N(SiMe_3)_2$

$-COOSiMe_3$

V.B-2 H. Eckert and I. Ugi, Angew. Chem. Int. Ed., 15, 681 (1976).

Use of Cobalt(I)-phthalocyanine to cleave β-halogenated urethane protecting groups, e.g.:

$$R\text{-}NH\text{-}\overset{\overset{\displaystyle O}{\|}}{C}\text{-}OCH_2CH_2X \longrightarrow R\text{-}NH_2$$

X = Cl, Br 70-91%

V.B-3 S. Brandänge and L. Lindblom, Acta Chem. Scand. B, 30, 93 (1976).

Use of N-vinyl as an N-H protecting group in the synthesis of myosmine:

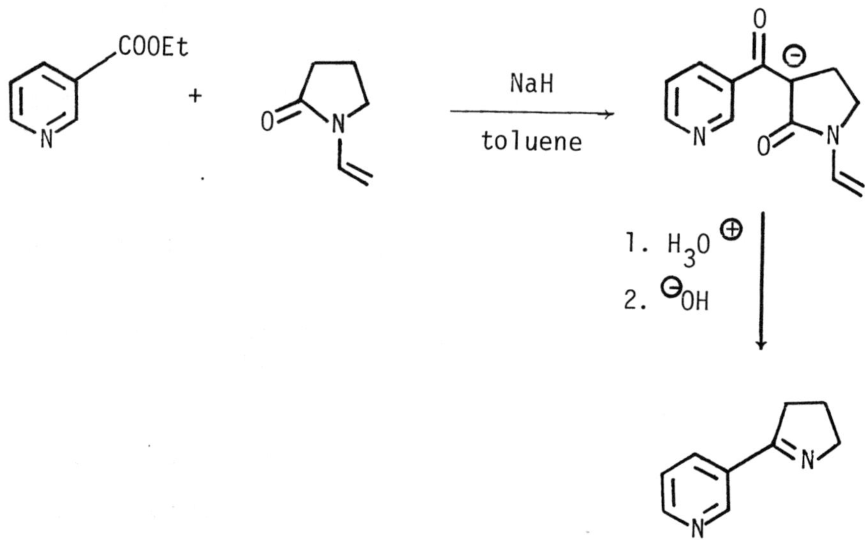

V.B-4 G. W. Kenner, G. A. Moore, and R. Ramage, Tetrahedron Lett., 3623 (1976).

Use of phosphinamides as amino protecting groups in peptide chemistry.

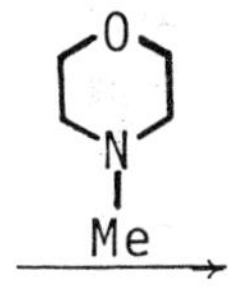

Ph_2POCl + $H_2NCHRCOOR'$ ⟶ $Ph_2PONHCHRCOOR'$

R' = Me, Bz

Cleaved with $HOAc/HCOOH/H_2O$.

V.B-5 A. Tun-Kyi and R. Schwyzer, Helv. Chim. Acta, 59, 1642 (1976).

Use of the (p-phenylazophenyl)-isopropyloxycarbonyl (AZOC) protecting group in peptide synthesis. Put on using AZOC-OPh or AZOC-N_3, and removed with mild acid.

V.B-6 A. P. Hope and B. Halpern, Aust. J. Chem., 29, 1591 (1976).

Use of the 2-hydroxyarylmethylene protecting group for N-terminii of amino acids in peptide synthesis. Removed under conditions (80% HOAc) which leave the t-butyloxycarbonyl group intact.

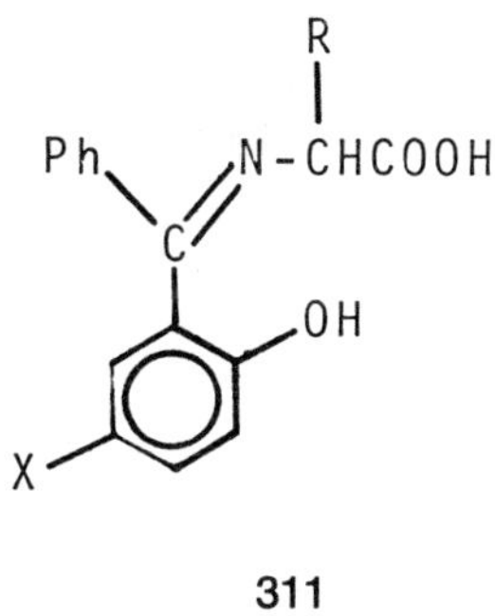

X = Cl, Me

V.B-7 B. Weinstein *et al.*, *Synth. Commun.*, **6**, 17 (1976).

Use of the Tac (*p*-tosylaminocarbonyl) N-protecting group in peptide synthesis:

$$H_2NCHRCO_2H + TsNCO \longrightarrow TsNHCONHCHRCO_2H$$

Stable to most acids and bases, and to H_2/Pd. Removed by refluxing in ethanol or isopropanol.

V.B-8 H. Kunz, *Liebigs Ann. Chem.*, 1674 (1976).

Use of the 2-phosphonioethoxycarbonyl function as an amine-protecting group in peptide synthesis, giving a higher water solubility than the corresponding triphenylphosphonium derivatives. Removed with .01 N NaOH.

V.B-9 H. Kunz, *Chem. Ber.*, **109**, 2670 (1976).

Use of the 2-(triphenylphosphonio)ethoxycarbonyl (Peoc) group as an amino protective function in peptide chemistry. Extremely stable to acids, but removed under mildly basic conditions.

V.B-10 H. Kunz, *Chem. Ber.*, **109**, 3693 (1976).

Use of the 2-(methylthio)ethoxycarbonyl (Mtc) group as an amino protecting group in peptide chemistry. Put on using Mtc-phenolate, and removed by methylation followed by basic hydrolysis.

V.B-11 M. Juillerat and J. P. Bargetzi, <u>Helv. Chem. Acta</u>, **59**, 855 (1976).

Use of NTBA or its methyl ester in cleaving the *o*-nitrophenylsulfenyl α-amino protecting group. Gives nearly quantitative yields under mild conditions.

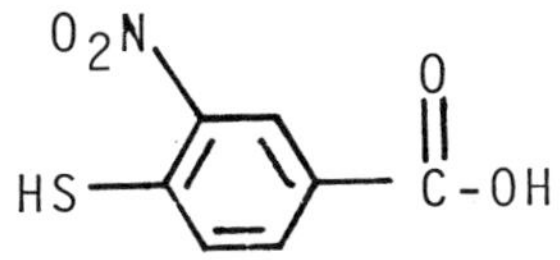

NTBA

V.B-12 O. Nishimura and M. Fujino, <u>Chem. Pharm. Bull.</u>, **24**, 1568 (1976).

Use of the *p*-methoxybenzenesulfonyl (MBS) group to protect the ω-guanidino function in arginine in peptide synthesis. More easily removed than tosyl group, using methanesulfonic acid and BTFA.

V.B-13 M. Chorev and Y. S. Klausner, <u>J. Chem. Soc., Chem. Commun.</u>, 596 (1976).

Use of the benzyloxycarbonyl group to protect the indole ring of tryptophan in peptide synthesis. Achieved by acylation with *p*-nitrophenyl benzyl carbonate in the presence of unsolvated fluoride anion, generated by crown ethers. Removed by cat. H_2, H_2NNH_2, or liquid HF.

V.C-1 C. G. Kruse, N.L.J.M. Broekhof, and A. Van der Gen, *Tetrahedron Lett.*, 1725 (1976).

Protection of thiol groups as 2'-S-tetrahydrofuranyl ethers. Similar to THP ethers, but removed faster and more easily (under mildly acidic conditions).

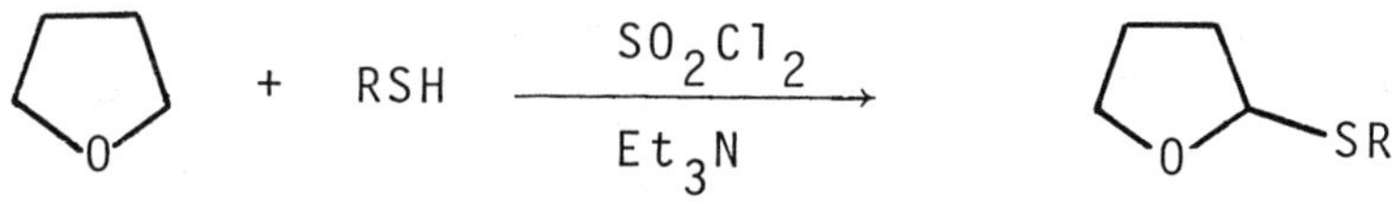

V.C-2 C. Seidel *et al.*, *J. Prakt. Chem.*, **318**, 229 (1976).

Use of the S-ethylthio protecting group for cysteine in solid-phase peptide synthesis. Relatively stable to base, but not stable to acid.

V.C-3 S. Coyle and G. T. Young, *J. Chem. Soc., Chem. Commun.*, 980 (1976).

Use of the diphenyl-4-pyridlymethyl group as a protecting group for the thiol group of cysteine and the imidazole nitrogen of histidine in peptide synthesis. Removed by electrolytic reduction, Zn/HOAc, $Hg(OAc)_2$, and I_2/HOAc.

V.D-1 J. San Filippo, Jr. *et al.*, *J. Org. Chem.*, **41**, 586 (1976).

$$RCO_2R' \xrightarrow[2.\ H_3O^{\oplus}]{1.\ KO_2\text{-crown ether}} RCO_2H + R'OH$$

R,R' = alkyl, aryl, lactone

yields generally 35-95% (GLC)

V.D-2 T.-L. Ho and G. A. Olah, <u>Angew. Chem. Int. Ed.</u>, **15**, 774 (1976).

$$\text{R-C(=O)-OR}' \xrightarrow[\text{2. } H_2O]{\text{1. } Me_3SiI} RCO_2H$$

R = Ar, Bz, t-Bu, cyclohexyl

∿ 60-90%

R' = Me, Et, Bz

V.D-3 H. Eckert and I. Ugi, <u>Angew. Chem. Int. Ed.</u>, **15**, 681 (1976).

Use of Cobalt(I)-phthalocyanine to cleave β-halogenated alkyl ester protecting groups, e.g.:

$$\text{R-C(=O)-}OCH_2CH_2X \longrightarrow \text{R-COOH}$$

X = Cl, Br

57-98%

V.D-4 J. W. Huffman and P. G. Harris, <u>Synth. Commun.</u>, **6**, 481 (1976).

$$R\text{-}COOCH_3 \xrightarrow{\text{KH, DMSO}} RCOOH$$

~50%

Useful for most hindered esters, but fails with methyl benzoate and methyl mesitoate.

V.D-5 G. Just and K. Grozinger, <u>Synthesis</u>, 457 (1976).

$$R\text{-}\overset{\overset{\displaystyle O}{\|}}{C}\text{-}O\text{-}CH_2CCl_3 \xrightarrow[\text{pH 4.2-7.2}]{\text{Zn, THF/}H_2O} RCOOH$$

R = alkyl, aryl, unsat., etc. ~80%

V.D-6 R. L. Prestidge, D.R.K. Harding, and W. S. Hancock, <u>J. Org. Chem.</u>, **41**, 2579 (1976).

Use of 

esters as carboxyl-protecting groups in solid-phase peptide synthesis.

V.E-1 E. J. Corey, E. J. Trybulski, and J. W. Suggs, Tetrahedron Lett., 4577 (1976).

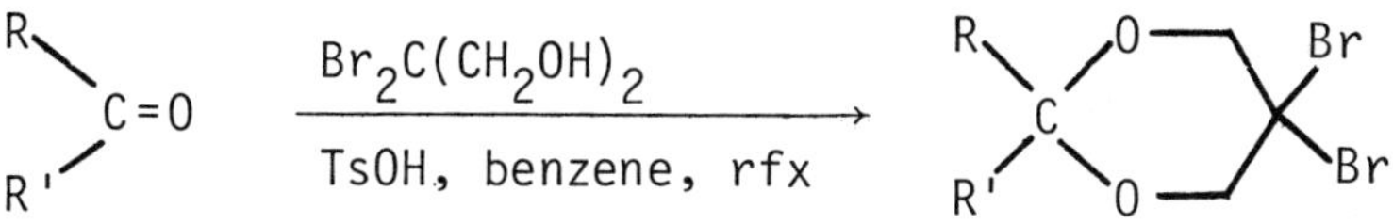

R,R' = H, alkyl

84-94%

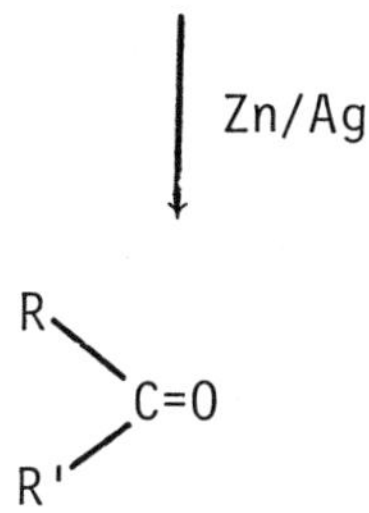

Stable to $NaBH_4$, RCO_3H, B_2H_6

>90%

V.E-2 J. B. Paine III, R. B. Woodward, and D. Dolphin, J. Org. Chem., 41, 2826 (1976).

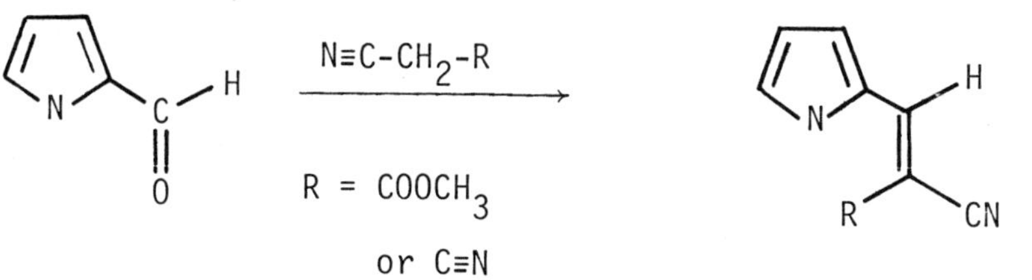

Stable to acids, most oxidants, reductants

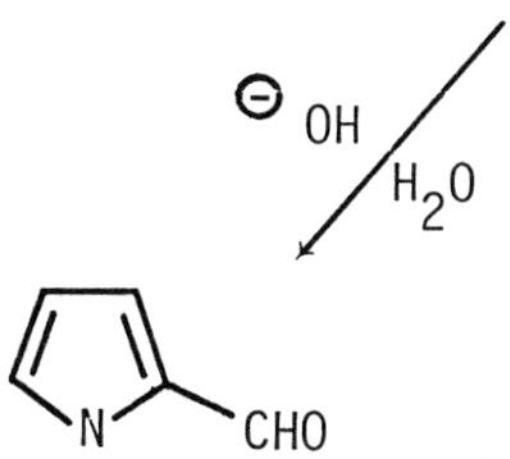

V.E-3 E. Nakamura *et al.*, *J. Am. Chem. Soc.*, **98**, 2346 (1976).

Use of ethyl trimethylsilylacetate-tetra-*n*-butylammonium fluoride (ETSA-TBAF) for silylation of ketones, e.g.:

V.E-4 E. J. Corey and S. Knapp, *Tetrahedron Lett.*, 3667 (1976).

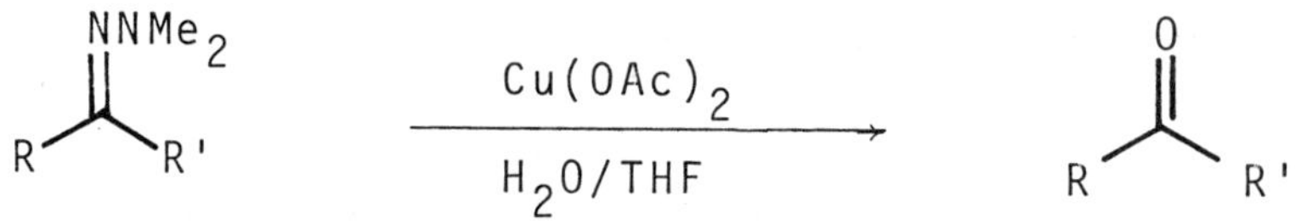

R,R' = H, alkyl, vinyl, etc.

generally >95%

V.E-5 E. J. Corey and D. Enders, *Tetrahedron Lett.*, 3 (1976).

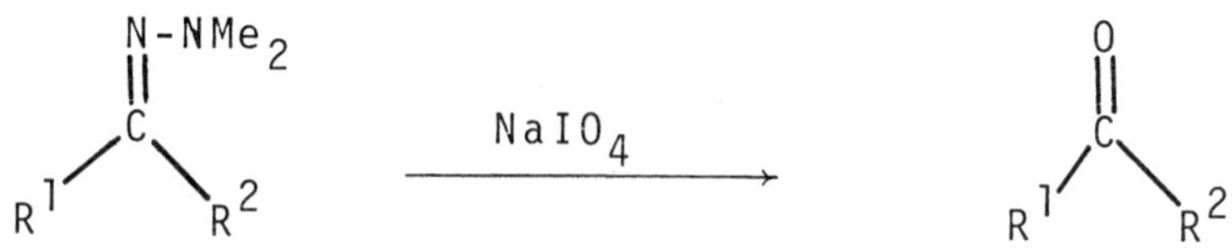

R^1,R^2 = H, alkyl

>90% isolated

V.E-6 G. A. Olah *et al.*, Synthesis, 808, 809 (1976).

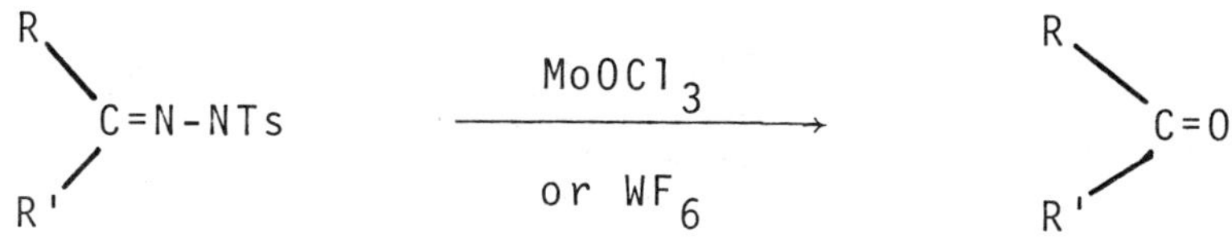

R,R' = alkyl, aryl ~80-90%

V.E-7 T.-L. Ho and G. A. Olah, Synthesis, 611 (1976).

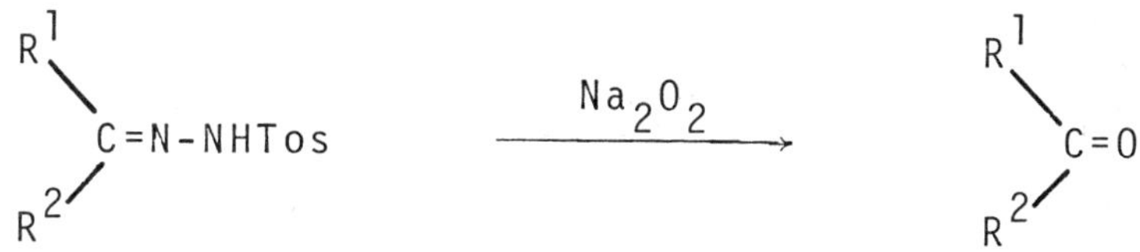

cyclohexanone, acetophenone, etc. 69-73%

V.E-8 C. E. Sacks and P. L. Fuchs, Synthesis, 456 (1976).

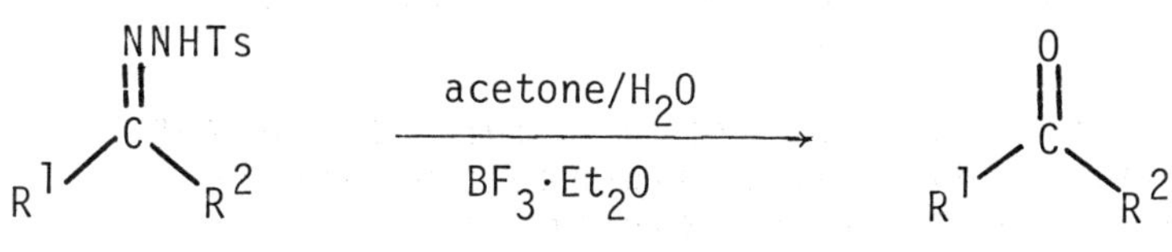

R^1, R^2 = alkyl, aryl, cyclic 80-97%

V.E-9 M. Hojo and R. Masuda, Synthesis, 678 (1976).

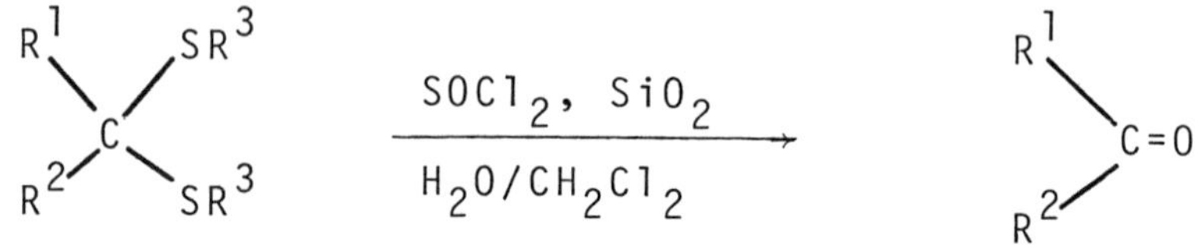

R^1,R^2 = H, alkyl, aryl quantitative by NMR

V.E-10 E. Fujita, Y. Nagao, and K. Kaneko, Chem. Pharm. Bull., 24, 1115 (1976).

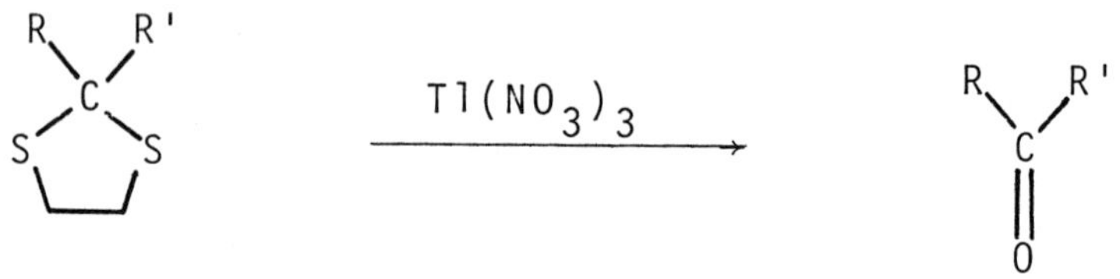

R,R' = H, aryl, 1°,2°,3° alkyl, cyclic 73-99%

V.F-1 K. K. Ogilvie, S. L. Beaucage, and D. W. Entwistle, Tetrahedron Lett., 1255 (1976).

Use of tetrabutylammonium fluoride in THF for the removal of phosphate-protecting groups in nucleotide synthesis.

V.G-1 G. Evans, B.F.G. Johnson and J. Lewis, <u>J. Organometal. Chem.</u>, <u>102</u>, 507 (1975).

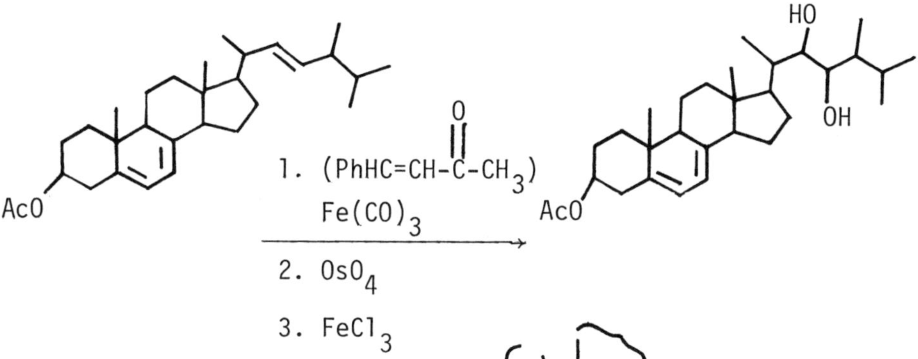

Fe protects the diene system as

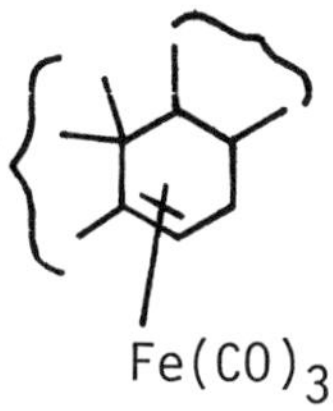

V.G-2 T. C. Jain, C. M. Banks, and J. E. McCloskey, <u>Tetrahedron</u>, <u>32</u>, 765 (1976).

Reversible dimethylamine addition as a protecting reaction for α,β-unsaturated methylene groups of γ-lactones.

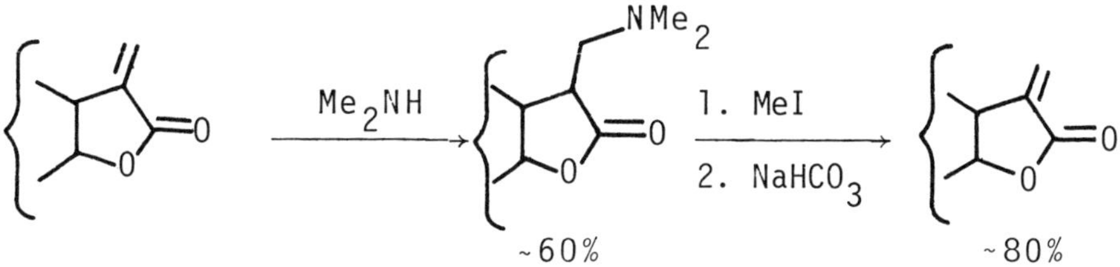

V.H-1 V. G. Mairanovsky, Angew. Chem. Int. Ed., **15**, 281 (1976).

Review: "Electro-Deprotection -- Electrochemical Removal of Protecting Groups"

Protecting groups discussed:

Tosyl, benzoyl, trityl, benzhydryl, benzyl, phenyl, cinnamyl, benzyloxycarbonyl, and benzylidene.

VI.A.1-1 A. I. Meyers *et al.*, *J. Am. Chem. Soc.*, **98**, 567 (1976).

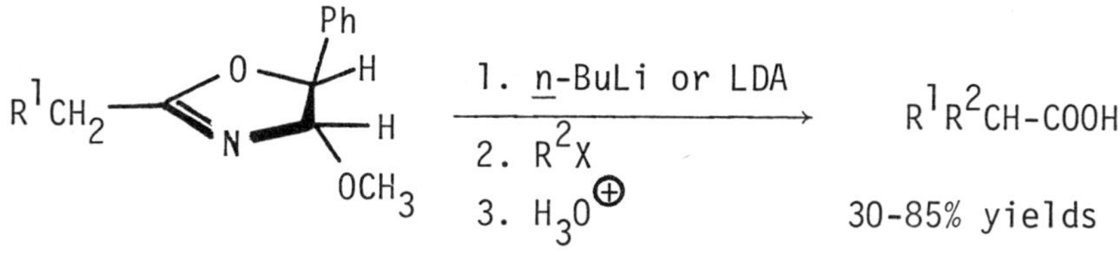

30-85% yields
50-90% ee

Full paper, many examples.

VI.A.1-2 A. I. Meyers and K. Kamata, *J. Am. Chem. Soc.*, **98**, 2290 (1976).

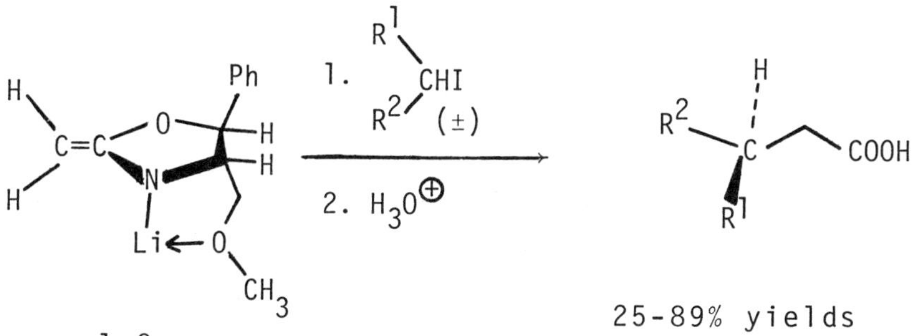

R^1R^2 = alkyl

25-89% yields
30-49% ee

VI.A.1-3 L. Cassar, M. Foa, and A. Gardano, *J. Organometal. Chem.*, **121**, C55 (1976).

$$\text{R-X} \xrightarrow[\textit{n}\text{-Bu}_4\text{NI, [Pd(PPh}_3)_2\text{Cl}_2]]{\text{PPh}_3\text{, NaOH, CO}} \text{R-COOH}$$

RX = "Aryl, benzyl, vinyl, and heterocyclic halides"

up to 83%
Few details.

VI.A.1-4 T. Mukaiyama and T. Tanaka, Chem. Lett., 303 (1976).

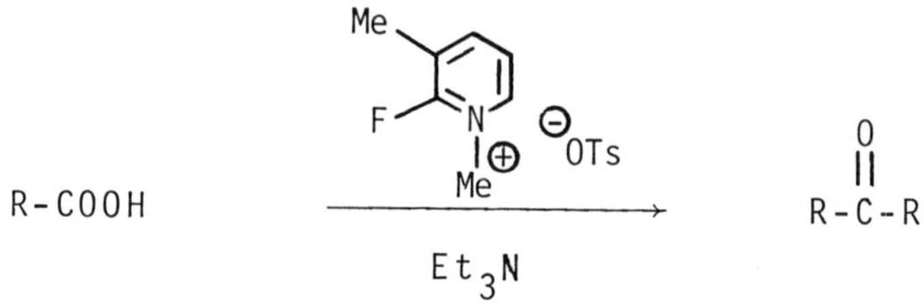

R = Ph, Bz, alkyl, cinnamyl, etc.

VI.A.2-1 Y. Yamamoto *et al.*, J. Chem. Soc., Chem. Commun., 672 (1976).

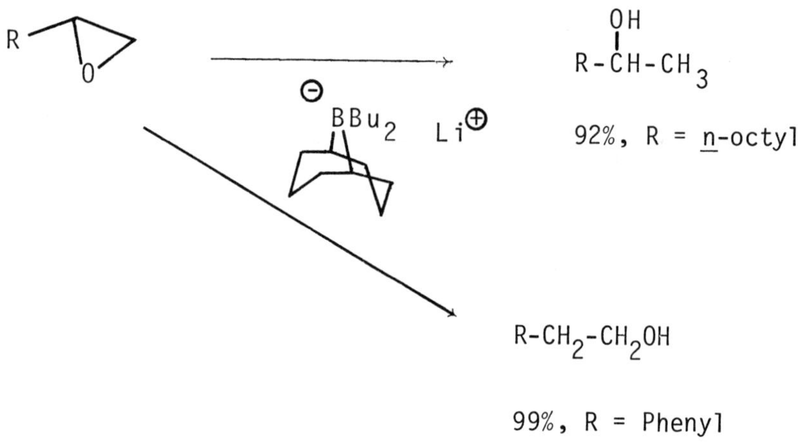

VI.A.2-2 B. Kokel. A. Lefebvre, and C. Lespagnol, Comptes Rendus (C), 282, 1125 (1976).

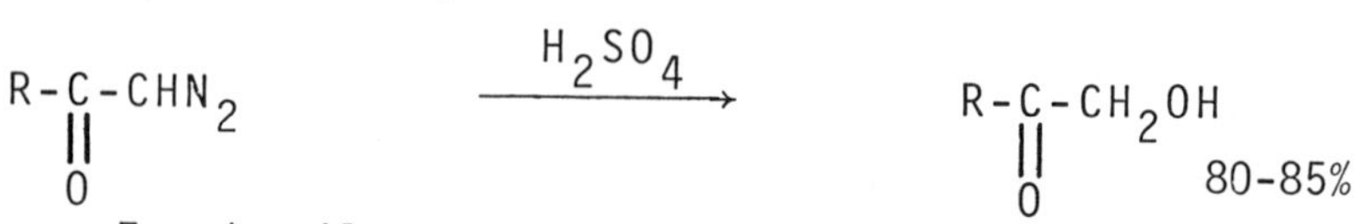

Few details; TFA esters may also be obtained.

VI.A.2-3 T. Kametani *et al.*, *J. Org. Chem.*, **41**, 2545 (1976).

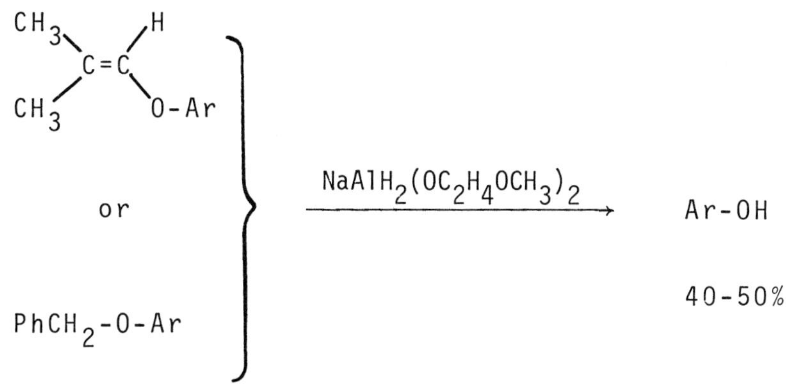

VI.A.2-4 E. Negishi, S. Baba, and A. O. King, *J. Chem. Soc., Chem. Commun.*, 17 (1976).

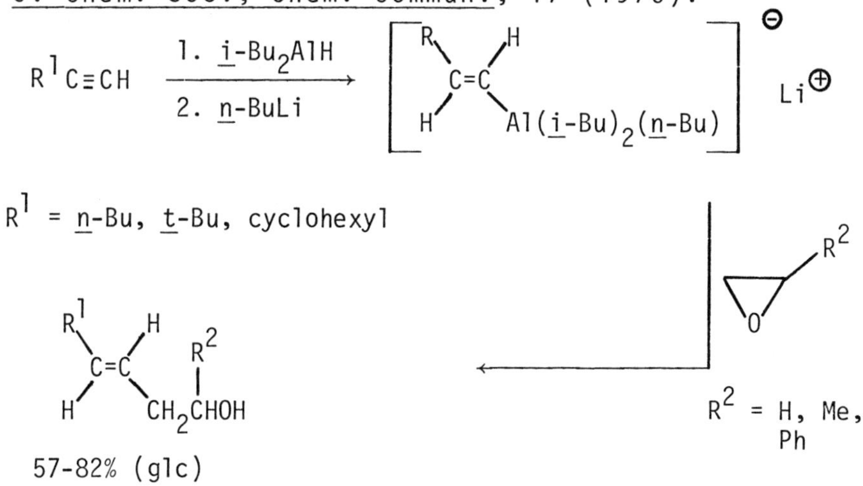

VI.A.2-5 G. Deleris, J. Dunoguès, and R. Calas, Tetrahedron Lett., 2449 (1976).

$$\text{R-CHO} + \text{R'-CH=CH-CH}_2\text{-SiMe}_3 \xrightarrow[\text{2. H}_2\text{O}]{\text{1. AlCl}_3} \text{R-CHOH-}\underset{\displaystyle\text{R'}}{\text{CH}}\text{-CH=CH}_2$$

$$\text{R-CHO} + \text{Me}_3\text{Si-C}\equiv\text{C-SiMe}_3 \xrightarrow[\text{2. H}_2\text{O}]{\text{1. AlCl}_3} \text{R-CHOH-C}\equiv\text{C-SiMe}_3$$

VI.A.2-6 G. N. Barber and R. A. Olofson, Tetrahedron Lett., 3783 (1976).

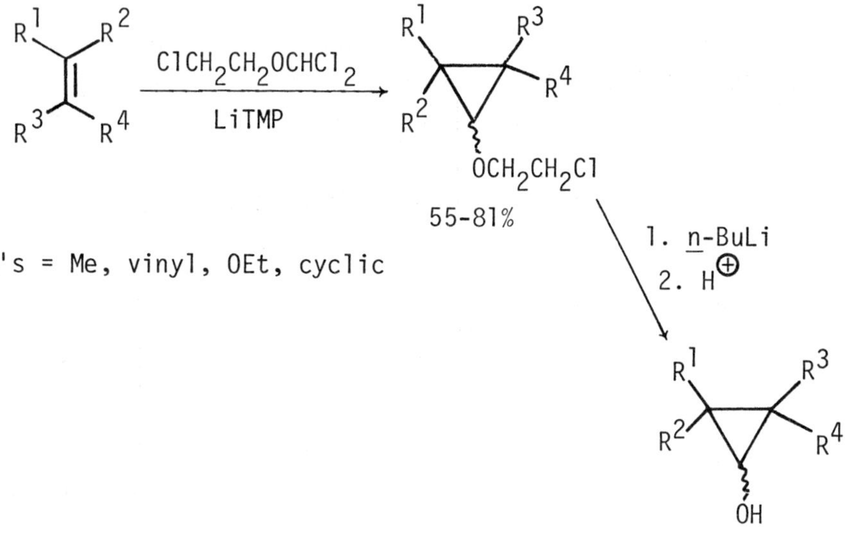

R's = Me, vinyl, OEt, cyclic

VI.A.2-7 J. F. Ruppert and J. D. White, J. Org. Chem., 41, 550 (1976).

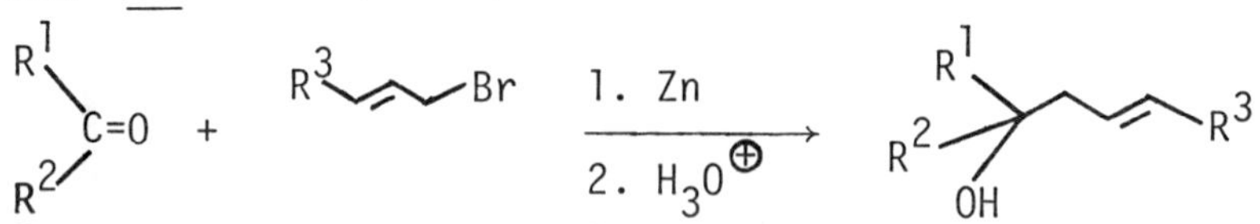

R^1,R^2= H,alkyl,aryl,cyclic; R^3=H, alkyl 74-97% (isolated)

VI.A.2-8   A. Debal, T. Cuvigny, and M. Larchevêque, *Synthesis*, 391 (1976).

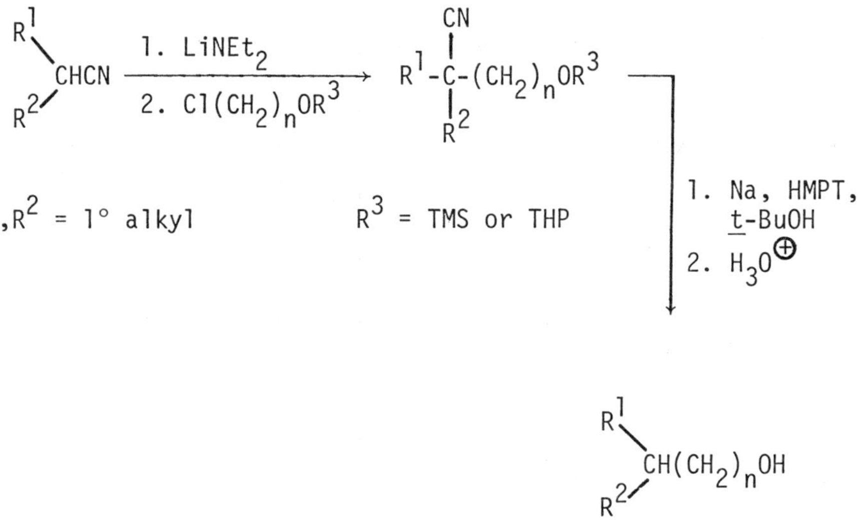

R^1,R^2 = 1° alkyl R^3 = TMS or THP

~50-80%

VI.A.2-9   A. B. Levy and S. J. Schwartz, *Tetrahedron Lett.*, 2201 (1976).

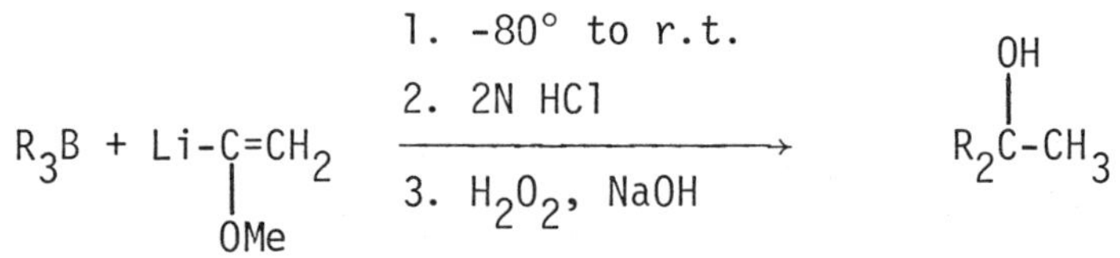

R = alkyl, cyclic   87-100% (VPC)

VI.A.3-1   A. G. Anderson, Jr. *et al.*, *Synthesis*, 398 (1976).

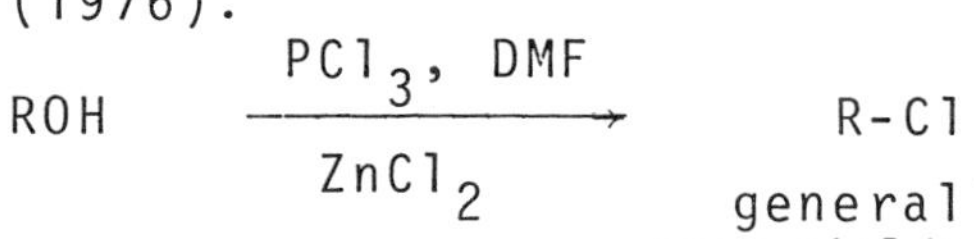

generally 60-87%; poor yields with hindered alcohols

R = 1°,2° alkyl

VI.A.3-2 R. M. Carman and I. M. Shaw, <u>Aust. J. Chem.</u>, <u>29</u>, 133 (1976).

$$R{-}OH \quad + \quad PCl_5 \xrightarrow[CHCl_3]{CaCO_3} R{-}Cl$$

R = 3° alkyl

Retention of configuration. "Good" yields.

VI.A.3-3 D. R. Hepburn and H. R. Hudson, <u>J. Chem. Soc. Perkin I</u>, 754 (1976).

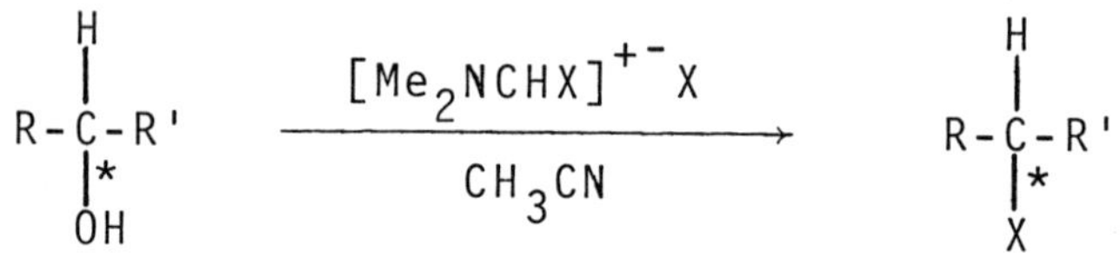

R,R' = alkyl X = Cl or Br

Chemical yields ~70-90%

~75% ee (retention)

VI.A.3-4 R. O. Hutchins, D. Masilamani, and C. A. Maryanoff, <u>J. Org. Chem.</u>, <u>41</u>, 1071 (1976).

Use of PBr_3 at low temperature (4°C) and long reaction times (1-2 days) to convert optically active alcohols to the corresponding bromides. Yields 60-75%, with 84-94% optical purity (on 1-phenylethanol, 2-octanol, and 2-pentanol).

VI.A.3-5 P. E. Sonnet, Synth. Commun., 6, 21 (1976).

$$\text{R-O-THP} \xrightarrow[\text{X = Cl, Br}]{PPh_3X_2} \text{RX}$$

R = 1° alkyl 80%

VI.A.3-6 P. Place, M.-L. Roumestant, and J. Gore, Bull. Soc. Chim. France, 169 (1976).

$$\text{R-OTs} \xrightarrow[MgX_2]{Et_2O} \text{R-X}$$

R = 1°,2° alkyl 27-98%

X = Br, I

VI.A.3-7 H. C. Brown et al., J. Am. Chem. Soc., 98, 1290 (1976).

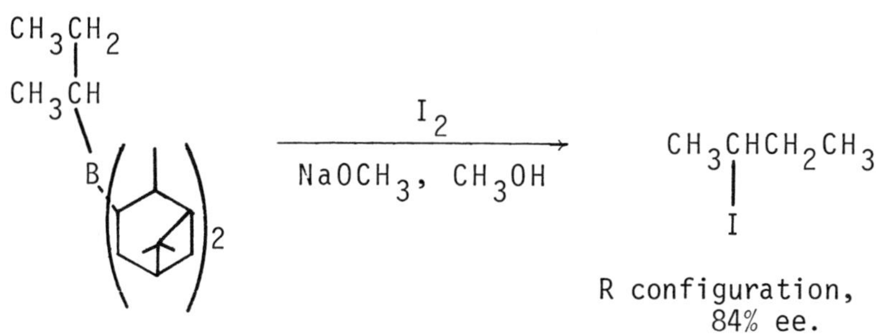

R configuration, 84% ee.

49% yield

VI.A.3-8 M. Zupan and A. Pollak, J. Org. Chem., 41, 4002 (1976).

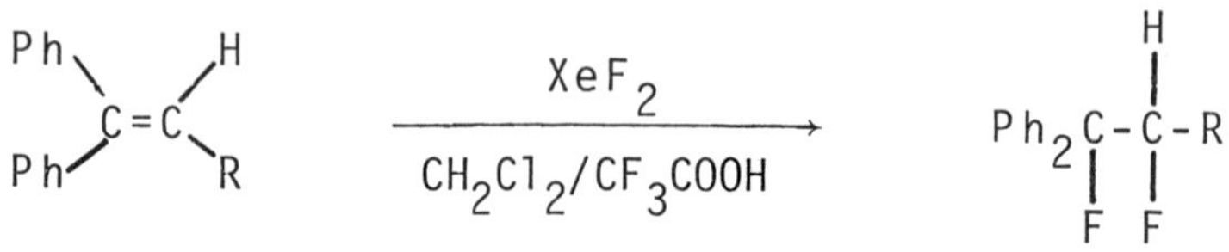

R = H, CH_3

95%

VI.A.3-9 N. R. DeLue and H. C. Brown, Synthesis, 114 (1976).

$$H_2C{=}CH(CH_2)_8COOCH_3 \xrightarrow[\text{2. } I_2\text{, NaOCH}_3]{\text{1. } B_6H_6\text{, THF}} I\text{-}CH_2(CH_2)_9\text{-}COOCH_3$$

68% isolated

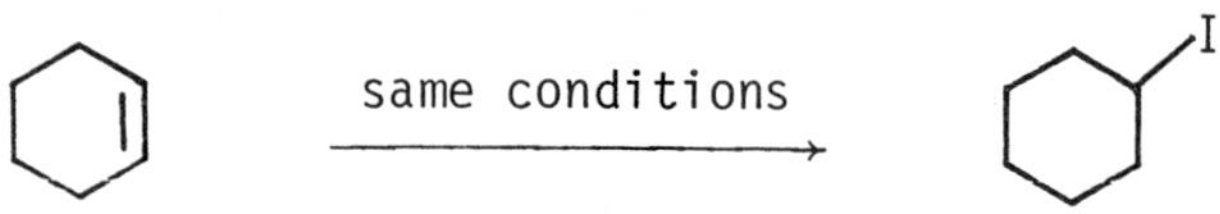

72% (VPC)

VI.A.3-10 J. Kollonitsch, S. Marburg, and L. M. Perkins, J. Org. Chem., 41, 3107 (1976).

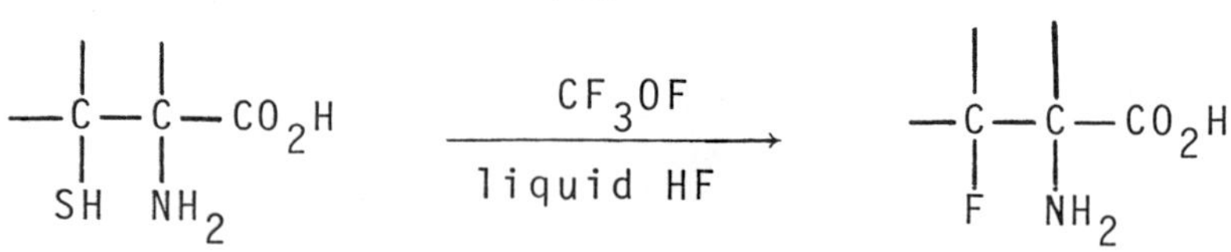

Widely varying yields.

VI.A.3-11 M. P. Doyle and B. Sigfried, J. Chem. Soc., Chem. Commun., 433 (1976).

$$RCH_2NH_2 \xrightarrow[CuX_2]{R'ONO} RCHX_2$$

R = unbranched alkyl — 38-79%

R' = isopentyl or _t_-butyl

X = Cl, Br

VI.A.3-12 M. P. Doyle, B. Siegfried, and J. J. Hammond, J. Am. Chem. Soc., 98, 1627 (1976).

$$RCH_2NH_2 \xrightarrow[2.\ (CuX_2 \cdot NO)_2,\ CH_3CN]{1.\ CuX_2} RCHX_2$$

R = Ph, Bz, alkyl, etc. — 14-58%

X = Cl, Br

VI.A.3-13 G. Cainelli and F. Manescalchi, Synthesis, 472 (1976).

$$R\text{-}X \xrightarrow{\text{Resin-}\overset{\oplus}{N}Me_3\overset{\ominus}{F}} R\text{-}F$$

X = Br, OTs, OMes — 60-90%

R = 1°,2° alkyl; may contain carbonyl

Reaction fails when R = cyclohexyl

VI.A.3-14 J. A. Miller and M. J. Nunn, <u>J. Chem. Soc. Perkin I</u>, 416 (1976).

$$\text{R-Cl} + \text{NaI} \xrightarrow[\text{ZnCl}_2]{\text{CS}_2} \text{R-I}$$

R = 3° alkyl, generally very hindered >90%

VI.A.3-15 W. E. Willy, D. R. McKean, and B. A. Garcia, <u>Bull. Chem. Soc. Japan</u>, <u>49</u>, 1989 (1976).

$$\text{R-Cl} \xrightarrow{\text{NaBr, EtBr}} \text{R-Br}$$

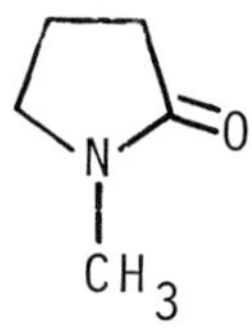

R = primary alkyl 70-93%

VI.A.3-16 A. L. Fridman and N. A. Kolobov, <u>J. Org. Chem. (USSR)</u>, <u>12</u>, 228 (1976).

$$\text{R-C(=O)-CHN}_2 \xrightarrow[(\text{X = Cl, Br})]{\text{X-I}} \text{R-C(=O)-CHIX}$$

R = alkyl, aryl ~70%

VI.A.3-17 P. E. Sonnet and J. E. Oliver, J. Org. Chem., 41, 3279 (1976).

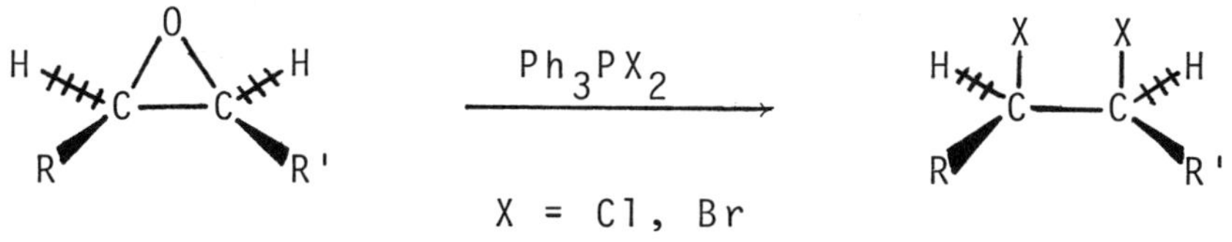

R,R' = alkyl

yields ~60-90%

VI.A.3-18 D.H.R. Barton et al., J. Am. Chem. Soc., 98, 3036 (1976).

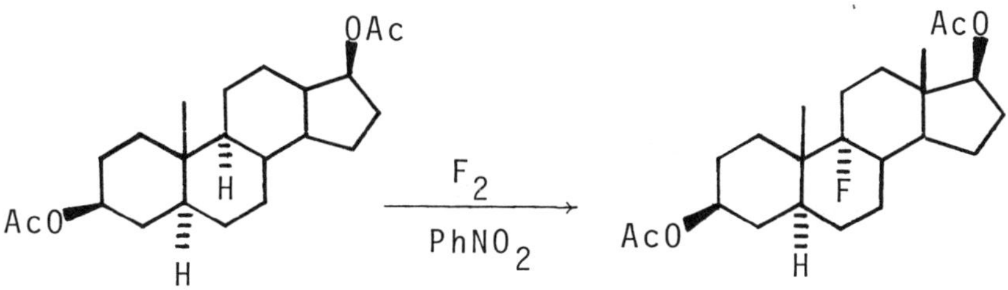

50%

VI.A.3-19 J. Pfab, Tetrahedron Lett., 943 (1976).

$$R^1R^2CClNO \xrightarrow{Br_2,\ h\nu} R^1R^2CClBr$$

R^1,R^2 = Me, Et, t-Bu, cyclic

75-92%

VI.A.3-20 F. de Reinach-Hirtzbach and T. Durst, Tetrahedron Lett., 3677 (1976).

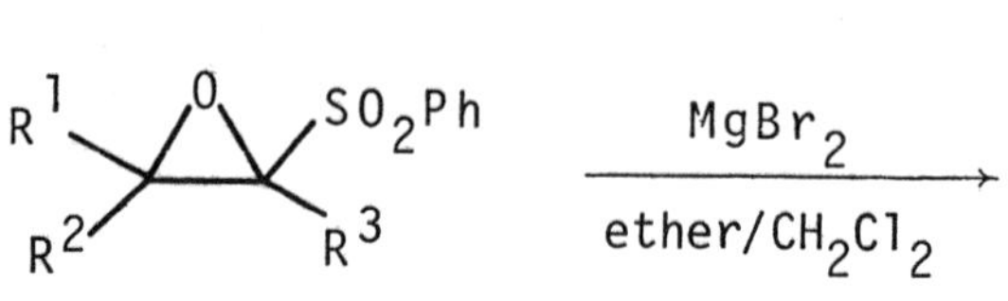

R^1,R^2 = H, Me, Ph, cyclic   80-95%

R^3 = H, Me, cyclic

VI.A.3-21 P. Wieland, Helv. Chim. Acta, 59, 1027 (1976).

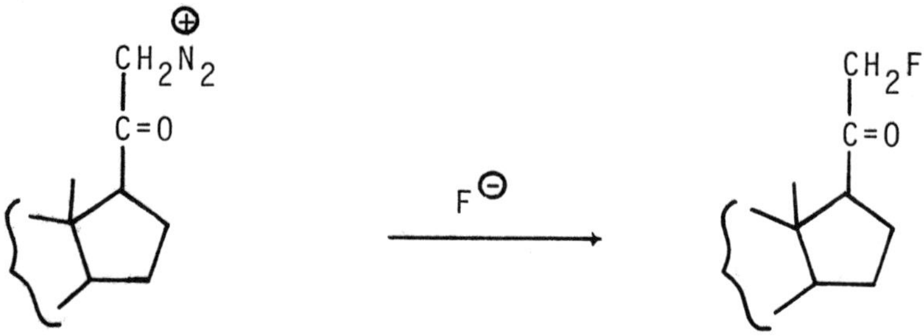

60%

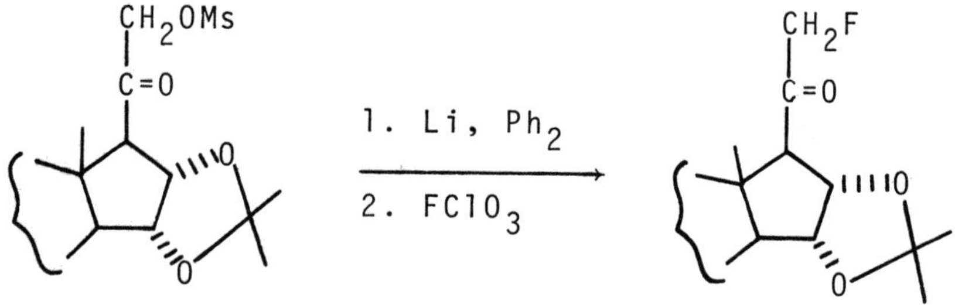

~70%

VI.A.3-22 V. E. Platonov and G. G. Yakobson, Synthesis, 374 (1976).

Review: "The Application of Thermolytic Reactions for the Syntheses of Fluoro-Organic Compounds"

VI.A.3-23 G. G. Yakobson and V. M. Vlasov, Synthesis, 652 (1976).

Review: "Recent Synthetic Methods for Polyfluoroaromatic Compounds"

VI.A.4-1 J. H. Hall and M. Gisler, J. Org. Chem., 41, 3769 (1976).

$$\text{R-C}\equiv\text{N} \xrightarrow{\text{KOH, } \underline{t}\text{-BuOH}} \text{R-}\overset{\text{O}}{\overset{\|}{\text{C}}}\text{-NH}_2$$

R = Bz, Et, Ph, etc. ~80-90%

VI.A.4-2 T. Mukaiyama, Y. Aikawa, and S. Kobayashi, Chem. Lett., 57 (1976).

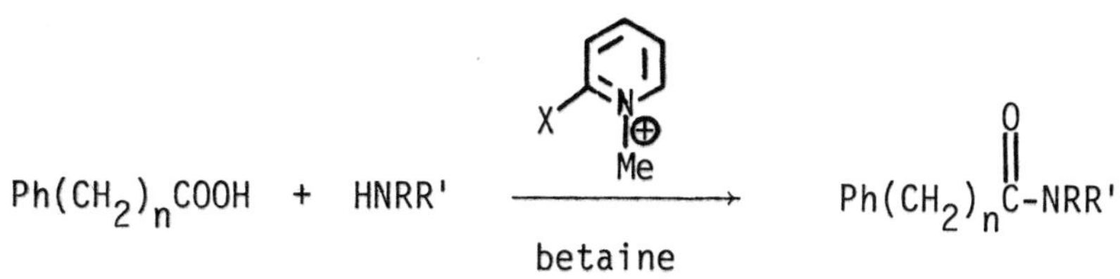

n = 0,1,2 X = F, Cl

R = H, Bu

R' = Bu, t-Bu, Ph, Bz

VI.A.4-3 K. Takeuchi, M. Nojima, and N. Tokura, J. Chem. Soc. Perkin I, 2205 (1976).

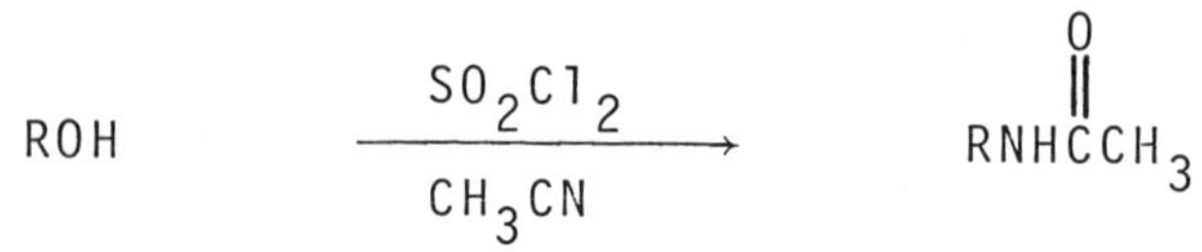

R = n-hexyl, bornyl, norbornyl, etc.

Widely varying yields

VI.A.4-4 H. Schuttenberg and R. C. Schulz, Angew. Chem. Int. Ed., 15, 777 (1976).

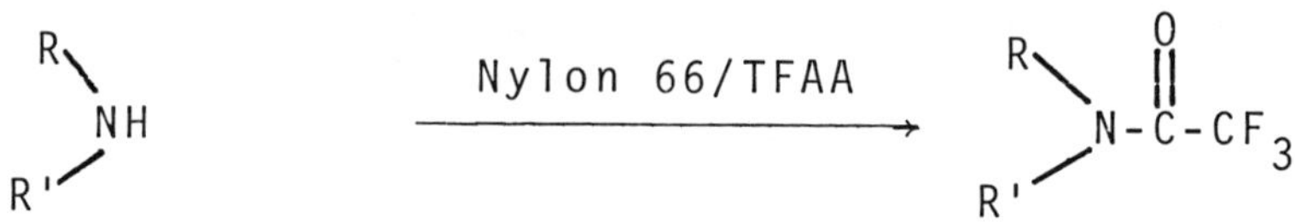

aryl amines, piperidine, hexylamine, hexamethylene diamine

~90% except for p-nitroaniline, 40%

VI.A.4-5 N. Ishikawa and S. Shin-Ya, Chem. Lett., 673 (1976).

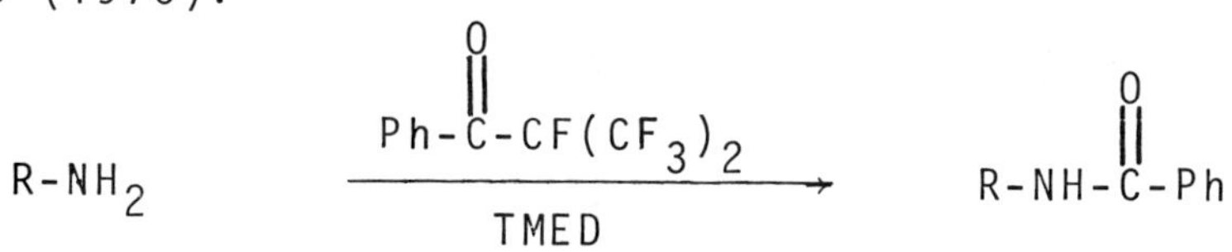

R = Bu, Ph

99%

VI.A.4-6 U. Schöllkopf and H. Beckhaus, <u>Angew. Chem. Int. Ed.</u>, <u>15</u>, 293 (1976).

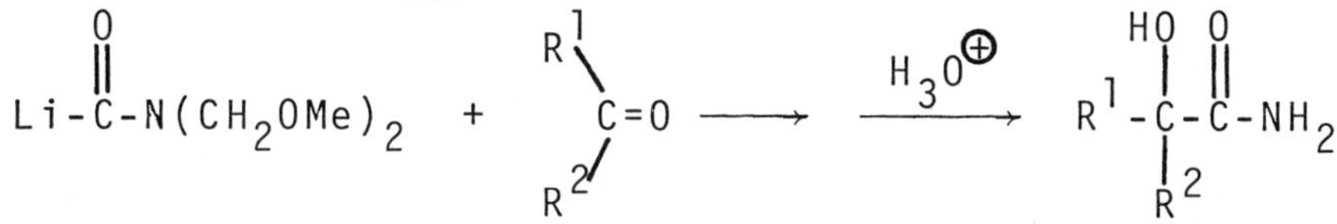

R^1, R^2 = H, alkyl, aryl 44-88%

VI.A.4-7 J. Auerbach, M. Zamore, and S. M. Weinreb, <u>J. Org. Chem.</u>, <u>41</u>, 725 (1976).

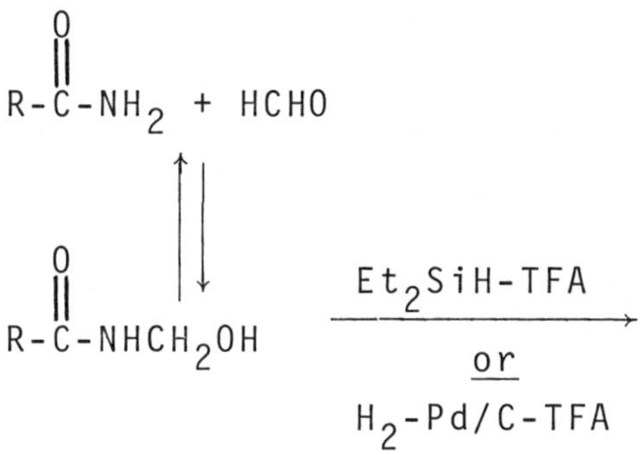

Yields generally 80-90%

VI.A.4-8 S. R. Jones and J. M. Mellor, <u>Synthesis</u>, 32 (1976).

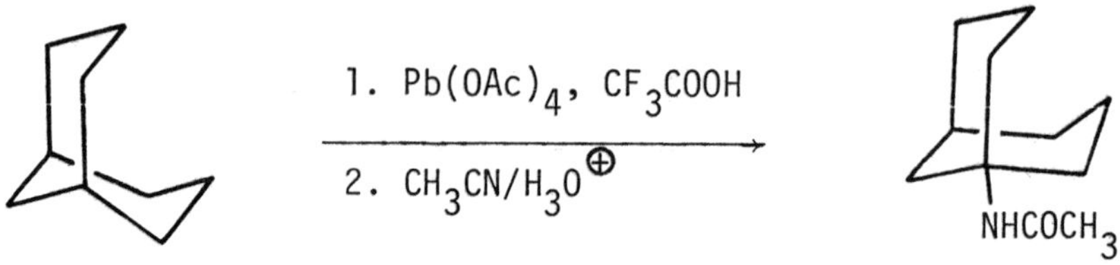

78%

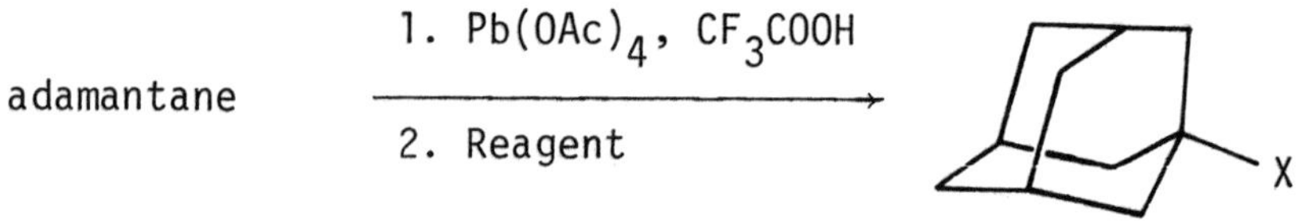

Reagent	X	Yield (%)
KCN	-NHCOH	82
CH_3CN	$-NHCOCH_3$	85
phenol	<u>p</u>-OHC_6H_4	81
ethyl acetoacetate	$-CH(COOEt)COOCH_3$ (-CHCOOCH3 with COOEt)	87
<u>n</u>-C_4H_9SH	$-S-C_4H_9$	97

VI.A.5-1 T.-L. Ho, M. Henninger and G. A. Olah, <u>Synthesis</u>, 815 (1976).

$$Ar\text{-}N_3 \xrightarrow[H_2O/THF]{VCl_2} Ar\text{-}NH_2$$

~70-90%

VI.A.5-2 Y. Watanabe *et al.*, *Bull. Chem. Soc. Japan*, **49**, 1378 (1976).

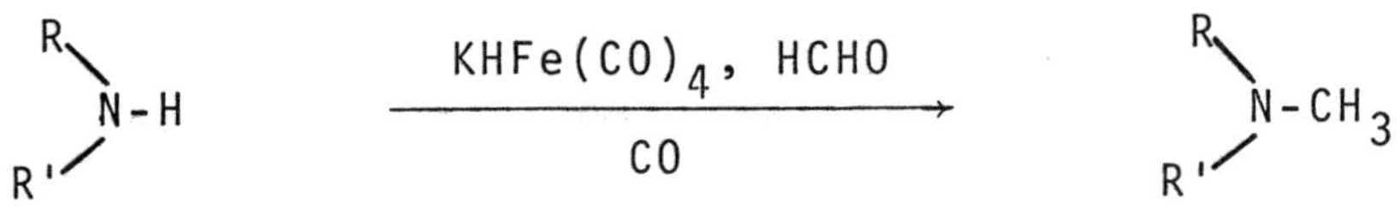

R,R' = alkyl, aryl, H

Widely varying yields.

VI.A.5-3 K. B. Sharpless *et al.*, *J. Am. Chem. Soc.*, **98**, 269 (1976). See also *J. Org. Chem.*, **41**, 176 (1976).

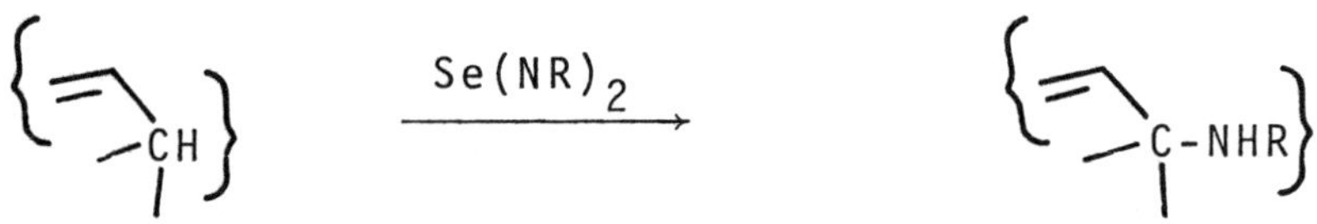

R = *t*-Bu, Ts

Isolated yields generally 40-60%

29 examples cited

VI.A.5-4 D. Enders *et al.*, *Synthesis*, 548 (1976).

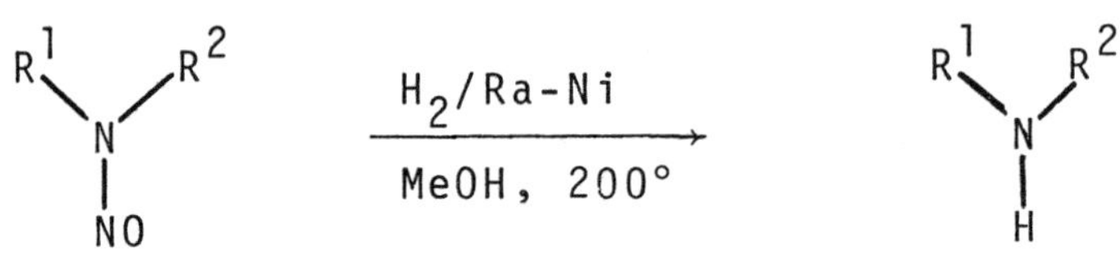

R^1,R^2 = Ph, 1°,2°,3° alkyl

70-95%

VI.A.5-5 F.C.M. Chen and N. L. Benoiton, Can. J. Chem., 54, 3310 (1976).

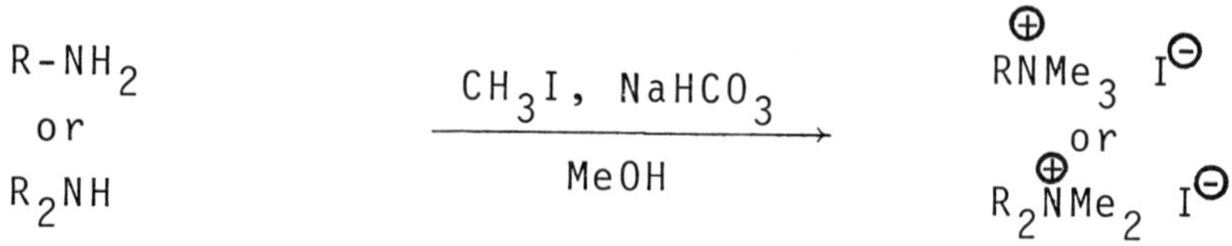

75-96%

VI.A.5-6 S. Coulton, G. A. Moore, and R. Ramage, Tetrahedron Lett., 4005 (1976).

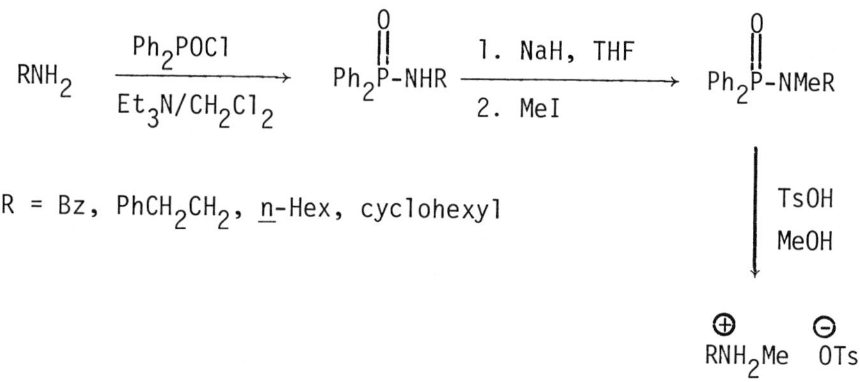

R = Bz, $PhCH_2CH_2$, n-Hex, cyclohexyl

Overall yields ~50%

VI.A.5-7 H. Berbalk, K. Eichinger, and E. Decker, Monat. für Chem., 107, 401 (1976).

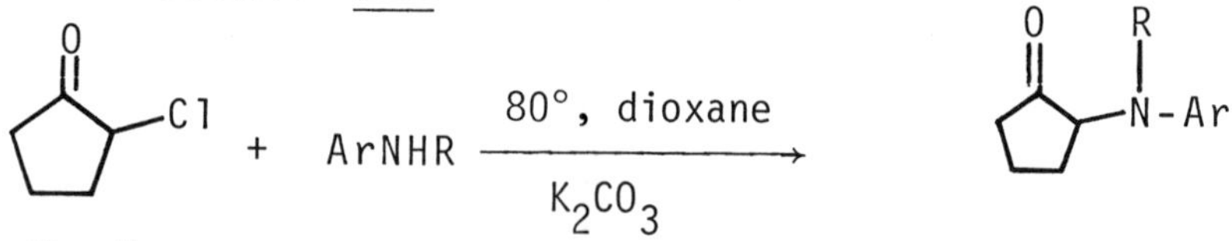

R = H, Me

Ar = Cl, CF_3, $C_6H_4OCF_3$

~50%

VI.A.5-8 S. Yamada, N. Ikota, and K. Achiwa, Tetrahedron Lett., 997, 1001 (1976).

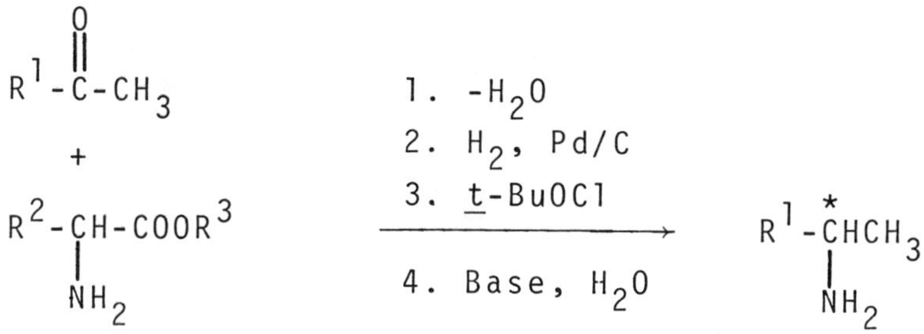

pure (L)

Predominantly (S)
Chemical yield up to 63%
ee up to 81%

R^1 = Ph, Bz

VI.A.5-9 L. E. Overman, J. Am. Chem. Soc., 98, 2901 (1976).

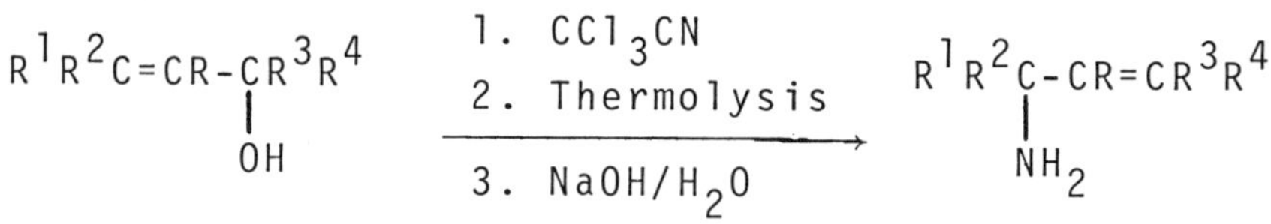

R's = H, alkyl, aryl 23-83% overall

Full paper, many examples cited.

VI.A.5-10 J. C. Arnould, J. Cossy and J. P. Pète, Tetrahedron Lett., 3919 (1976).

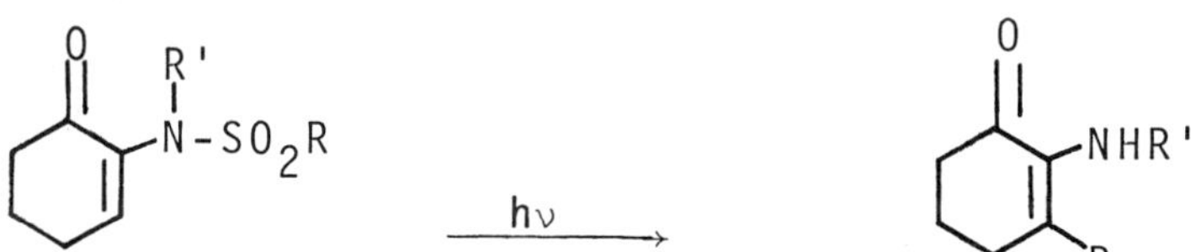

benzene

R = Ts, α,β-naphthyl
R'= Et, Bz, i-Pr, allyl

30-70%

VI.A.5-11 R. Chaabouni, A. Laurent, and B. Marquet, Tetrahedron Lett., 757 (1976).

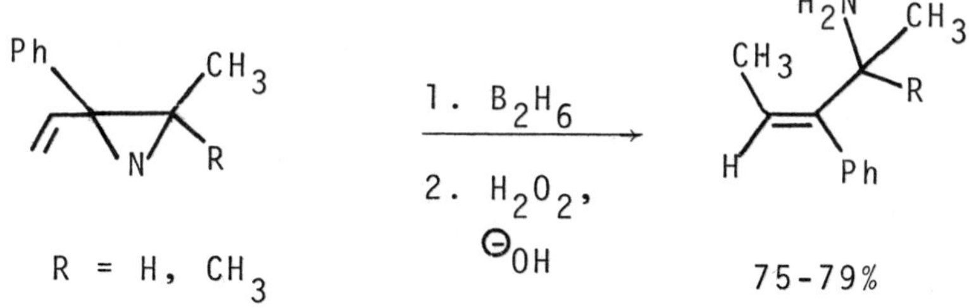

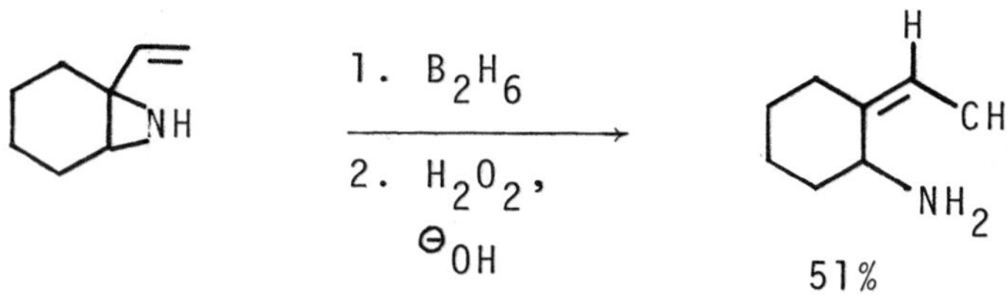

VI.A.5-12 E. Schmitz, Russ. Chem. Rev., 45, 16 (1976).

Review: "Electrophilic Amination"

VI.A.6-1 M. Ali, N. H. Khan, and A. A. Siddiqui, Synth. Commun., 6, 227 (1976).

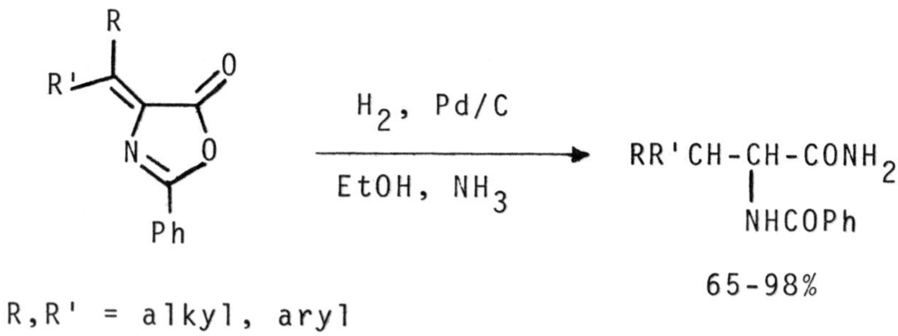

VI.A.6-2 I. Tabushi, Y. Yabushita, and T Nakajima, Tetrahedron Lett., 4343 (1976).

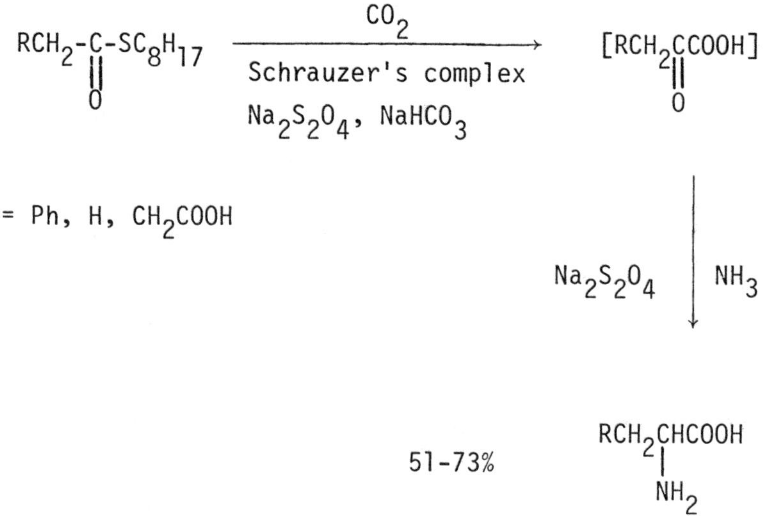

R = Ph, H, CH_2COOH

VI.A.6-3 D. Ben-Ishai, I. Sataty, and Z. Bernstein, Tetrahedron, 32, 1571 (1976).

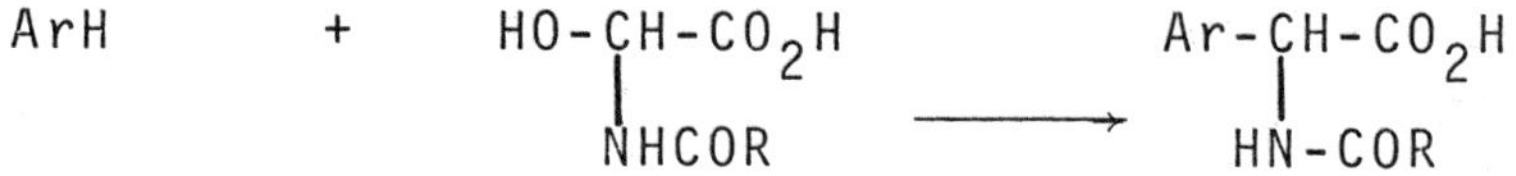

Ar = Ph, activated Ph, furoyl, thiophenyl   R = Ph, Bz, OMe   41-92%

VI.A.6-4 S. Yamada, T Oguri, and T. Shioiri, J. Chem. Soc., Chem. Commun., 136 (1976).

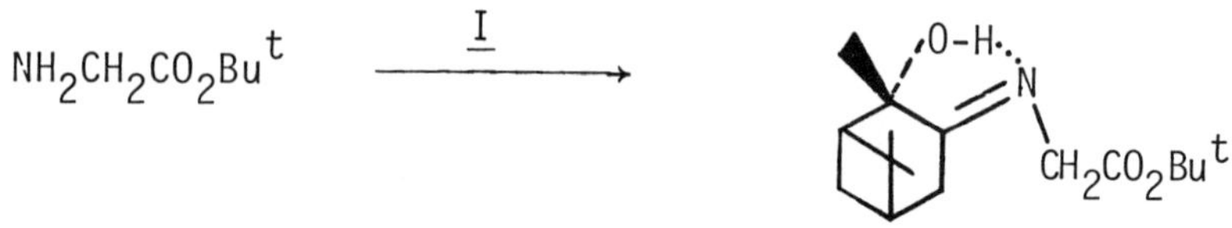

1. $LiNPr^i_2$
2. RX
3. $H_3O^{\oplus}$

I = 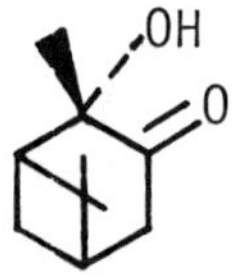

RX	Overall Yield	Optical Yield
MeI	52	83
Bu^iI	50	83
$PhCH_2Br$	79	72

I + $H_2NCH(R)CO_2Bu^t$

(predominantly D isomer)

VI.A.6-5 N. Takaishi et al., J. Am. Chem. Soc., 98, 5400 (1976).

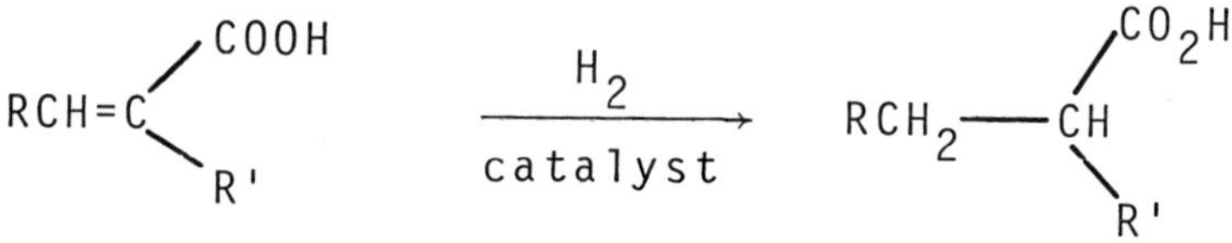

R = H, Ph

R' = Ph, $NHCOCH_3$

Yields ~100% (NMR)

52-86% ee

catalyst = polymer-supported Rh-DIOP

VI.A.6-6 H. Poisel and U. Schmidt, Angew. Chem. Int. Ed., 15, 294 (1976).

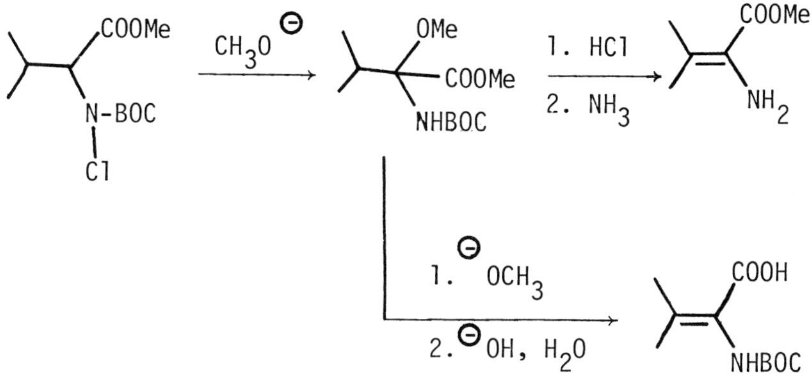

yields ~60% overall

VI.A.7-1 G. W. Griffin, R. L. Smith, and A. Manmade, J. Org. Chem., 41, 338 (1976).

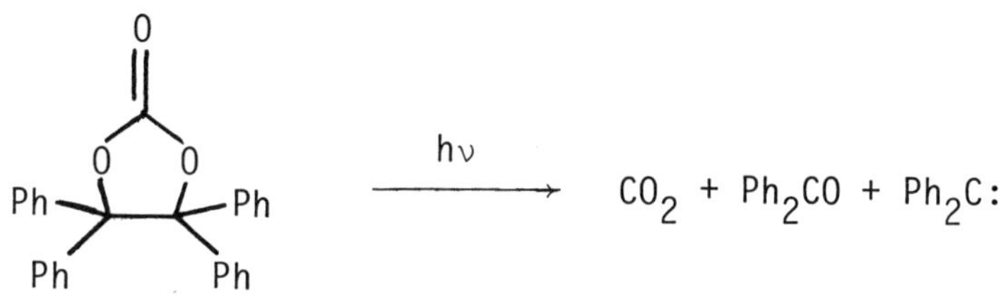

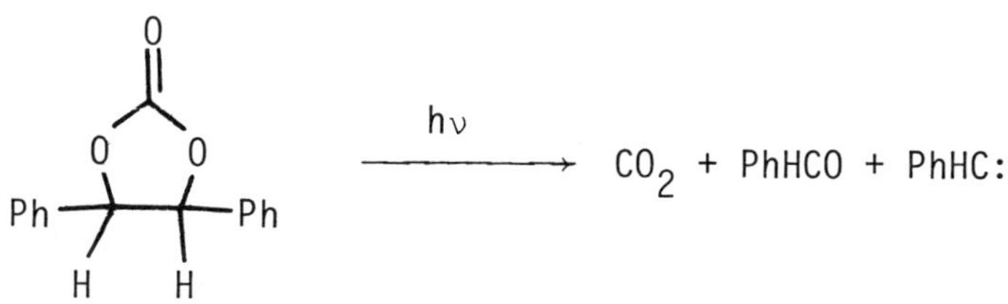

VI.A.7-2 M. Sekiya *et al.*, *Chem. Pharm. Bull.*, **24**, 369 (1976).

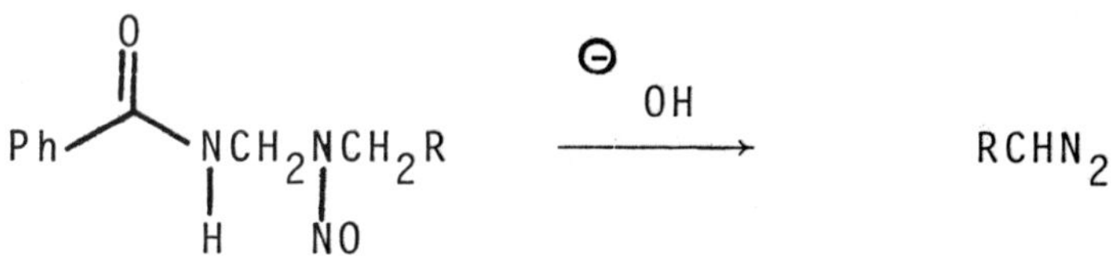

R = Me, Et, *n*-Bu 75%, 65%, 44%, respectively

VI.A.7-3 R. A. Olofson, K. D. Lotts, and G. N. Barber, *Tetrahedron Lett.*, 3779 (1976).

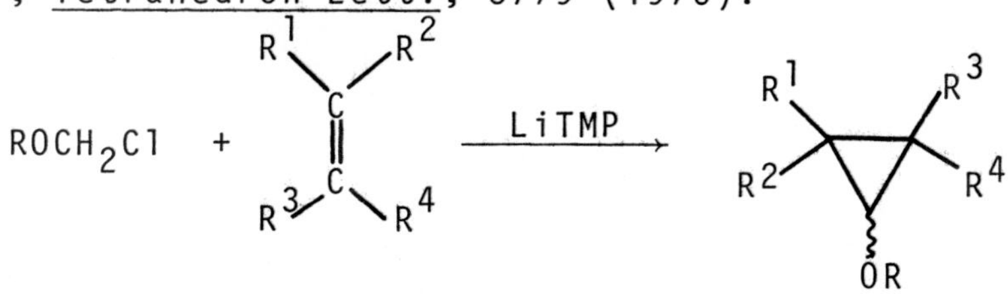

R = Me, Et, *t*-Bu, *i*-Pr ~40-60%
$R^1, \ldots, R^4$ = Me, vinyl, cyclic, etc.

VI.A.7-4 T. L. Gilchrist and D. J. Pearson, *J. Chem. Soc. Perkin I*, 1257 (1976).

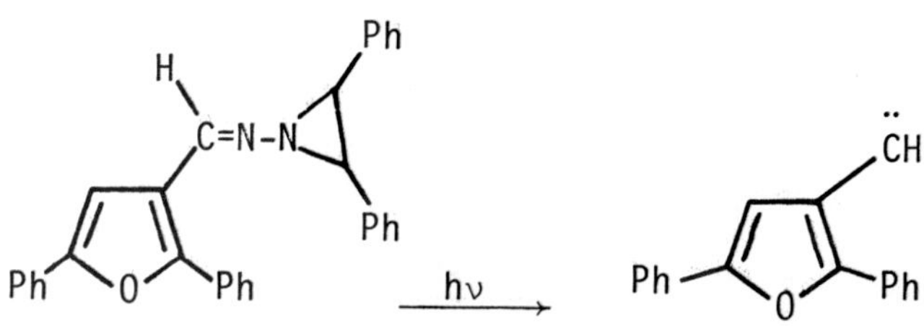

VI.A.7-5 M. Makosza, Pure and Appl. Chem., 43, 439 (1975).

Review: "Two-phase reactions in the chemistry of carbanions and halocarbenes. A useful tool in organic synthesis."

VI.A.7-6 C. Wentrup, Fortschritte der Chem. Forschung, 62, 173 (1976).

Review: "Rearrangements and Interconversions of Carbenes and Nitrenes"

VI.A.8-1 K. Nyberg, Synthesis, 545 (1976).

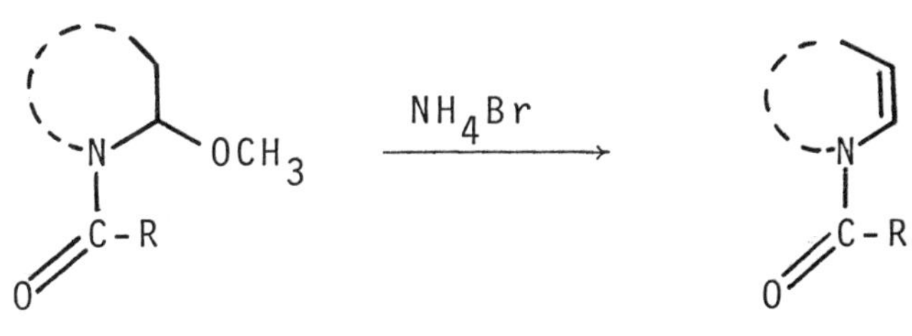

~70-90%

VI.A.8-2 H. Bohme and K.-H. Weisel, Chem. Ber., 109, 2908 (1976).

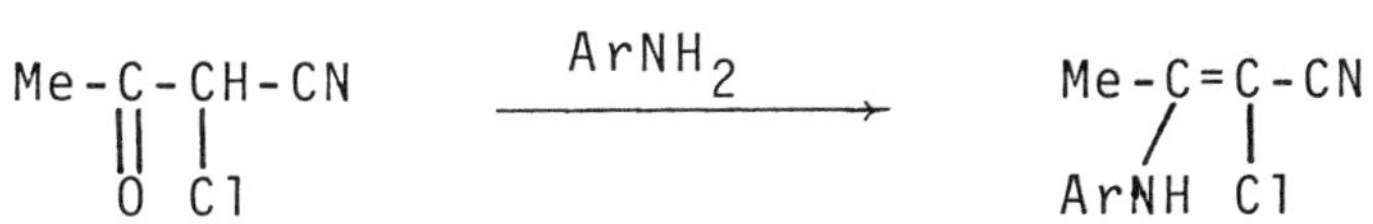

~50-70%

VI.A.8-3 E. Elkik and M. Imbeaux-Oudotte, Comptes Rendus (C), 282, 935 (1976).

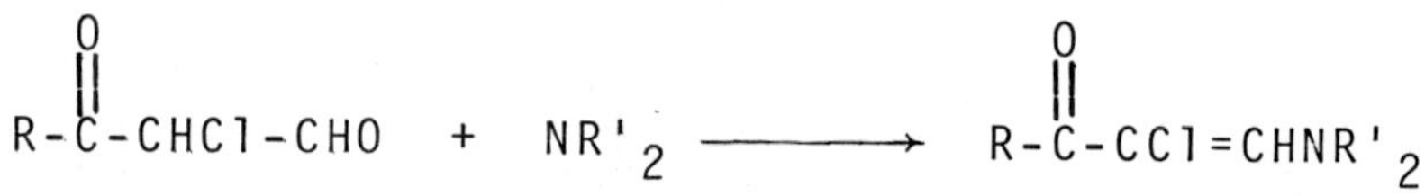

R = Ph, OEt $R'_2 = Me_2$, Et_2, $(CH_2)_5$ 35-48%

VI.A.8-4 J. Ficini, Tetrahedron, 32, 1449 (1976).

Review: "Ynamine: a versatile tool in organic synthesis."

VI.A.9-1 Y. Tamura et al., Synthesis, 35 (1976).

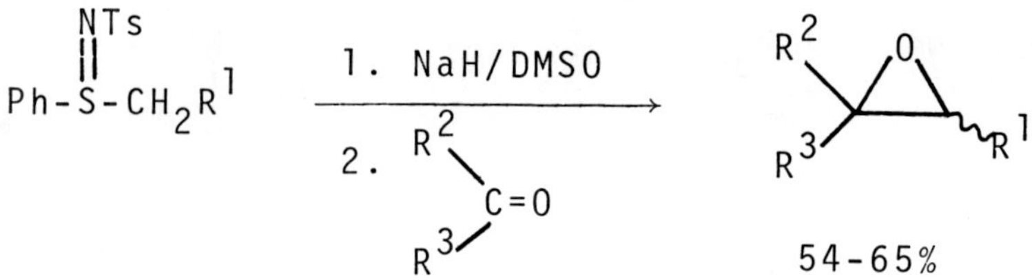

54-65%

R^1 = H, CH_3

R^2, R^3 = H, alkyl, aryl, cyclic

VI.A.9-2 D. Van Ende and A. Krief, Tetrahedron Lett., 457 (1976).

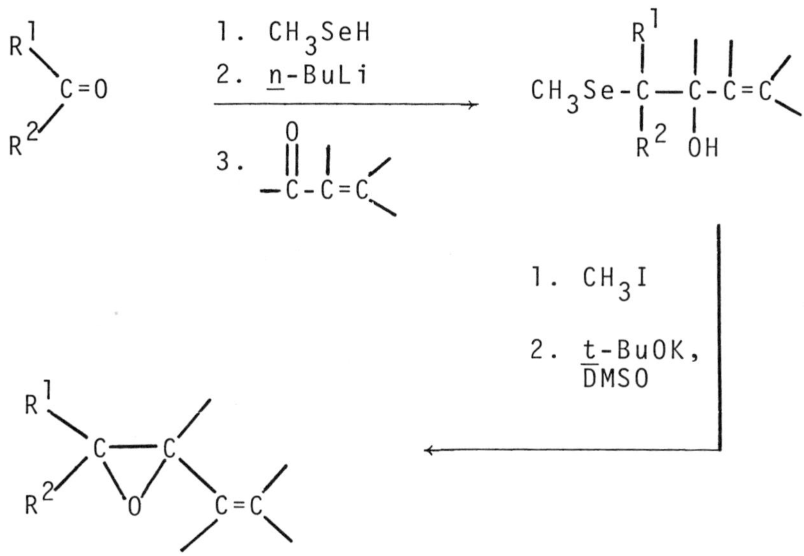

Yields up to 80% (isolated)

VI.A.10-1 J. Fairhurst and D. C. Horwell, Synth. Commun., 6, 89 (1976).

$$\text{R-COOH} \xrightarrow{\text{Et}_2\text{SO}_4\text{, DBN}} \text{R-COOEt}$$

76-95%

R = β-keto, β-cyano alkyl, etc.

Particularly useful for thermally unstable acids.

VI.A.10-2 H. Normant and C. Laurenco, Comptes Rendus (C), 283, 483 (1976).

$$\text{R-COOK} + \text{R'X} \xrightarrow{\text{CH}_3\text{CN, TMEDA}} \text{R-COOR'}$$

R= hindered alkyl R'X = BzCl, n-BuBr ~80-90%

VI.A.10-3 T. Mukaiyama, H. Toda, and S. Kobayashi, Chem. Lett., 13 (1976).

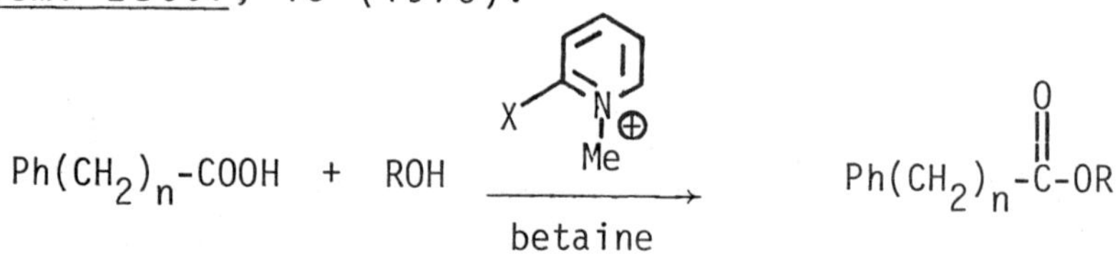

n = 1, 2 X = F, Cl

R = Ph, alkyl, Bz, cinnamyl, crotyl

VI.A.10-4 U. Weber, Z. Naturforsch (B), 31, 1157 (1976).

$$\text{R-COOH} \xrightarrow[\text{BuOH, pyridine}]{\text{R'OH, DCC}} \text{R-COOR'}$$

(N-protected amino acid) R' = Me, Bz ~80-90%

VI.A.10-5 S. Takimoto et al., Bull. Chem. Soc. Japan, 49, 2335 (1976).

$$\text{R-}\overset{\text{O}}{\overset{\|}{\text{C}}}\text{-Cl} \quad + \quad \text{R'-OH} \xrightarrow{\text{AgCN}} \text{R-}\overset{\text{O}}{\overset{\|}{\text{C}}}\text{-OR'}$$

R = hindered alkyl or aryl R' = 1°,2°, or 3° alcohol 57-100% (VPC)

VI.A.10-6 B. M. Trost *et al.*, *Tetrahedron Lett.*, 3477 (1976).

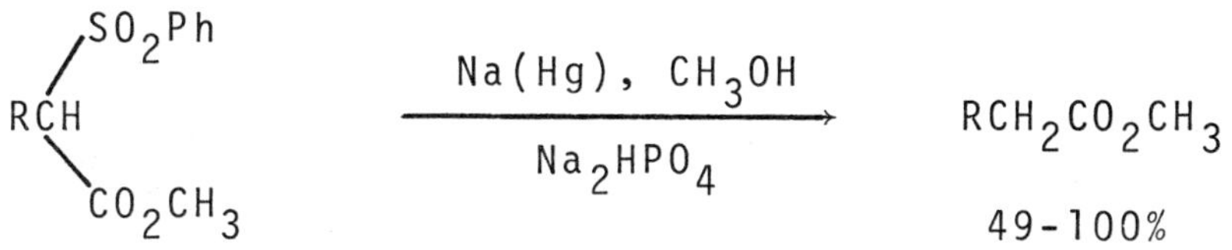

R = complicated alkyl

VI.A.10-7 K. R. Fountain and M. Pierschbacher, *J. Org. Chem.*, **41**, 2039 (1976).

A phosgeneless synthesis of diaryl carbonates, e.g.:

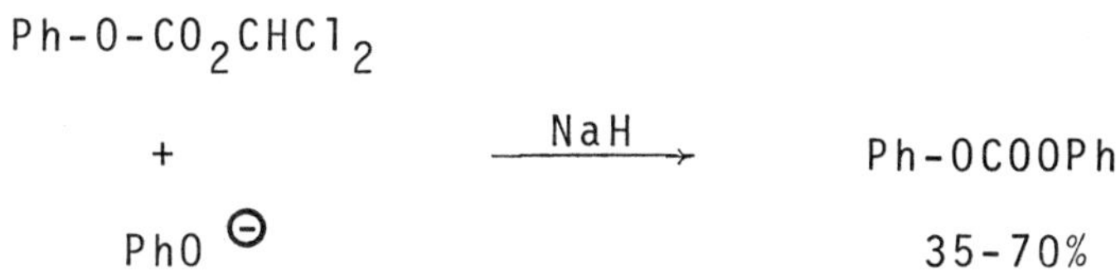

VI.A.10-8 T. Ogawa and M. Matsui, *J. Am. Chem. Soc.*, **98**, 1629 (1976).

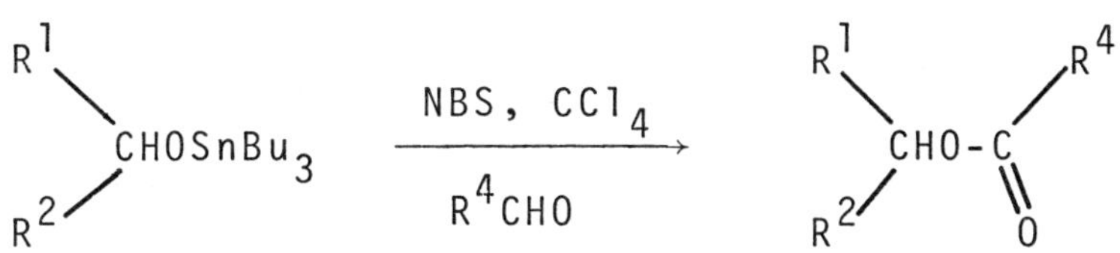

R's = H, alkyl, aryl yields up to 95%

VI.A.10-9 P. Catsoulacos, Bull. Soc. Chim. France, 642 (1976).

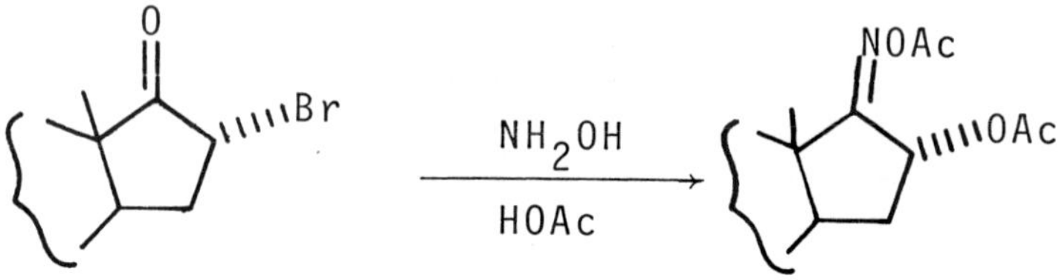

~50-85%

VI.A.10-10 T. Francis and M. P. Thorne, Can. J. Chem., 54, 24 (1976).

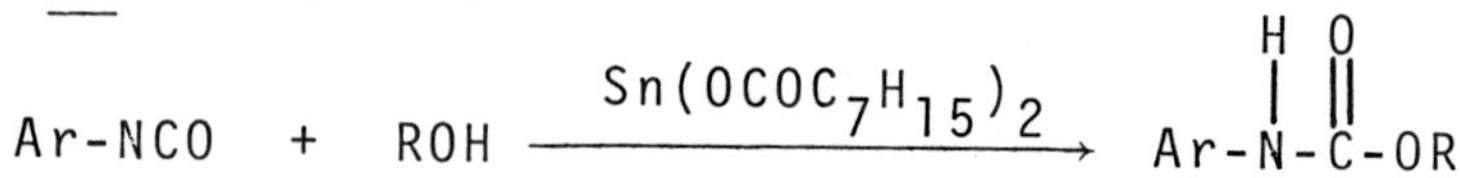

R = 3° alkyl

~70-95%

VI.A.10-11 M. Julia and B. Badet, Bull. Soc. Chim. France, 525 (1976).

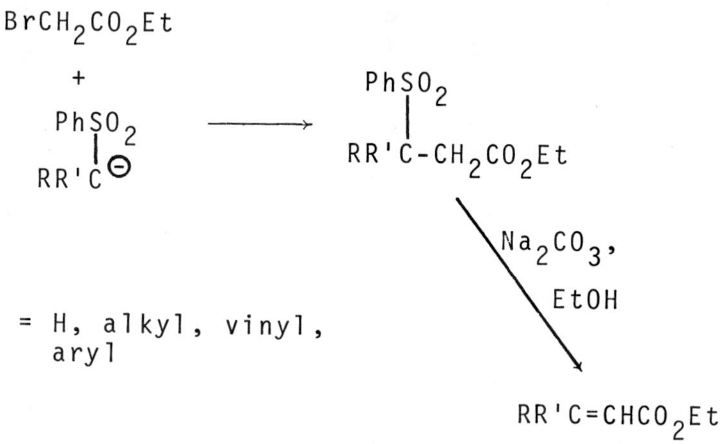

R,R' = H, alkyl, vinyl, aryl

30-90%

VI.A.10-12 E. Yoshii, T. Koizumi, and T. Kawazoe, *Chem. Pharm. Bull.*, **24**, 1957 (1976).

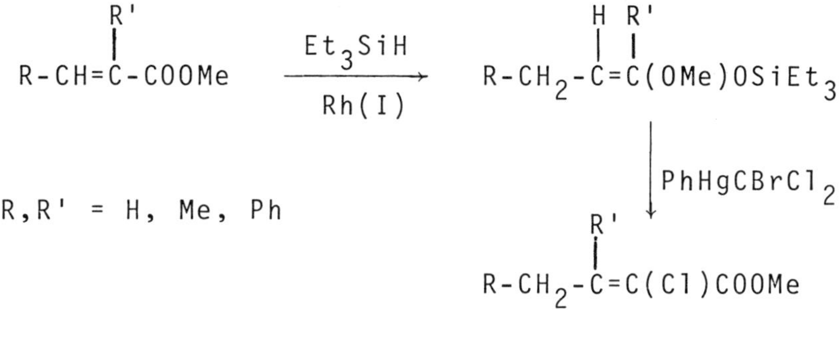

52-92%

VI.A.10-13 G. M. Rubottom, J. M. Gruber, and K. Kincaid, *Synth. Commun.*, **6**, 59 (1976).

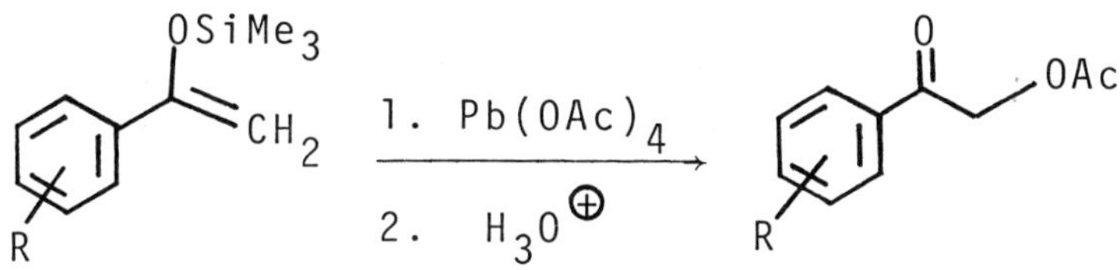

>90%

VI.A.11-1 J. D. Daley, J. M. Rosenfeld, and E. V. Younglai, *Steroids*, **27**, 481 (1976).

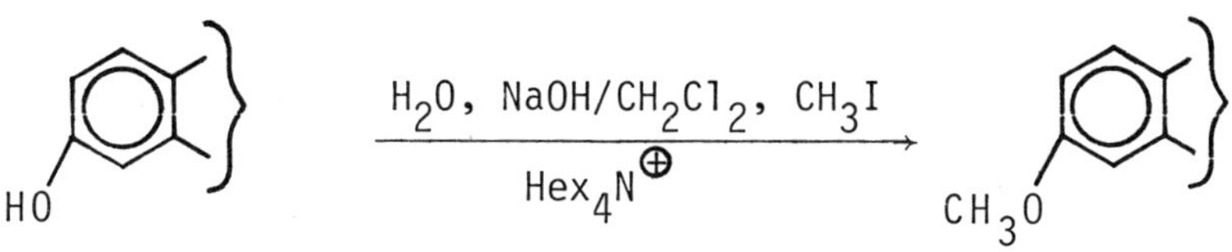

quantitative

VI.A.11-2 J. E. Shaw, D. C. Kunerth, and S. B. Swanson, J. Org. Chem., 41, 732 (1976).

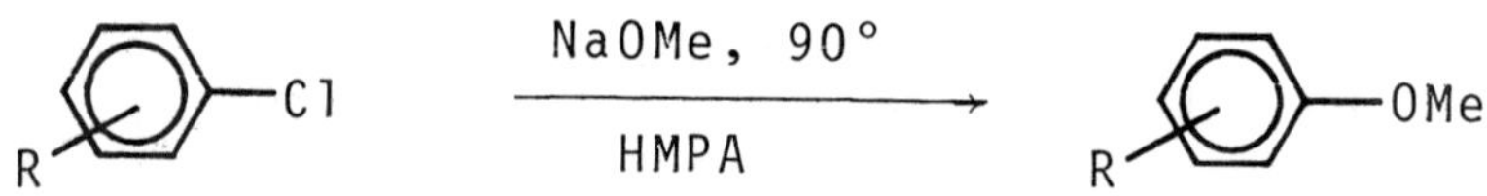

R = H, Cl

50-80%

13% for R = p-CH_3

VI.A.12-1 A. W. Burgstahler, L. O. Weigel, and C. G. Shaefer, Synthesis, 6, 767 (1976).

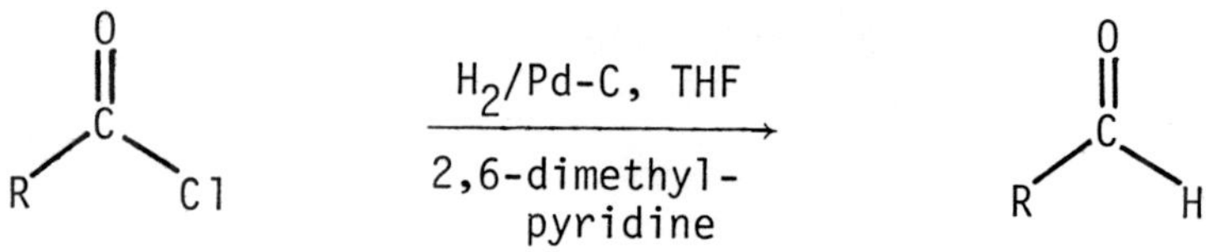

R = alkyl, aryl

77-96% isolated

VI.A.12-2 G. Cardillo, M. Orena, and S. Sandri, J. Chem. Soc., Chem. Commun., 190 (1976).

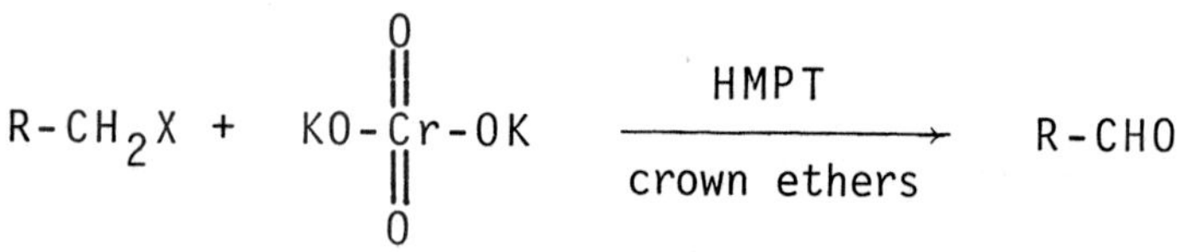

halide	Yield (%, by glc)
γ,γ-dimethylallyl Br	78
Geranyl Br	82
Farnesyl Br	80
Benzyl Br	80
Octyl Br	20

VI.A.12-3 I. Degani and R. Fochi, J. Chem. Soc. Perkin I, 323 (1976).

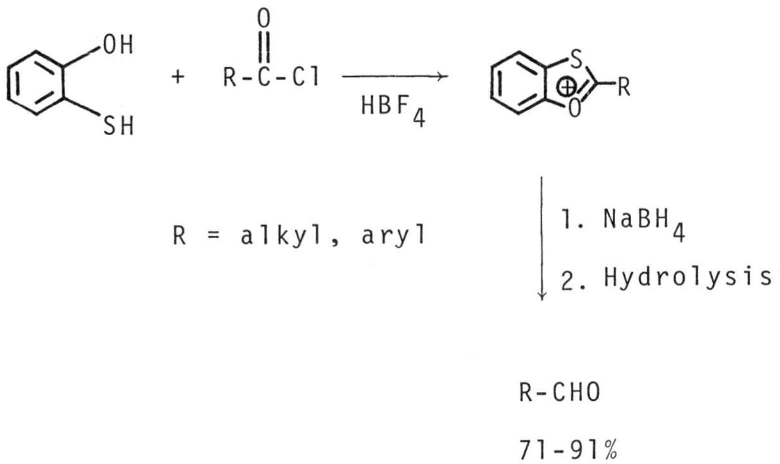

R = alkyl, aryl

VI.A.12-4 G. Doleschall, Tetrahedron, 32, 2549 (1976).

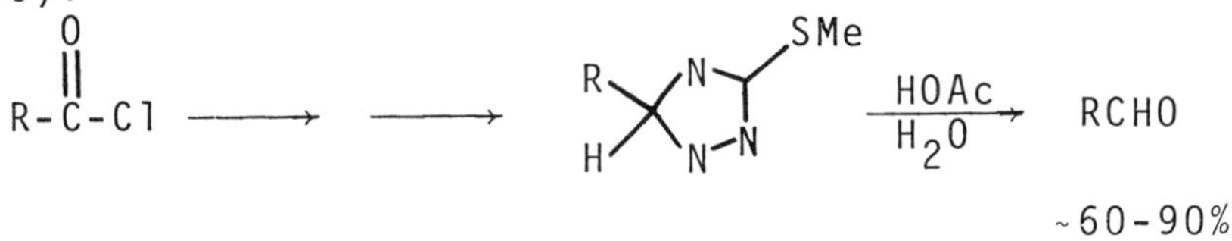

R = 1°,2°,3° alkyl, aryl, cinnamoyl

VI.A.12-5 C. A Scott, D. G. Smith, and D.J.H. Smith, Synth. Commun., 6, 135 (1976).

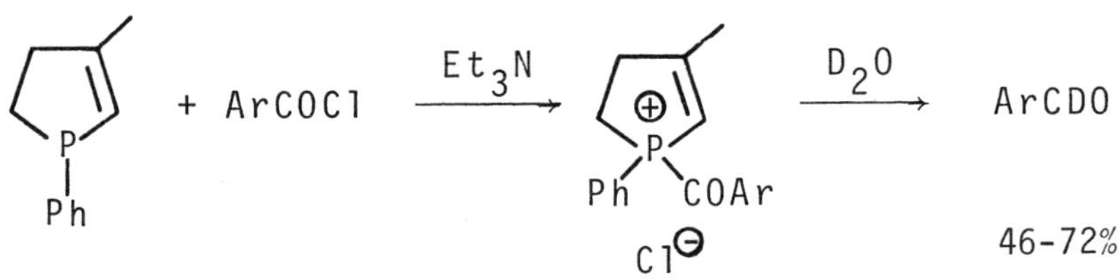

VI.A.12-6 K. Sachdev and H. S. Sachdev. <u>Tetrahedron Lett.</u>, 4223 (1976).

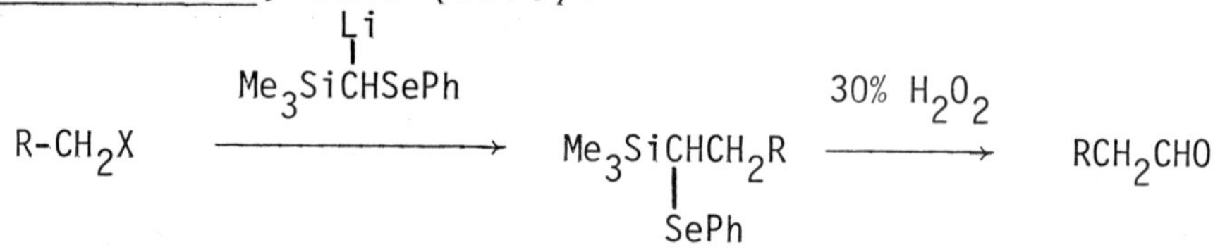

X = Br, I

R = 1° alkyl

>66%

VI.A.12-7 J. Nakayama, <u>J. Chem. Soc. Perkin I</u>, 540 (1976).

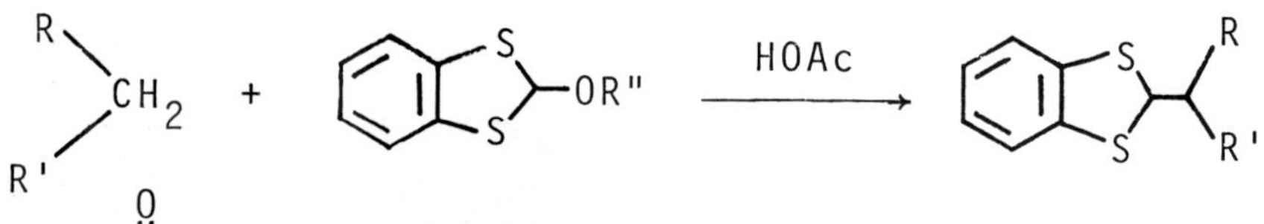

R,R' = $-\overset{O}{\overset{\|}{C}}-R$, CN, SPh   R" = alkyl

~50-80%

VI.A.12-8 P. Bakuzis *et al.*, <u>J. Org. Chem.</u>, **41**, 2769 (1976).

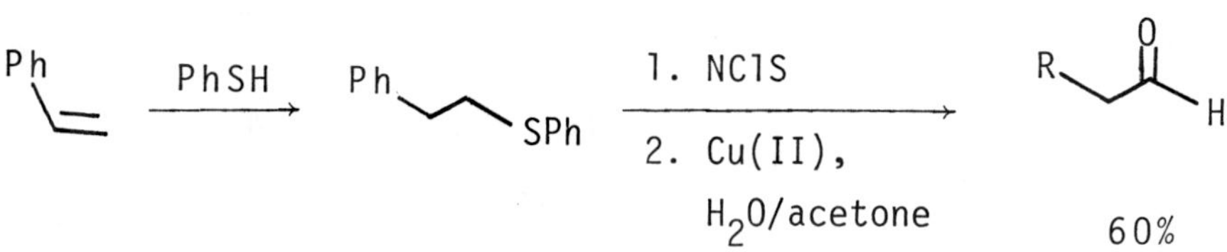

60%

VI.A.12-9 H. Pauling, D. A. Andrews, and N. C. Hindley, Helv. Chim. Acta, 59, 1233 (1976).

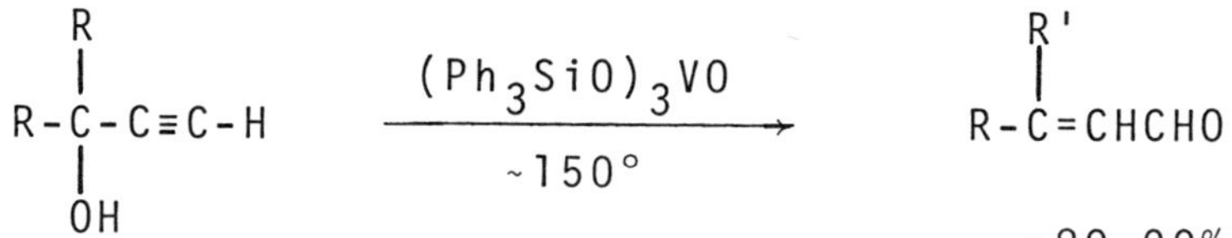

~80-90%

R' = H, CH_3

R = alkyl, aryl

VI.A.12-10 J. H. Babler and M. J. Coghlan, Synth. Commun., 6, 469 (1976).

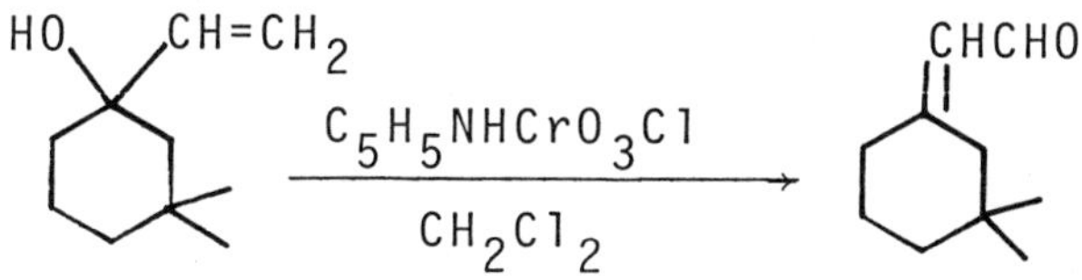

85%

(mixture of E and Z)

VI.A.12-11 J. Hocker, H. Giesecke, and R. Merten, Angew. Chem. Int. Ed., 15, 169 (1976).

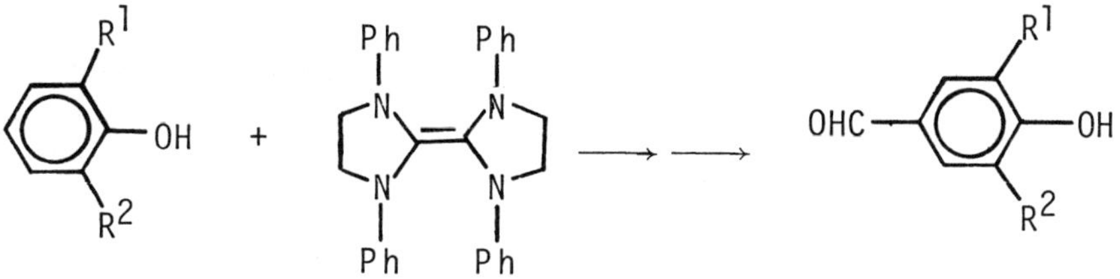

26-84%

R^1,R^2 = H, Cl, CH_3, t-Bu

VI.A.12-12 A. Rahman and A. Basha, J. Chem. Soc., Chem. Commun., 594 (1976).

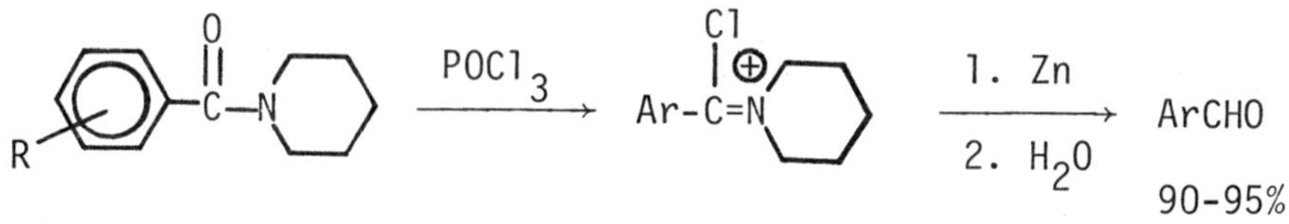

R = H, Cl, Br

VI.A.12-13 C. A. Bertelo and J. Schwartz, J. Am. Chem. Soc., 98, 262 (1976).

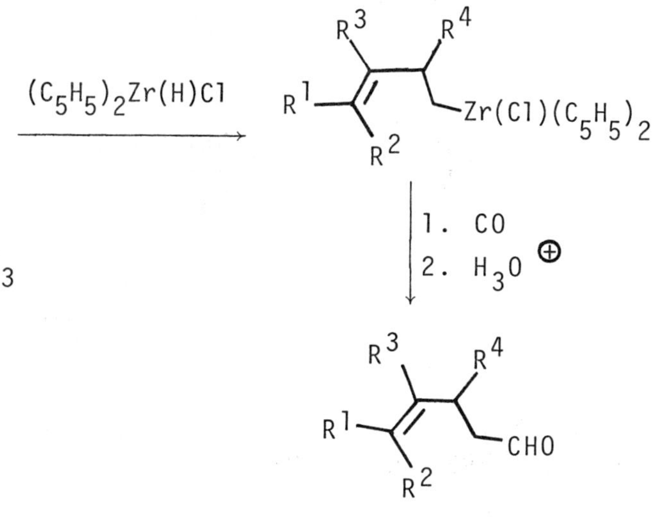

R's = H, CH_3

59-98% (VPC, based on Zr complex)
Five examples cited.

VI.A.12-14 G. Cahiez, *et al.*, Tetrahedron Lett., 3155 (1976).

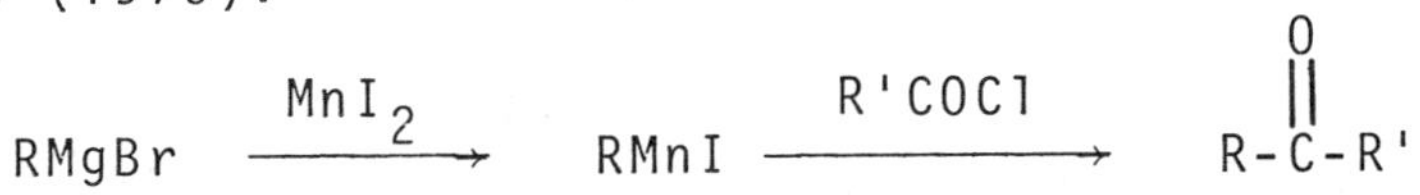

R,R' = Me, *n*-alkyl, *i*-Pr, *t*-Bu 60-90%

VI.A.12-15 G. Cardillo, M. Orena, and S. Sandri, Tetrahedron Lett., 3985 (1976).

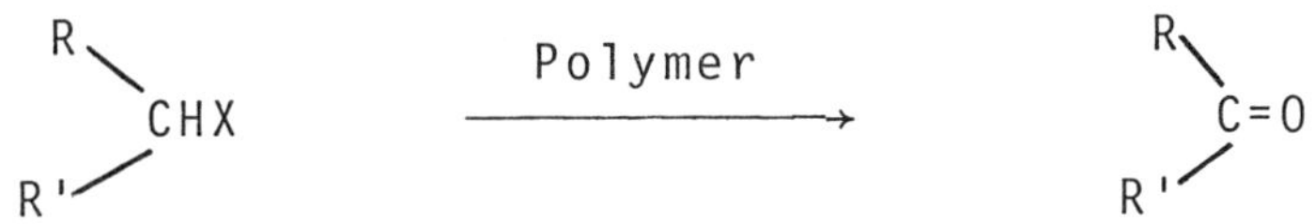

BzCl, geranyl Br, ~95%

Ph_2CHBr, etc.

Polymer =

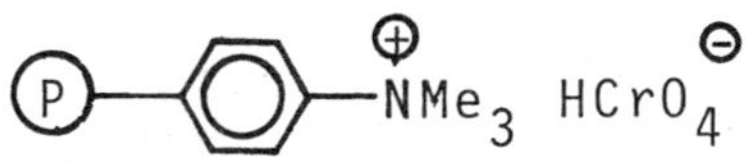

VI.A.12-16 R. Kirchoff, Tetrahedron Lett., 2533 (1976).

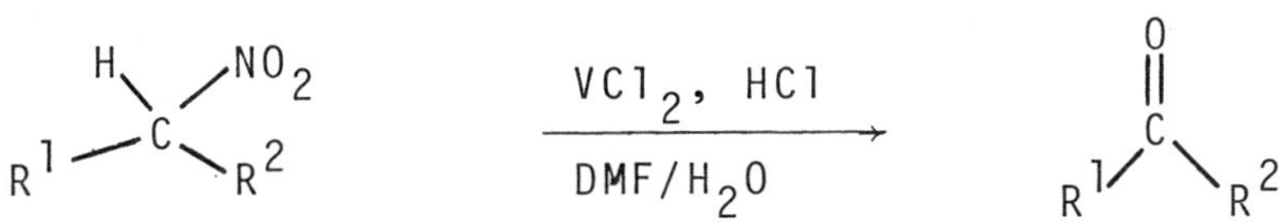

R^1 = H, alkyl 24% (heptaldehyde)

R^2 = alkyl, benzyl 53-71% (ketones)

VI.A.12-17 Y. Masuyama, Y. Ueno and M. Okawara, <u>Tetrahedron Lett</u>., 2967 (1976).

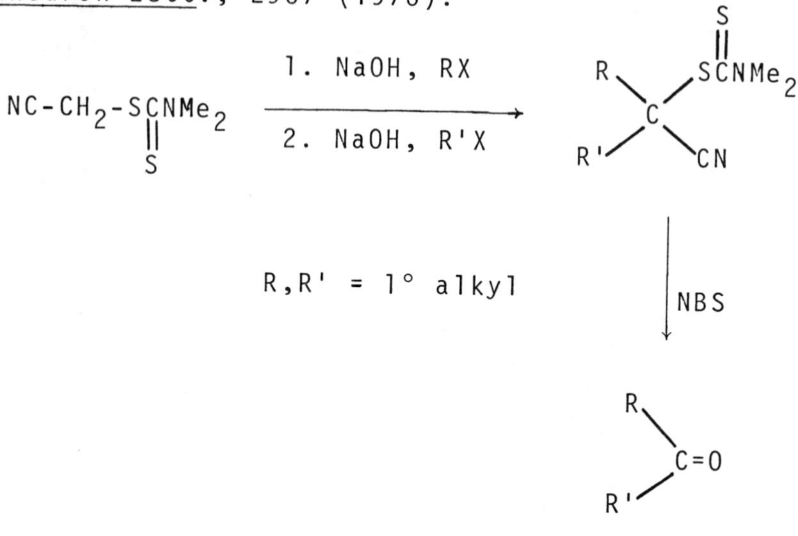

R,R' = 1° alkyl

overall ~60%

VI.A.12-18 J. Tsuji, I. Shimizu, and K. Yamamoto, <u>Tetrahedron Lett.</u>, 2975 (1976).

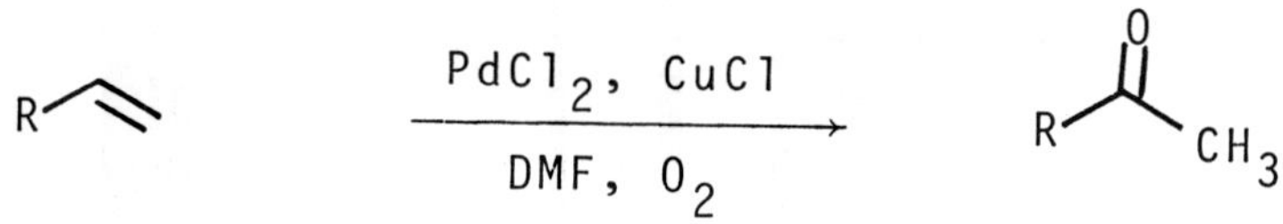

R = alkyl, often γ-keto

58-77%

VI.A.12-19 P. C. Traas, H. Boelens, and H. J. Takken, Tetrahedron Lett., 2287 (1976).

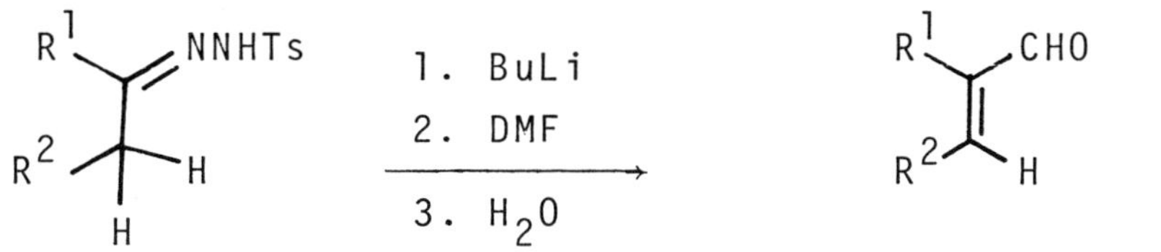

Yields generally 50-60%

VI.A.12-20 L. M. Stephenson and L. C. Falk, J. Org. Chem., 41, 2928 (1976).

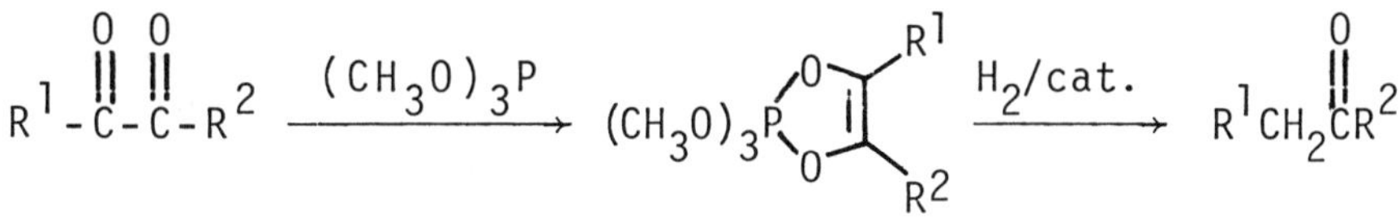

R^1, R^2 = alkyl, aryl

VI.A.12-21 R. C. Larock and J. C. Bernhardt, Tetrahedron Lett., 3097 (1976).

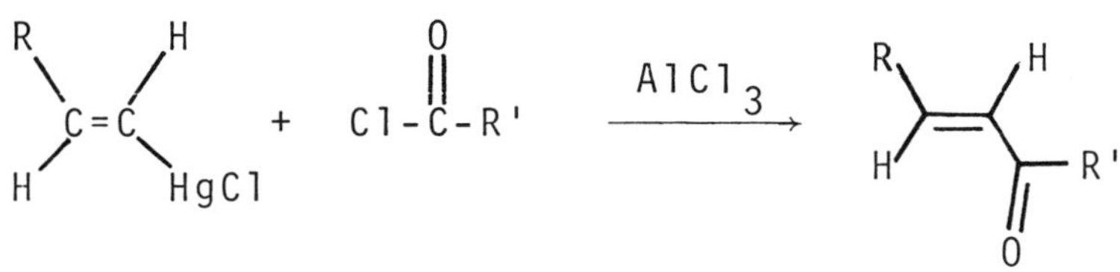

R = alkyl, aryl
R' = alkyl, vinyl

~90-100%

VI.A.12-22 L. Barsky *et al.*, *J. Org. Chem.*, **41**, 3651 (1976).

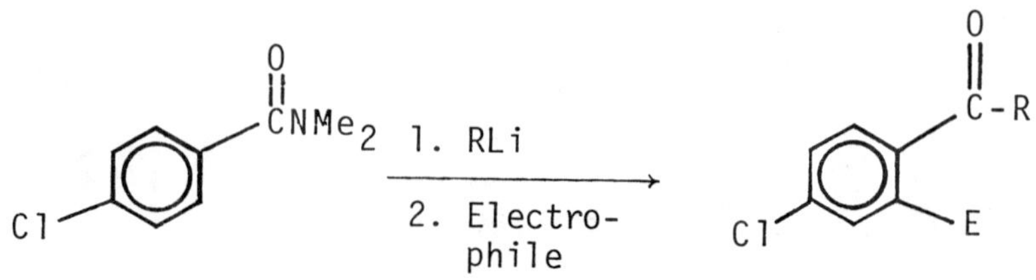

R	Electrophile	E	Yield (%)
Bu	$(CH_3S)_2$	CH_3S	56
Ph	$(CH_3S)_2$	CH_3S	53
Me	Ph_2CO	Ph_2COH	47
Me	DMF	CHO	56
Me	MeI	CH_3	81 (NMR)

Several additional examples.

VI.A.12-23 P. T. Lansbury and R. W. Britt, *J. Am. Chem. Soc.*, **98**, 4577 (1976).

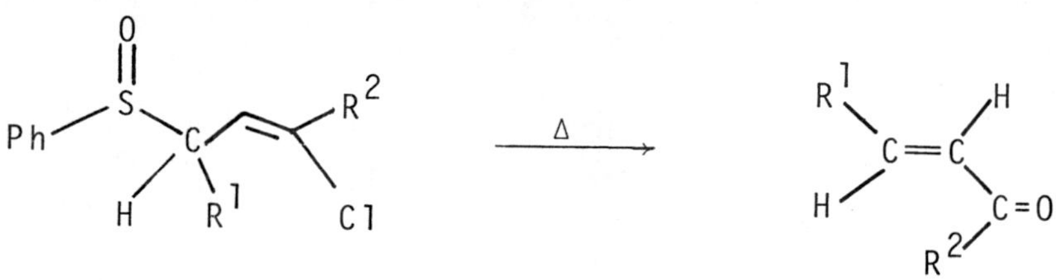

R^1, R^2 = H, alkyl

VI.A.12-24 B. Corbel *et al.*, *Tetrahedron Lett.*, 835 (1976).

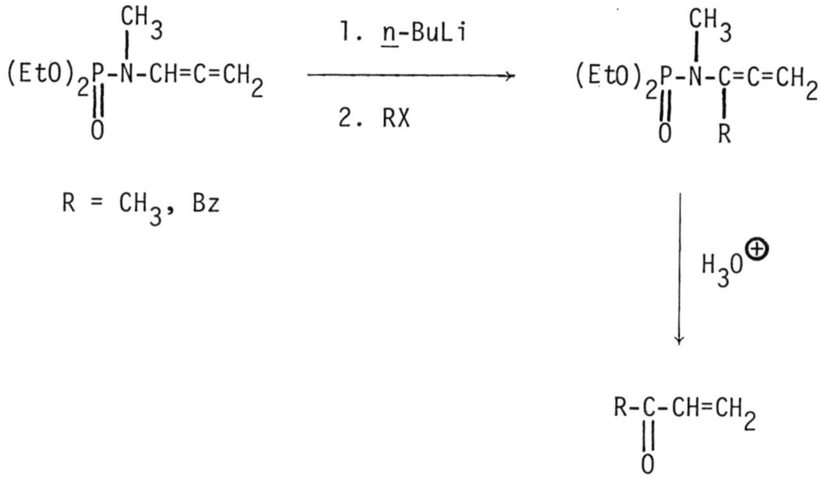

$R = CH_3$, Bz

yields generally 50-70%

VI.A.12-25 J. S. Walia and A. S. Walia, *J. Org. Chem.*, **41**, 3765 (1976).

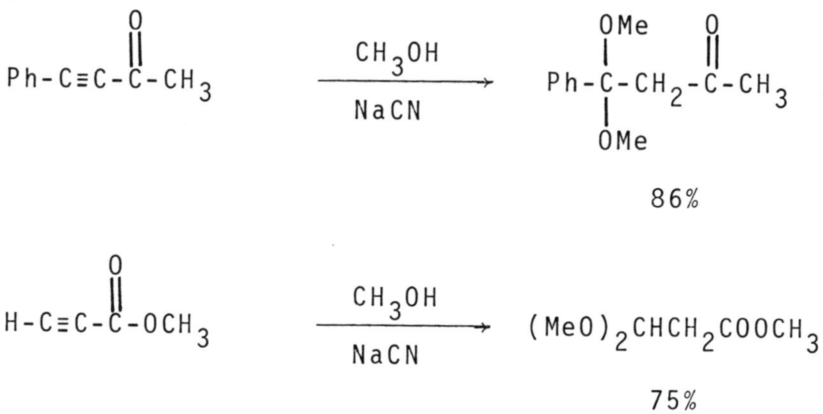

VI.A.12-26 J. I. Okogun, J. Chem. Soc. Perkin I, 2241 (1976).

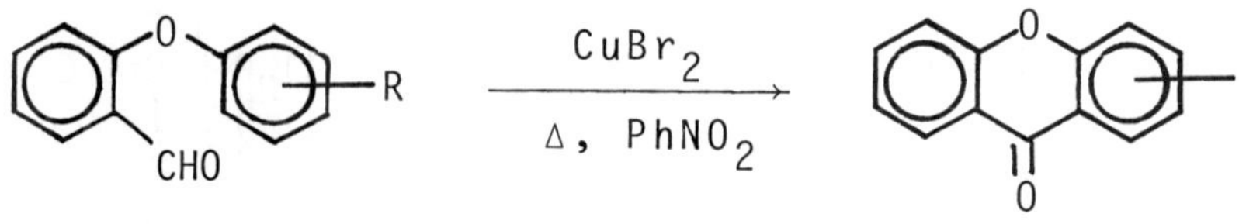

R = H, OMe, Me, NO_2 low yields (<50%)

VI.A.12-27 S. Nimgirawath, E. Ritchie, and W. C. Taylor, Aust. J. Chem., 29, 339 (1976).

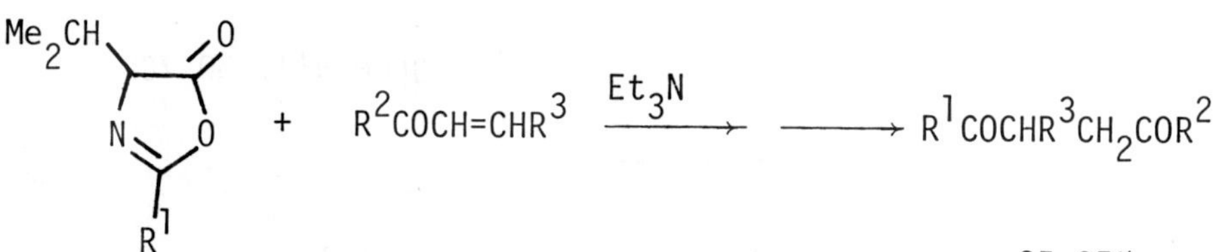

25-35%

R^1, R^2 = alkyl, aryl

R^3 = H, Ph

VI.A.12-28 R. C. Cookson and R. M. Lane, J. Chem. Soc., Chem. Commun., 804 (1976).

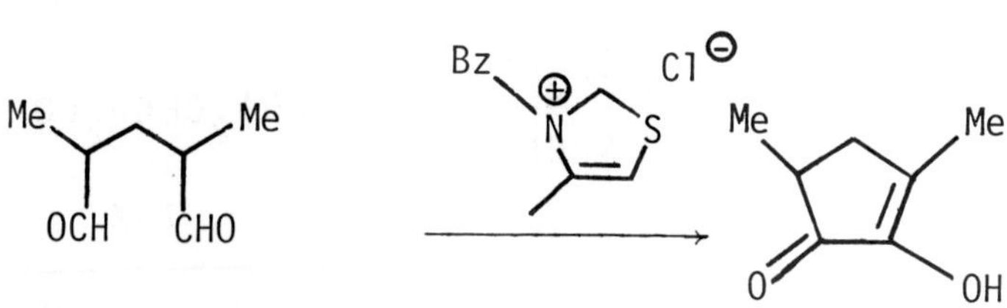

88%

VI.A.12-29 H. D. Scharf, P. Friedrich, and A. Linckens, Synthesis, 256 (1976).

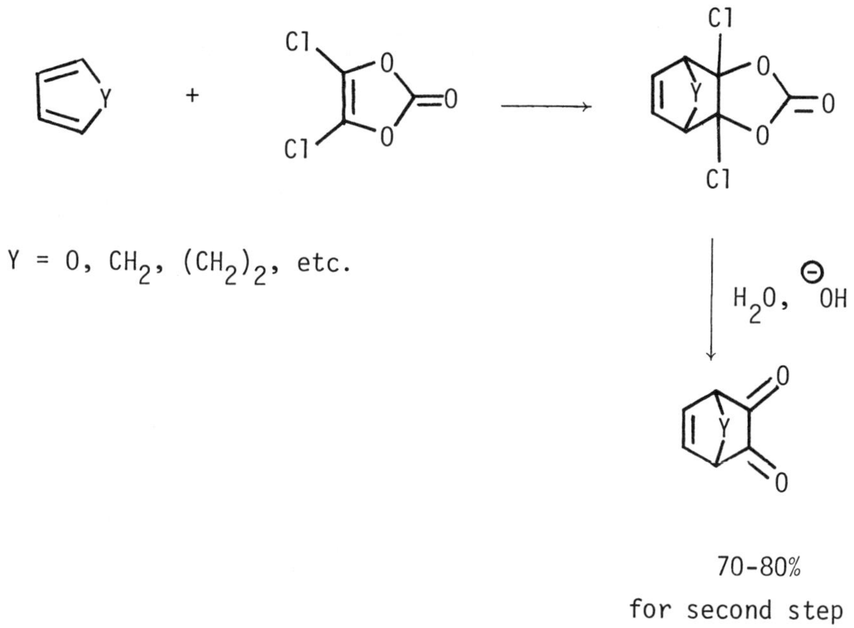

VI.A.12-30 K. Oguar, M. Yamashita, and G. Tsuchihashi, Tetrahedron Lett., 759 (1976).

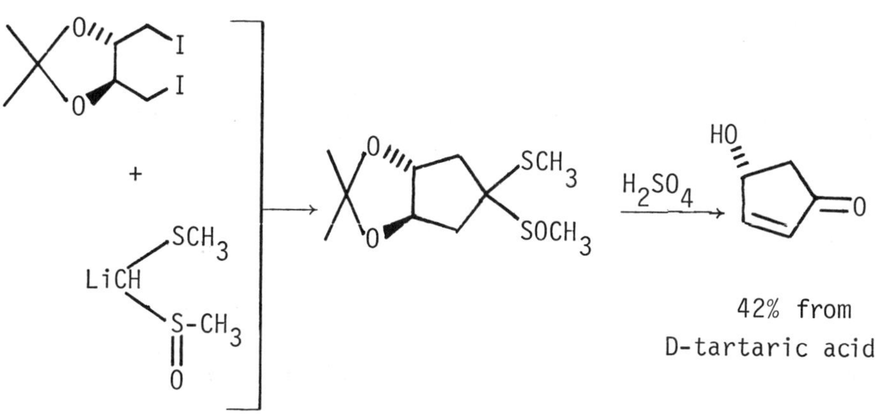

VI.A.12-31 J. M. Conia, Pure and Appl. Chem., **43**, 317 (1975).

Review: "The Cyclopropanation of Silyl Enol Ethers. A Powerful Synthetic Tool."

Particularly useful in the formation of alkylated ketones, e.g.:

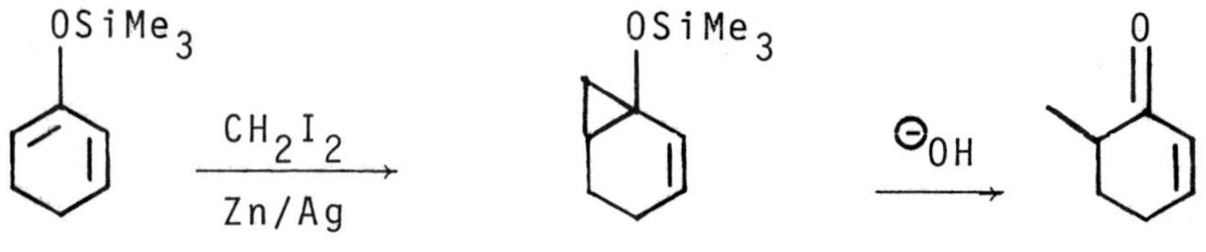

~80% overall

VI.A.12-32 G. Rio and A. Lecas-Nawracka, Bull. Soc. Chim. France, 317 (1976).

Review: "Preparation of Saturated and Unsaturated 1,4-Diketones"

VI.A.13-1 J. Streith, C. Fizet, and H. Fritz, Helv. Chem. Acta, **59**, 2786 (1976).

$$R{-}CHO \xrightarrow[H_2O]{NH_2OSO_3H} R{-}C{\equiv}N$$

~70-90%

R = alkyl, cyclic alkyl, aryl, heterocyclic

VI.A.13-2 J. B. Hendrickson, K. W. Bair, and P. M. Keehn, Tetrahedron Lett., 603 (1976).

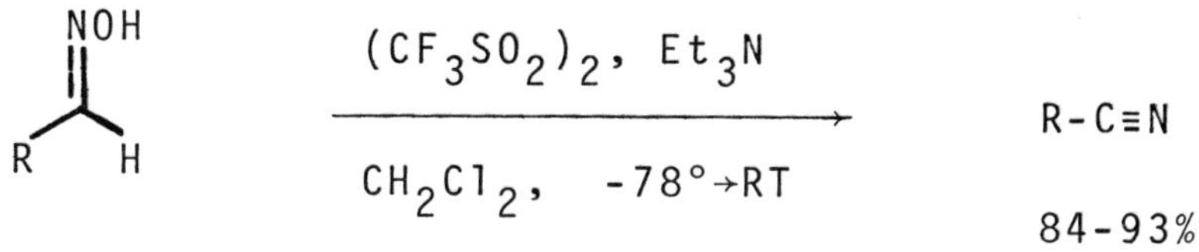

R = alkyl, aryl, cinnamyl, etc.

VI.A.13-3 M. R. Czarny, Synth. Commun., 6, 285 (1976).

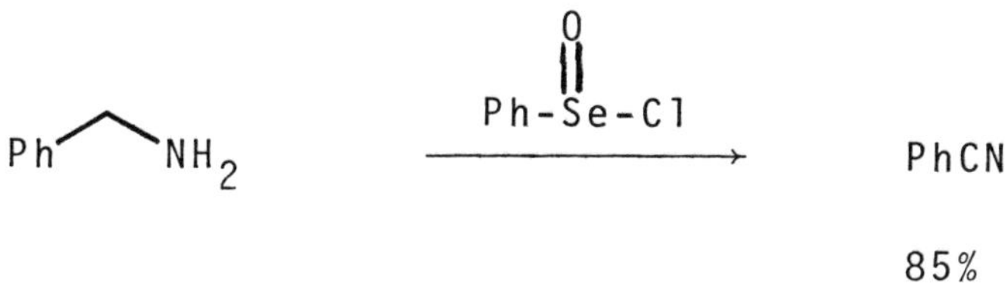

VI.A.13-4 W. P. Reeves and M. R. White, Synth. Commun., 6, 193 (1976).

$$R\text{-}Br \quad + \quad NaCN \xrightarrow{Bu_3N} R\text{-}CN$$

R = unhindered 1° alkyl >90%

VI.A.13-5 J. B. Bapat et al., Tetrahedron Lett., 2691 (1976).

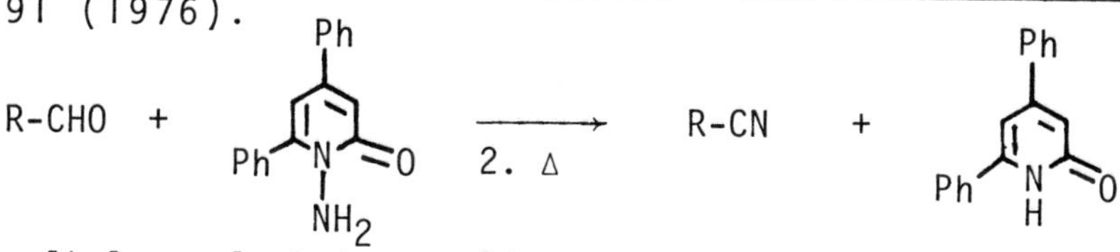

R = alkyl, aryl, heterocyclic 50-90%

VI.A.13-6 A. Ahmad, Synthesis, 418 (1976).

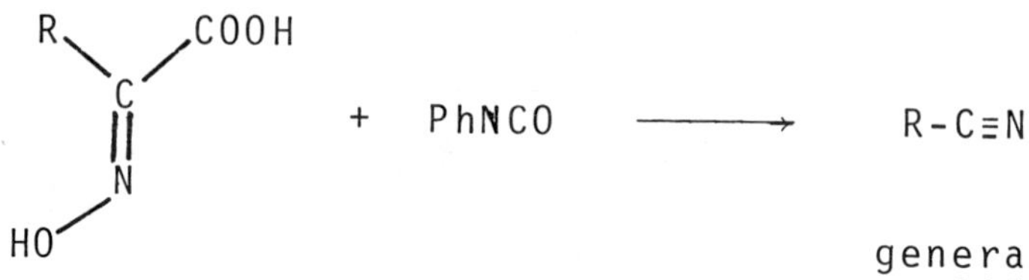

generally >90%

R = 1°,2° alkyl, aryl

VI.A.13-7 T. Cuvigny et al., Synthesis, 237, 238 (1976).

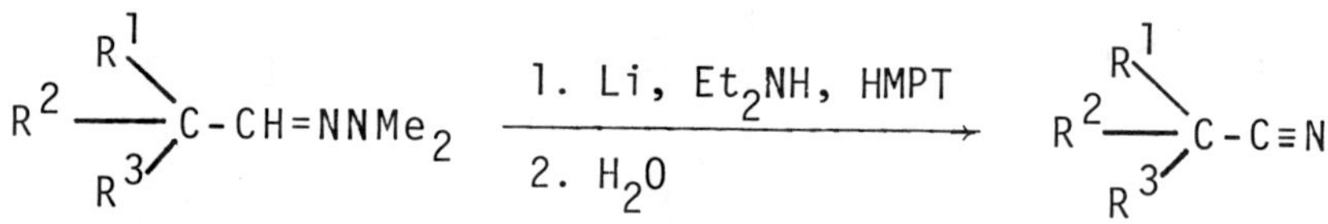

71-96%

R's = H, n-alkyl

VI.A.13-8 A. Antonowa and S. Hauptmann, Z. Chem., 16, 17 (1976).

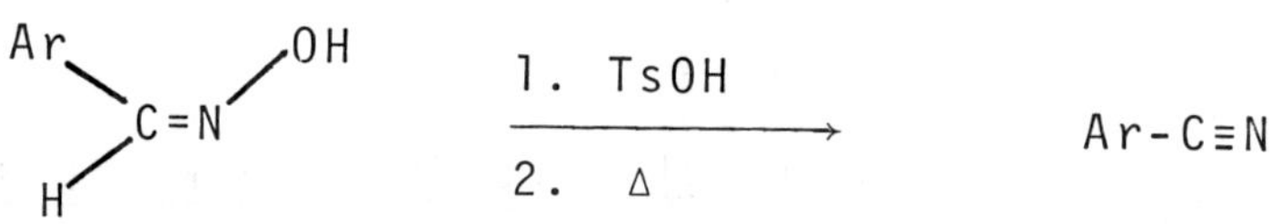

Widely varying yields,
~50-80% for subst. Ph

VI.A.13-9 R. S. Glass and R. C. Hoy, Tetrahedron Lett., 1781 (1976).

$$\text{Ar-CHO} \longrightarrow \text{ArCH=NTs} \xrightarrow[\text{HMPA}]{\text{NaCN}} \text{ArC}\equiv\text{N}$$

generally 70-90%

VI.A.13-10 N. Suzuki *et al.*, J. Chem. Soc. Perkin I, 1901 (1976).

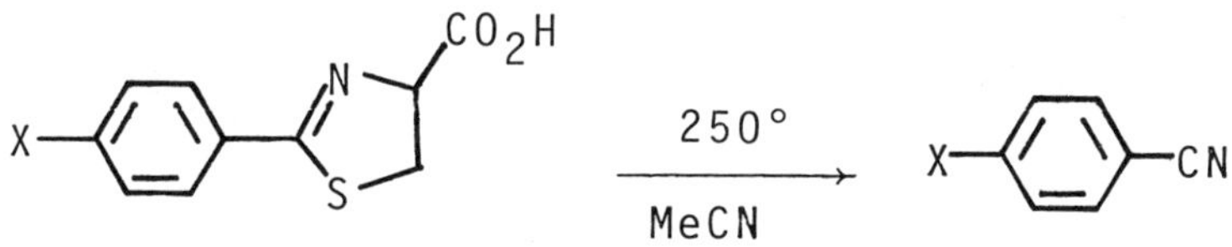

X = H, Cl

~100%

VI.A.13-11 M. A. Schwartz *et al.*, J. Org. Chem., 41, 2502 (1976).

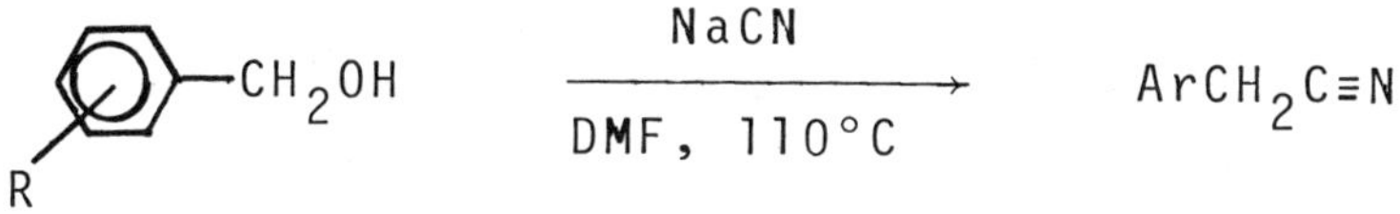

R = OH, OCH_3

0-79%

VI.A.13-12 G. W. Gokel, S. A. DiBiase, and B. A. Lipisko, Tetrahedron Lett., 3495 (1976).

$$CH_3CN + R\text{-}\overset{O}{\overset{\|}{C}}\text{-}R' \xrightarrow[\text{crown ether}]{\text{KOH}} RR'C{=}CH\text{-}CN$$

R,R' = alkyl, aryl, H 50-86% isolated

VI.A.13-13 D. S. Watt, J. Am. Chem. Soc., 98, 271 (1976).

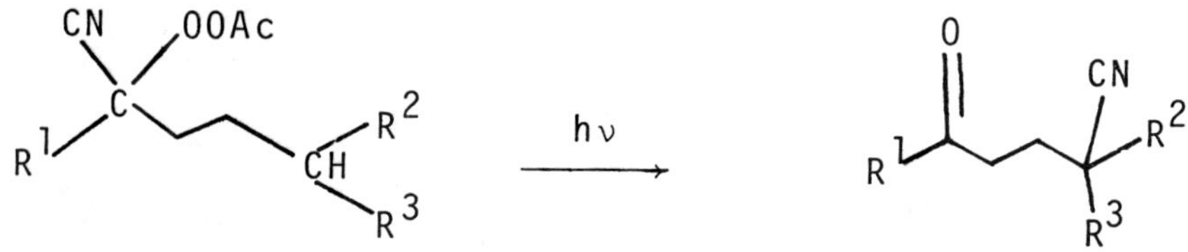

R^1, R^2, R^3 = alkyl, aryl, etc. 9-50%

VI.A.13-14 J. Tsuji, H. Takayanagi and Y. Toshida, Chem. Lett., 147 (1976).

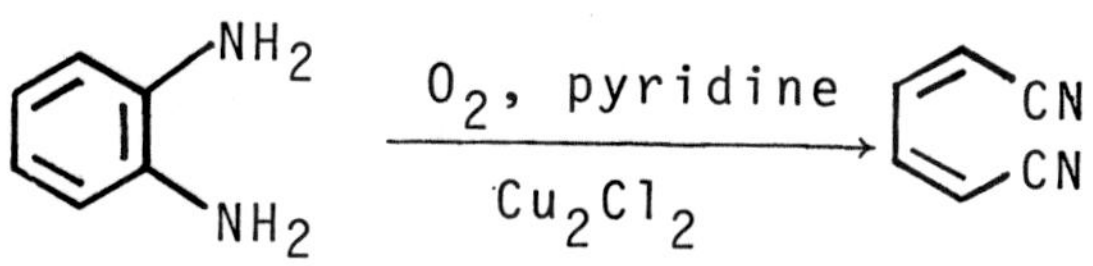

90%

VI.A.13-15 G. Höfle and B. Lange, Angew. Chem. Int. Ed., 15, 113 (1976).

R = Ph, n-Bu, Bz

VI.A.14-1 Y. Takahashi et al., Synthesis, 616 (1976).

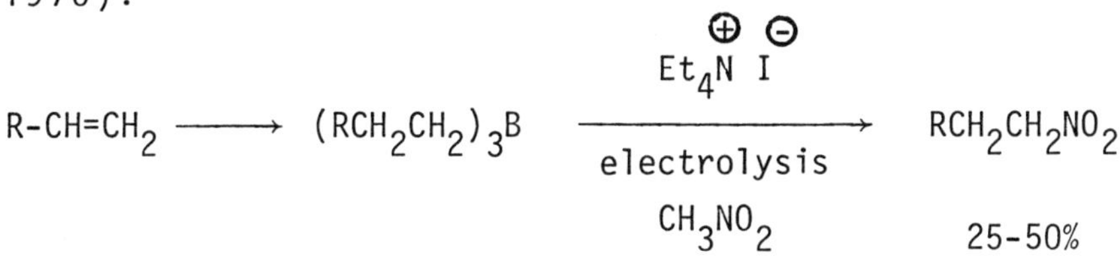

VI.A.14-2 M. W. Barnes and J. M. Patterson, J. Org. Chem., 41, 733 (1976).

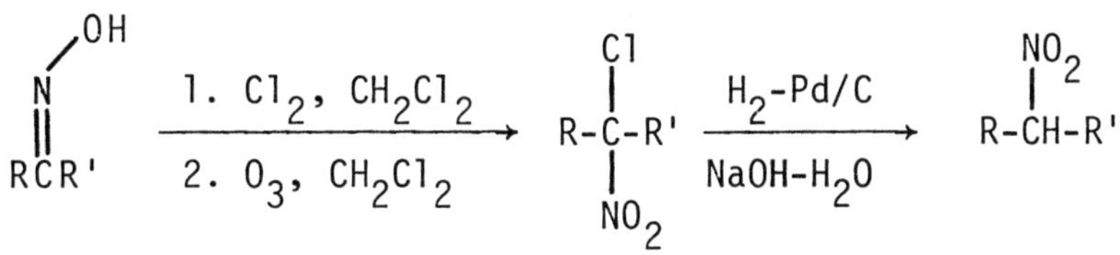

R,R' = alkyl

Yields ~50-80%

VI.A.14-3 M. E. Kurz, and R.T.Y. Chen, <u>J. Chem. Soc., Chem. Commun.</u>, 968 (1976).

$$\text{X-Ph-H} + CH_3NO_2 \xrightarrow{Mn(OAc)_3} \text{X-Ph-}CH_2NO_2$$

~77% for X = H, Me, OMe

0-20% for X = NO_2, Cl

VI.A.14-4 N. Kornblum <u>et</u> <u>al</u>., <u>J. Org. Chem.</u>, <u>41</u>, 1560 (1976).

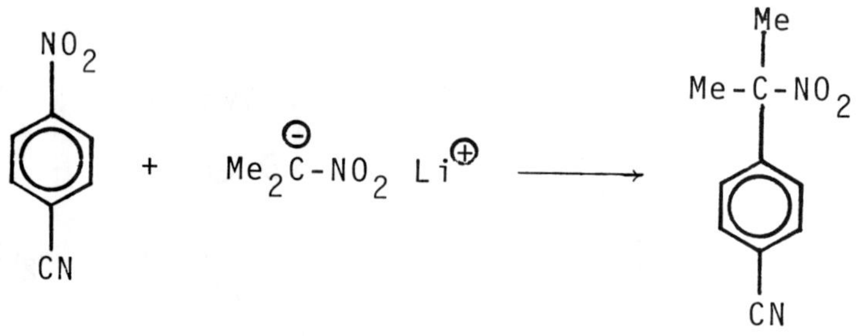

82%

VI.A.15-1 M. J. Robins <u>et</u> <u>al</u>., <u>J. Am. Chem. Soc.</u>, <u>98</u>, 7381 (1976).

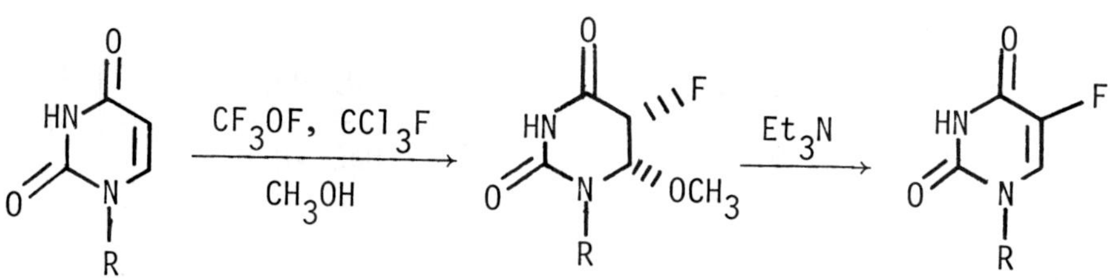

R = H, CH_3, sugar

~80%

VI.A.15-2 U. Reichman *et al.*, *Synthesis*, 533 (1976).

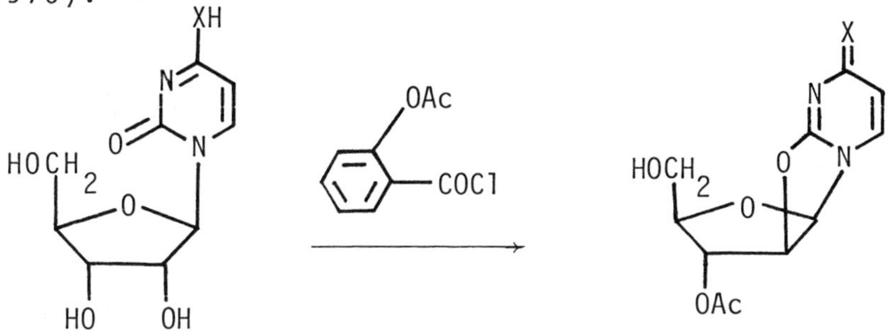

X = NH, O ~65%

VI.A.15-3 D. E. Bergstrom and J. L. Ruth, *J. Am. Chem. Soc.*, **98**, 1587 (1976).

5-chloromercuriuridine → 5-ethyluridine 68%

1. Li_2PdCl_4/MeOH/ethylene
2. $NaBH_4$
3. H_2/Pd

Reactions with allyl chloride and methyl acrylate are also described. (Yields 78% and 57%).

VI.A.15-4 M. Sekiya, J. Suzuki, and Y. Terao, *Chem. Pharm. Bull.*, **24**, 1331 (1976).

NCCHCN(N=NPh) —[H_2, Ra-Ni, 150°; NH_3, $HCONH_2$]→

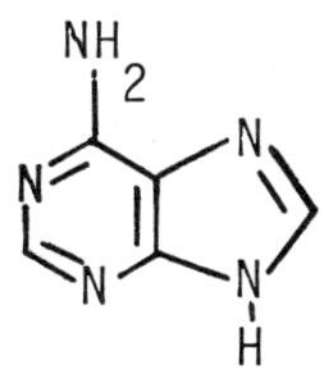

Similar procedures can be used to form xanthine and guanine.

VI.A.15-5 F. Yoneda and T. Nagamatsu, J. Chem. Soc. Perkin I, 1547 (1976).

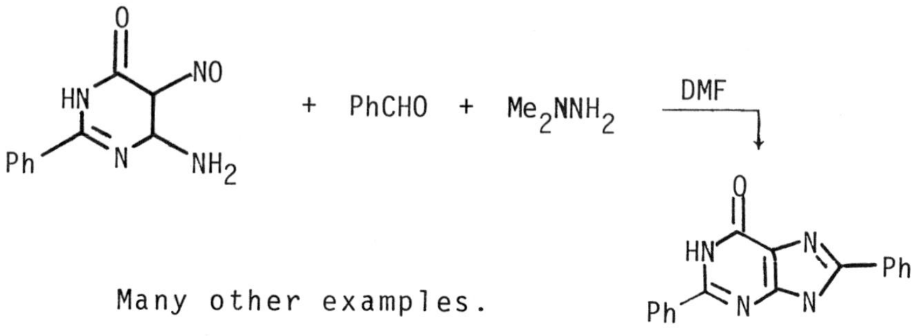

Many other examples.

49%

VI.A.15-6 G. Zvilichovsky and J. Feingers, J. Chem. Soc. Perkin I, 1507 (1976).

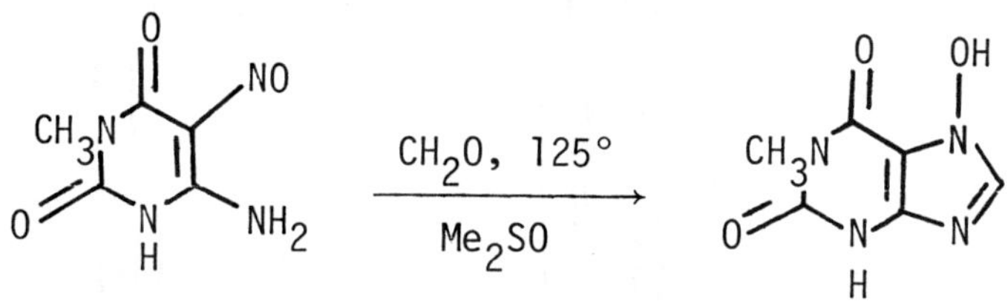

25%

VI.A.15-7 K. Yamauchi, T. Tanabe, and M. Kinoshita, J. Org. Chem., 41, 3691 (1976).

Use of trimethyl phosphate for the direct N-alkylation of cytosine (N-1), thiamine (N-1 and N-3), uracil (N-1 and N-3), adenine (N-9 and N-3), and guanine (N-1).

VI.A.15-8 A. D. Broom, J. L. Shim, and G. L. Anderson, J. Org. Chem., 41, 1095 (1976).

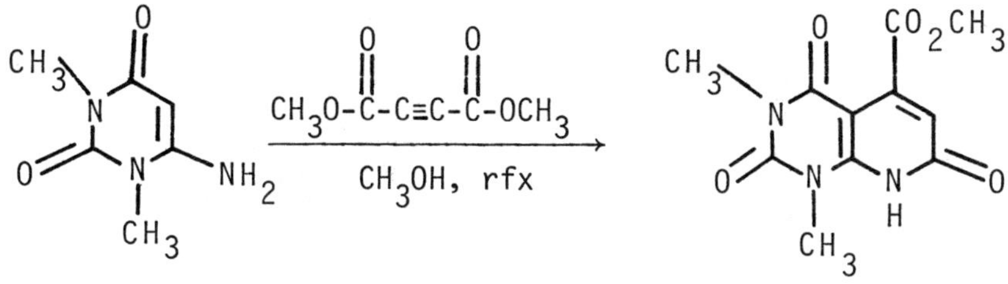

64%

VI.A.15-9 T. Fujii *et al.*, Chem. Pharm. Bull., 24, 655 (1976).

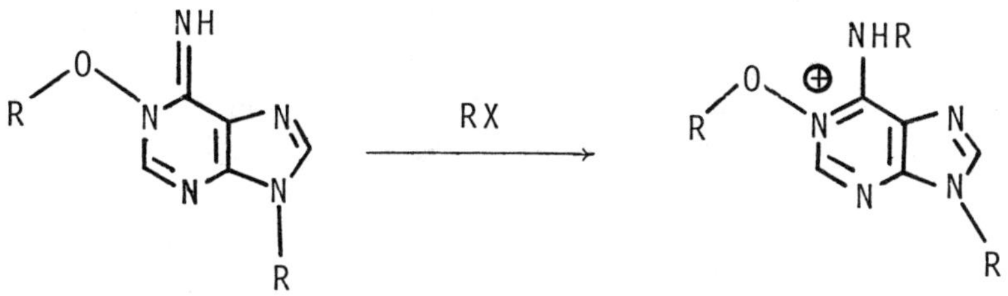

R = Me, Et, Bz

40-80%

VI.A.15-10 J. Herscovici, M. Bessodes, and K. Antonakis, J. Org. Chem., 41, 3827 (1976).

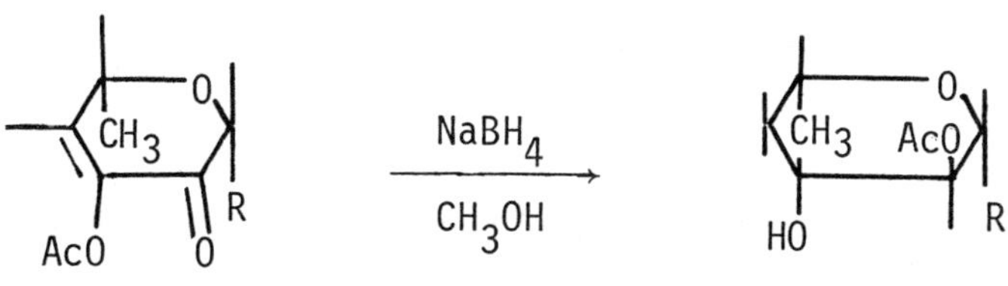

VI.A.15-11 K. Senga, H. Kanazawa, and S. Nishigaki, J. Chem. Soc., Chem. Commun., 155 (1976).

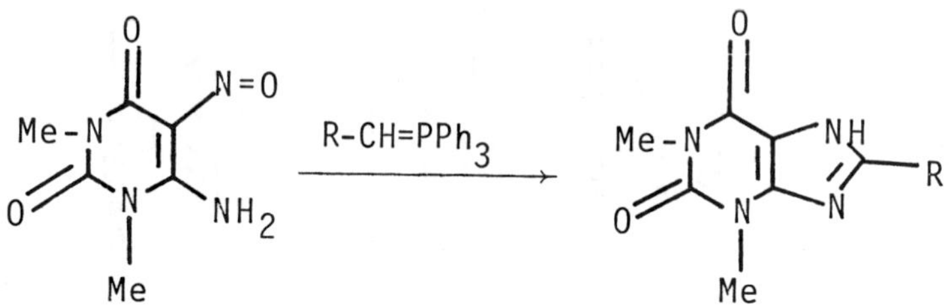

R = aryl

48-75%

VI.A.15-12 J. H. van Boom, P.M.J. Burgers, and P. H. van Deursen, Tetrahedron Lett., 869 (1976).

Use of I for the formation of internucleotide linkages.

I =

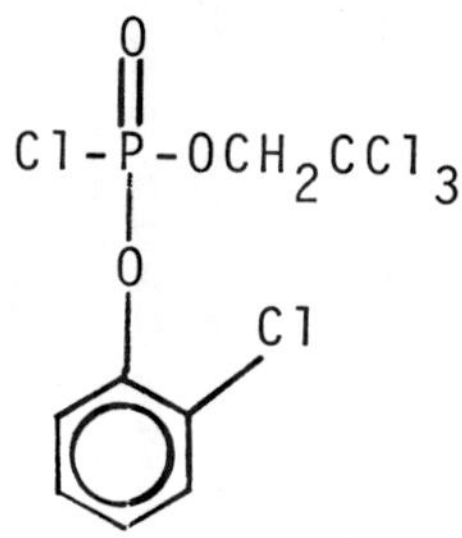

VI.A.15-13 I. Ohtsuka, S. Morioka, and M. Ikehara, Chem. Phar. Bull., 24, 560 (1976).

Synthesis of deoxyribooligonucleotide blocks by an extraction method.

VI.A.15-14 J. Stawinski, T. Hozumi, and S. A. Narang, Can. J. Chem., 54, 670 (1976).

Use of various arylsulfonyltetrazoles to form the phosphotriester bond in polynucleotide synthesis.

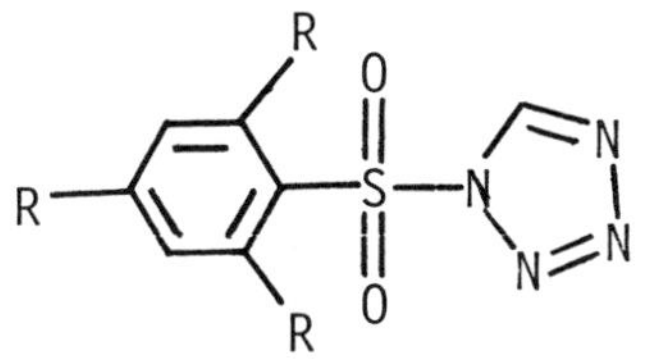

R = H, Me, *i*-Pr

VI.A.15-15 K. Furusawa, M. Sekine, and T. Hata, J. Chem. Soc. Perkin I, 1711 (1976).

Use of nucleoside phosphoric di-*n*-butylphosphinothioic mixed anhydrides as precursors to nucleoside di- and tri-phosphates and nucleotide coenzymes. Stable to H_2O, but react with phosphoric acid, pyrophosphoric acid, or nucleotides in the presence of Ag^+.

VI.A.15-16 M. J. Gait and R. C. Sheppard, J. Am. Chem. Soc., 98, 8516 (1976).

Use of a polydimethylacrylamide resin as a solid-phase support for oligonucleotide synthesis.

VI.A.16-1 S. Baba, D. E. VanHorn and E. Negishi, <u>Tetrahedron Lett.</u>, 1927 (1976).

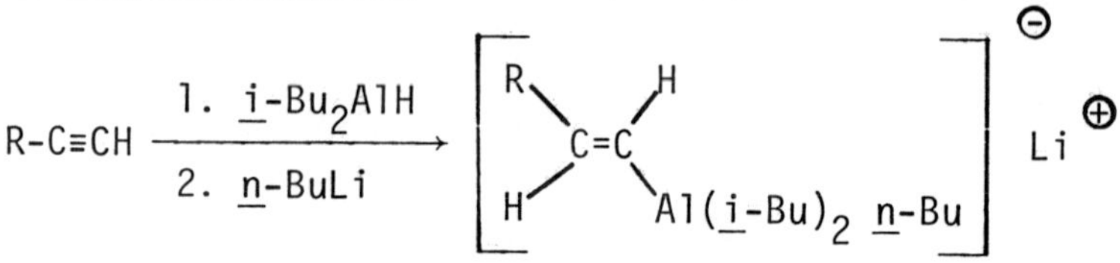

R = n-butyl, n-pentyl

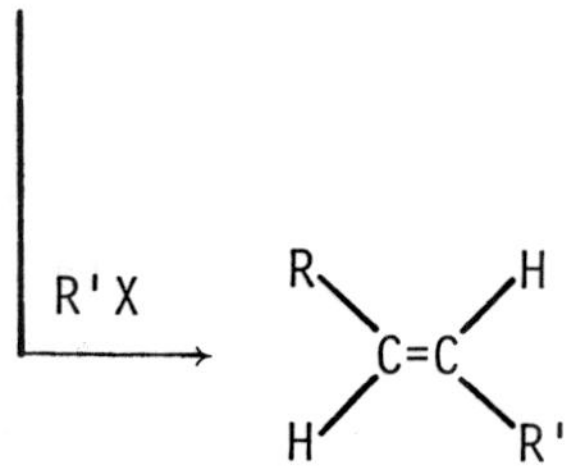

41-73% (GLC)

R' = alkyl, propargyl, benzyl

VI.A.16-2 J. Hooz and R. Mortimer, <u>Tetrahedron Lett.</u>, 805 (1976).

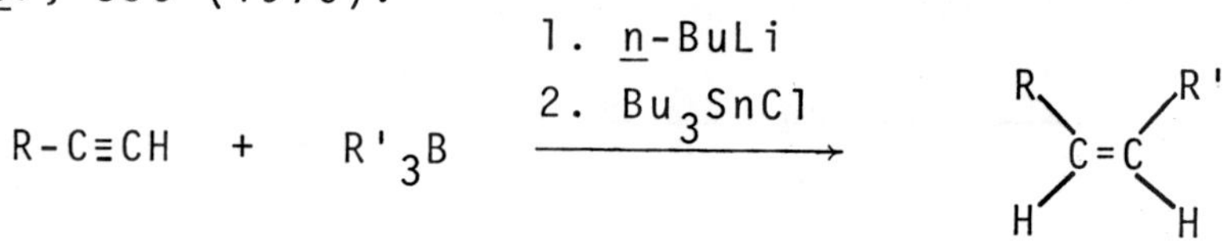

R = Ph, n-butyl, n-pentyl 45-71% isolated

R' = Et, Bu, Hex, $-(CH_2)_4-$ 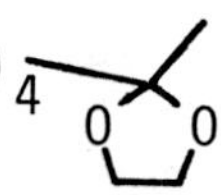

VI.A.16-3 C. W. Spangler and T. W. Hartford, <u>Synthesis</u>, 108 (1976).

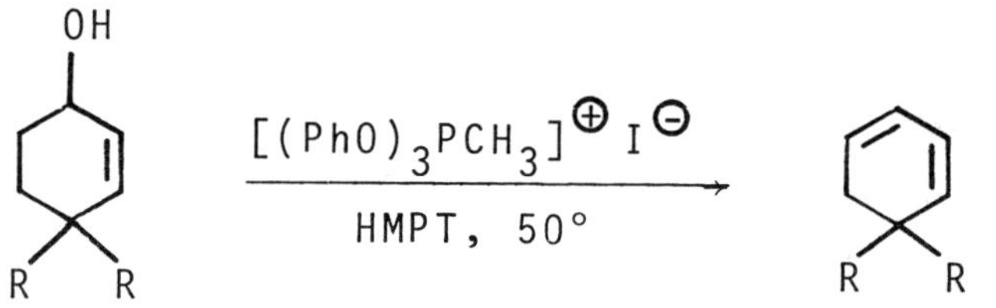

R = H, alkyl 58-78%

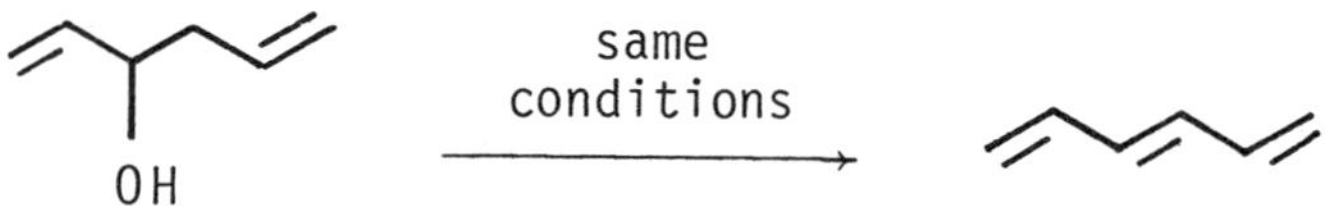

56%

VI.A.16-4 J. Rémion and A. Krief, <u>Tetrahedron Lett.</u>, 3743 (1976).

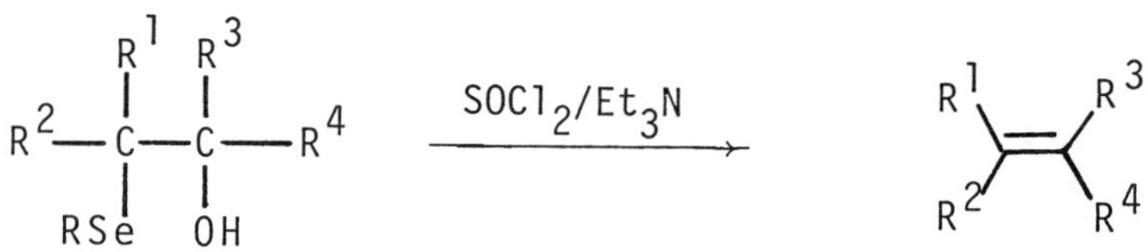

R's = H, 1° alkyl, Ph 50-80%

VI.A.16-5 S. M. Makin, <u>Pure and Appl. Chem.</u>, <u>47</u>, 173 (1976).

Review: "The Enol Ether Synthesis of Polyenes"

VI.A.16-6 R. C. Larock, J. Org. Chem., 41, 2241 (1976).

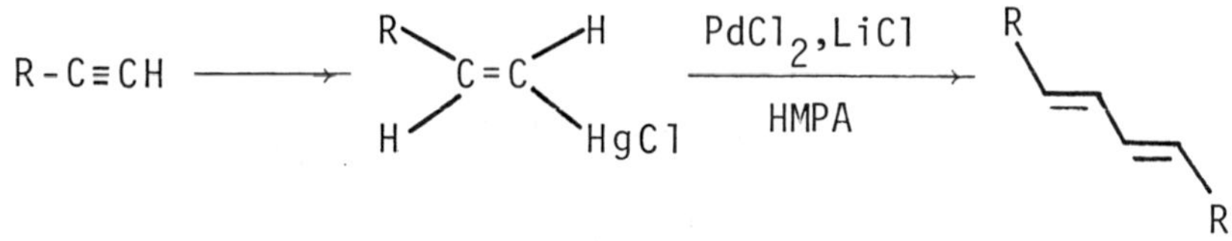

R = Ph, n-Bu >90% (GLC)

VI.A.17-1 H. R. Kricheldorf, M. Fehrle, and J. Kaschig, Angew. Chem. Int. Ed., 15, 305 (1976).

Use of the following compounds as condensation agents in polypeptide synthesis:

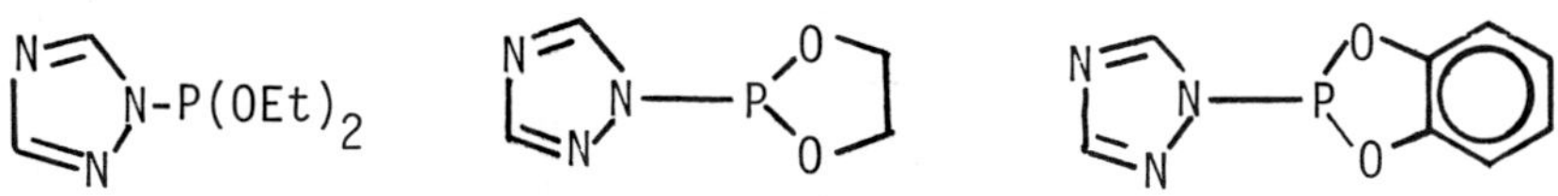

VI.A.17-2 R. Appel, G. Bäumer, and W. Strüver, Chem. Ber., 109, 801 (1976).

Use of PPh_3/CCl_4 as a condensation agent in peptide synthesis.

VI.A.17-3 A. C. Veronese et al., Gazz. Chim. Ital., 106, 179 (1976).

Use of hydroximino esters (from acetylacetone, benzoylacetophenone, dimethyl malonate, etc.) as coupling agents in peptide synthesis. Amino acids such as serine, threonine, and tyrosine do not require protection of the OH group.

VI.A.17-4 K. Horiki, *Tetrahedron Lett.*, 4106 (1976).

Use of a polymeric triphenylphosphine/2,2'-dipyridyl disulfide resin as a reagent for anti-Merrifield solid-phase peptide synthesis. In this case, the coupling reagent is bound to the resin (and can be recycled), and the peptide is in solution.

VI.A.17-5 D. Yamashiro, J. Blake, and C. H. Li, *Tetrahedron Lett.*, 1469 (1976).

Use of trifluoroethanol to increase the reactivity of the α-amino group in solid-phase peptide synthesis.

VI.A.17-6 A. G. Jackson *et al.*, *Tetrahedron Lett.*, 3627 (1976).

Activation of carboxylic acids as diphenylphosphinic mixed anhydrides: Application to peptide chemistry.

VI.A.17-7 Yu. A. Davidovich and V. Gut, *Coll. Czech. Chem. Commun.*, **41**, 1805 (1976).

Use of polymeric *o*-nitrobenzenesulfenylaminoacyl N-hydroxysuccinimide esters as active esters in the synthesis of model protected peptides.

VI.A.17-8 S. Matsuura, C.-H. Niu, and J. S. Cohen, J. Chem. Soc., Chem. Commun., 451 (1976).

Use of HF in pyridine to remove completed peptides from the Merrifield resin. Also cleaves BOC protecting groups.

VI.A.17-9 A. E. Jackson and R.A.W. Johnstone, Synthesis, 685 (1976).

Removal of benzyloxycarbonyl groups from peptides by catalytic transfer hydrogenation.

VI.A.17-10 S. A. Kahn and K. M. Sivanandaiah, Synthesis, 614 (1976).

The use of a toluene medium and 1-hydroxybenzotriazole to accelerate coupling rates in solid-phase peptide synthesis.

VI.A.17-11 M. Akiyama, K. Shimizu, and M. Narita, Tetrahedron Lett., 1015 (1976).

Use of N-hydroxysuccinimide ester polymers as polymer reagents in peptide synthesis.

VI.A.17-12 R. C. Orlowski, R. Walter, and D. Winkler, J. Org. Chem., **41**, 3701 (1976).

Synthesis and use of *p*-methoxybenzhydrylamine resin in solid-phase peptide synthesis.

VI.A.17-13 J. K. Chang, M. Shimizu, and S.-S. Wang, J. Org. Chem., **41**, 3255, 3258 (1976).

Use of hydroxymethyl resin in a fully automated solid-phase peptide synthesis; also use of an α-methylphenacyl anchoring bond, which may be cleaved by irradiation at 350 nm.

VI.A.17-14 J. T. Sparrow, J. Org. Chem., **41**, 1350 (1976).

Use of a long spacer chain between the first amino acid and the polystyrene support to improve yields in solid phase peptide synthesis.

VI.A.17-15 A. R. Mitchell *et al.*, J. Am. Chem. Soc., **98**, 7357 (1976).

Use of a phenylacetamidomethyl (Pam) bridge between the peptide and the resin in solid-phase peptide synthesis. Provides for greater stability toward acid, minimizing acidolysis during peptide elongation.

VI.A.17-16 J. Blake and C. H. Li, J. Chem. Soc., Chem. Commun., 504 (1976).

Use of a *p*-carbamoylmethylbenzyl ester linkage between the polystyrene resin and the peptide in solid-phase peptide synthesis. Relatively stable toward TFA and liquid HF.

VI.A.17-17 R. W. Roeske and P. D. Gesellchen. *Tetrahedron Lett.*, 3369 (1976).

Use of 18-crown-6 to catalyze the esterification of the potassium salts of *Boc*-amino acids with chloromethyl polystyrene.

VI.A.17-18 Yu. P. Shvachkin *et al.*, *J. Gen. Chem. (USSR)*, **46**, 717 (1976).

A method for the recycling of unreacted N-protected amino acid derivatives in solid-phase peptide synthesis. The reagent is allowed to react twice with portions of the polymer-bound amino component, and finally with another portion of the amino component in solution.

VI.A.17-19 H. Eckstein, *Liebigs Ann. Chem.*, 2145 (1976).

"Synthesis of Tryptophan-Containing Peptides"

VI.A.18-1 P. Savignac and P. Coutrot, *Synthesis*, 197 (1976).

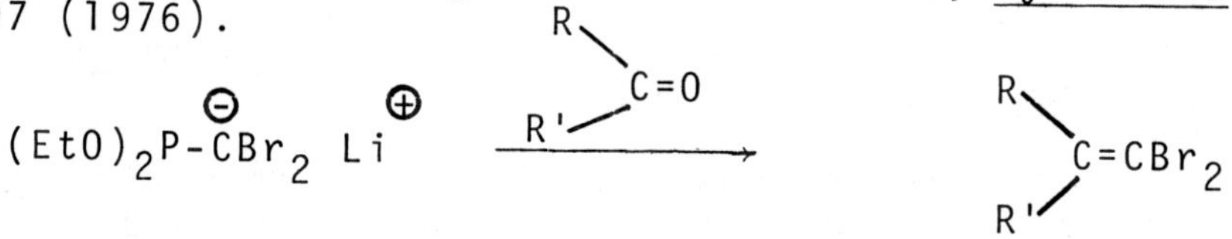

R,R' = H, alkyl, aryl, cyclic 40-70%

VI.A.18-2 G. A. Wheaton and D. J. Burton, Tetrahedron Lett., 895 (1976).

$$Ph_3\overset{\oplus}{P}\text{-}\overset{\ominus}{C}R^1R^2 + HCF_2Cl \longrightarrow R^1R^2C{=}CF_2$$

R^1,R^2 = H, alkyl, $-C_5H_{10}-$, etc.

Maximum initial yield 50%, but reagent is recycled to give yields of 80-100% (VPC).

VI.A.18-3 R. D. Clark and C. H. Heathcock, J. Org. Chem., 41, 636 (1976).

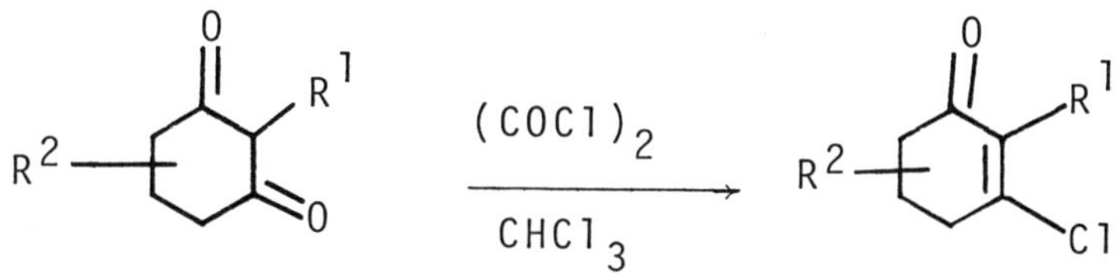

R^1 = H, alkyl

R^2 = H, alkyl, dialkyl

76-91%

VI.A.18-4 K. Tamao, M. Zembayashi, and M. Kumada, Chem. Lett., 1237 (1976).

$$BrCH{=}CHOEt + RMgBr \xrightarrow{[Ni(dppp)Cl_2]} RCH{=}CHOEt$$

R = alkyl, Ph, thiophenyl 53-77%

VI.A.18-5 L. Blanco, P. Amice, and J. M. Conia, Synthesis, 194, 196 (1976).

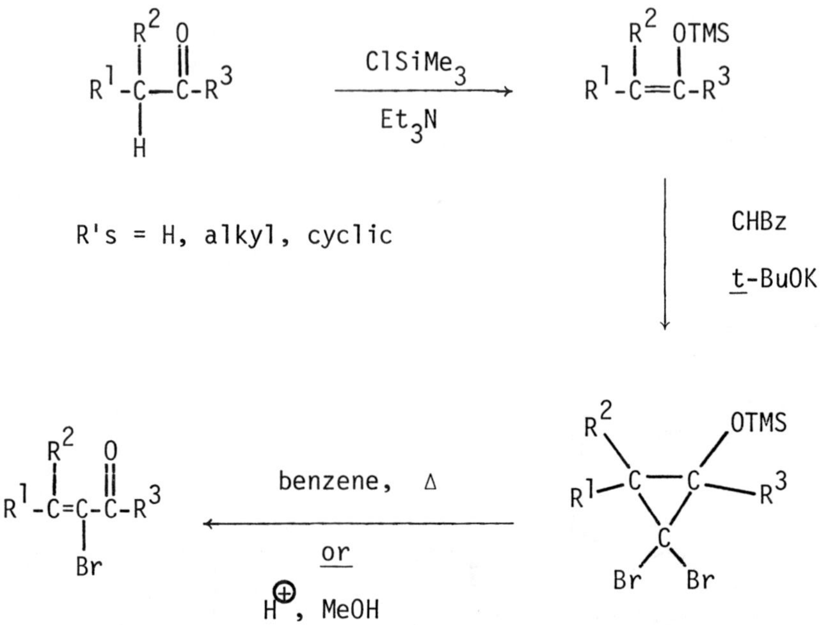

R's = H, alkyl, cyclic

16-80% overall

VI.A.18-6 G. Zweifel and R. A. Lynd, Synthesis, 816 (1976).

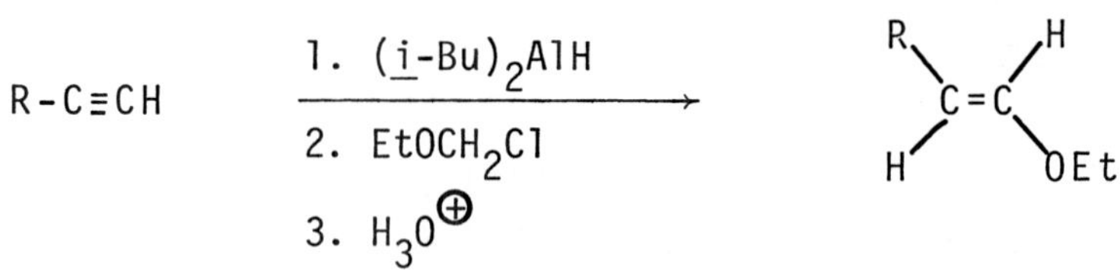

72-80%

R = n-Bu, t-Bu, cyclohexyl

VI.A.18-7 H. Klein *et al.*, *Tetrahedron Lett.*, 947 (1976).

$$[RCu(Br\ or\ R)]MgX + H_2C{=}C{=}CHOCH_3 \xrightarrow[2.\ H_3O^{\oplus}]{} R\text{-}CH_2CH{=}CHOCH_3$$

80-90%

R = alkyl, phenyl (mixture of E and Z)

VI.A.18-8 V. Subramanyam, E. H. Silver and A. H. Soloway, *J. Org. Chem.*, **41**, 1272 (1976).

$$Ph_3P{=}CHPh \quad + \quad H\text{-}\overset{O}{\overset{\|}{C}}\text{-}OR \quad \longrightarrow \quad Ph\text{-}CH{=}CH\text{-}OR$$

R = Et, *s*-Bu, cyclohexyl

VI.A.19-1 T. Mukaiyama, T. Masui, and T. Izawa, *Chem. Lett.*, 1177 (1976).

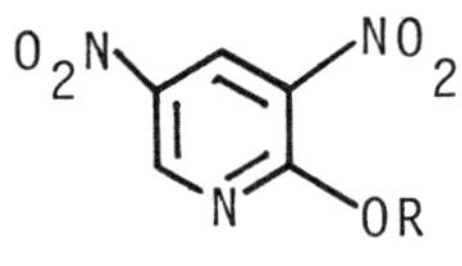

$$+ \quad R'SH \quad \xrightarrow[\text{catalyst}]{CH_2Cl_2} \quad R\text{-}S\text{-}R'$$

60-91%

R = 1°,2° alkyl, containing double bonds, halides, esters

catalyst = $BF_3 \cdot Et_2O$ or $TiCl_4$

R'= Ph, Bz, *t*-Bu

VI.A.19-2 A.J.H. Labuschagne *et al.*, Tetrahedron Lett., 3571 (1976).

$$\text{R-CH}_2\text{-Br} + 2\ \text{Me}_2\text{S} \xrightarrow{\Delta} \text{R-CH}_2\text{-S-Me}$$

R = vinyl, Ph, COOR, COPh — 63-100%

$$\text{RCH}_2\text{SMe} + \text{RCH}_2\text{Br} \xrightarrow{\Delta} \text{R-CH}_2\text{-S-CH}_2\text{-R}$$

R = vinyl, Ph, COOR, COPh — 43-76%

VI.A.19-3 R. S. Glass, Synth. Commun., **6**, 47 (1976).

$$\text{RCHO} + \text{R'SSiMe}_3 \longrightarrow \text{R-CH(OSiMe}_3\text{)SR'} \xrightarrow[\text{AlCl}_3]{\text{LiAlH}_4} \text{RCH}_2\text{SR'}$$

R,R' = Ph, alkyl, cycloalkyl — generally 50-80%

VI.A.19-4 F. Pochat and E. Levas, Tetrahedron Lett., 1491 (1976).

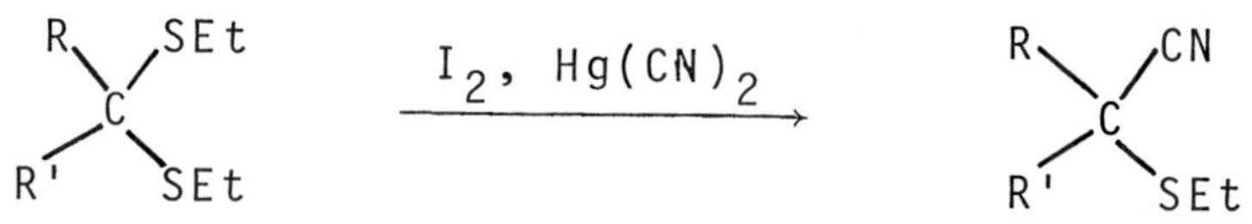

R,R' = H, alkyl, aryl

Yields generally 40-70%

VI.A.19-5 I. Kuwajima, T. Murofushi, and E. Nakamura, Synthesis, 602 (1976).

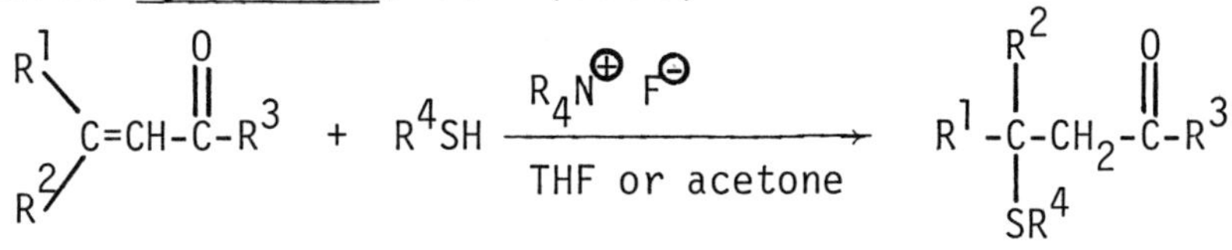

R^1,R^2 = H, alkyl, vinyl, cyclic — generally 80-90%

R^3 = H, Me, -COOR, cyclic

R^4 = Ph, Bz, $-CH_2COOR$

VI.A.19-6 I. I. Lapkin, G. G. Abashev, and F. G. Saitkulova, J. Org. Chem. (USSR), 12, 975 (1976).

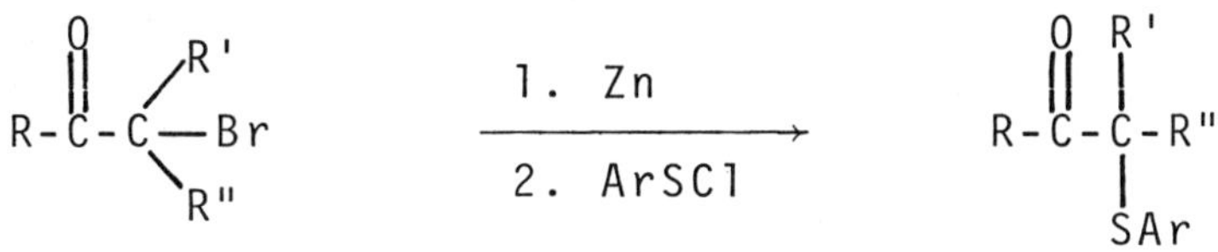

R's = H, 1° alkyl — ~70%

few details given

VI.A.19-7 D. Seebach and M. Teschner, Chem. Ber., 109, 1601 (1976).

R = Me, Ph, $CSNMe_2$

R^1 = Me, Ph, i-Pr, cyclic — ~60-90%
R^2 = Me, Ph, cyclic

VI.A.19-8 W. T. Flowers *et al.*, *J. Chem. Soc. Perkin I*, 2394 (1976).

$$ArSCN + ROH \xrightarrow{Ph_3P} Ar\text{-}S\text{-}R$$

Ar = Ph, *p*-MeOPh, *p*-NO_2Ph ~70-90%

R = *n*-alkyl, allyl, Bz

VI.A.19-9 B. B. Snider, N. J. Hrib, and L. Fuzesi, *J. Am. Chem. Soc.*, **98**, 7115 (1976).

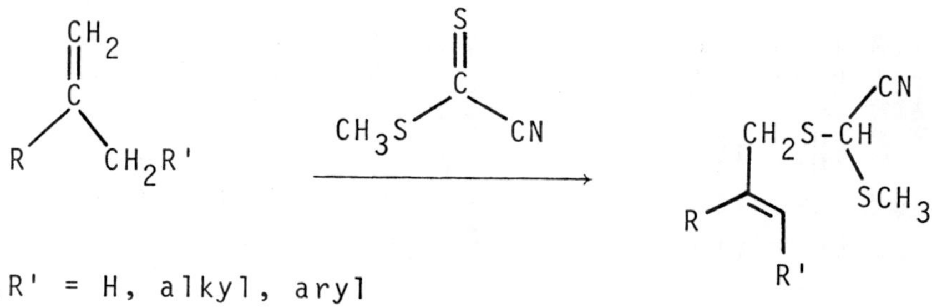

R,R' = H, alkyl, aryl

22-83%

VI.A.19-10 N. Kornblum *et al.*, *J. Org. Chem.*, **41**, 1560 (1976).

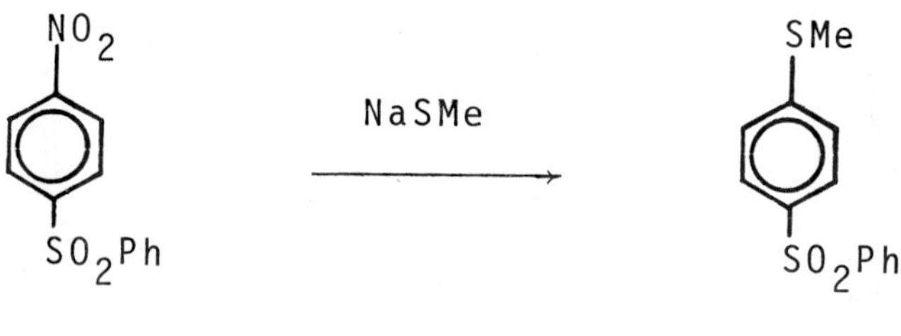

61%

VI.A.19-11 K. Ogura, S. Furukawa, and G. Tsuchihashi, Synthesis, 202 (1976).

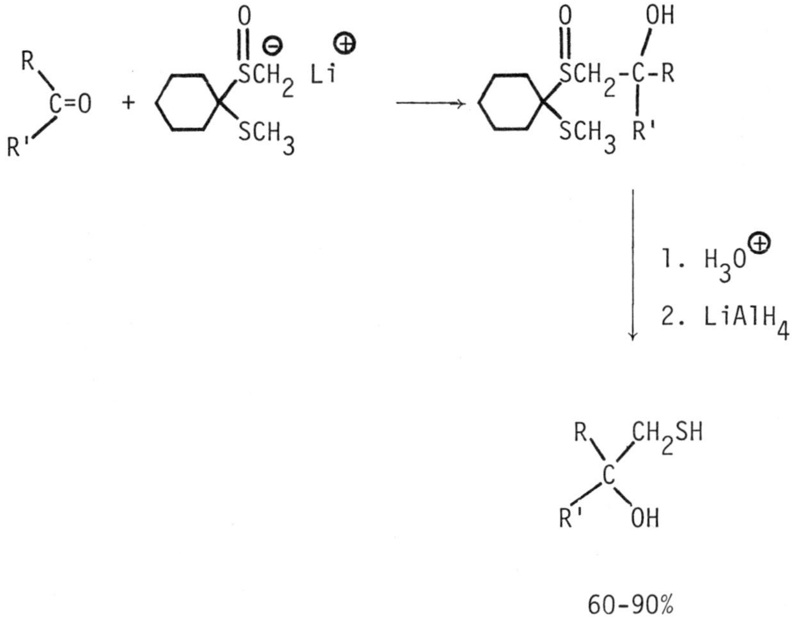

VI.A.19-12 P. Vermeer et al., Rec. Trav. Chim. Pays-Bas, 95, 25 (1976).

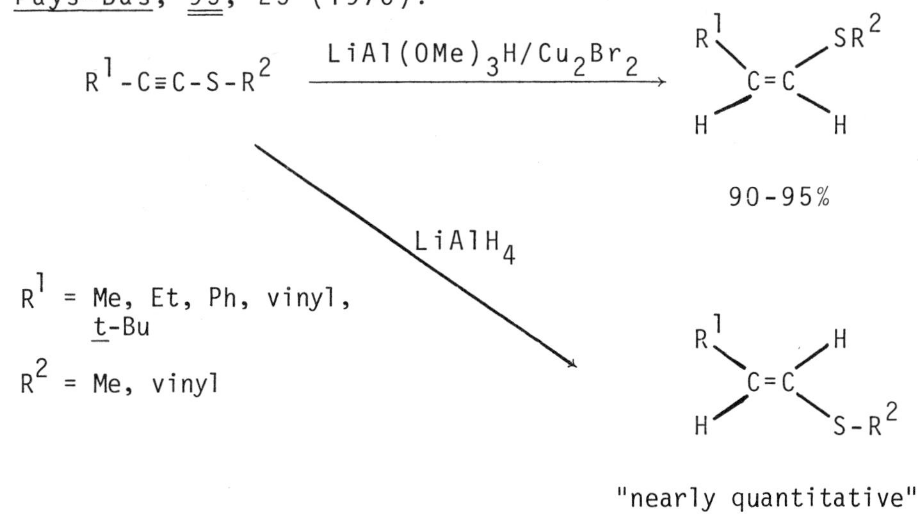

VI.A.19-13 P. Coutrot *et al.*, *Synthesis*, 107 (1976).

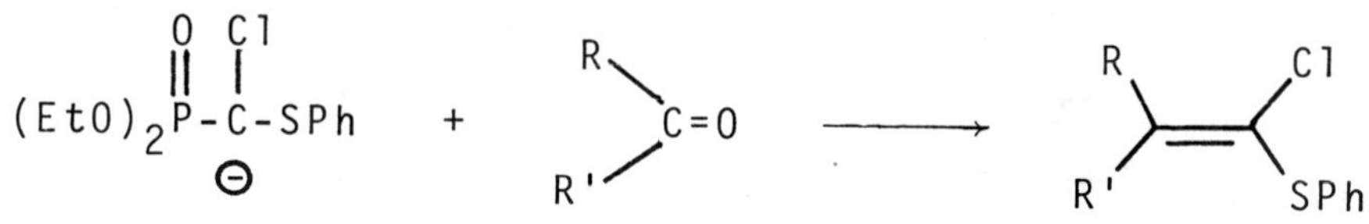

R,R' = H, alkyl, aryl, cyclic

46-68%

VI.A.19-14 H. Yoshida, T. Ogata, and S. Inokawa, Synthesis, 552 (1976).

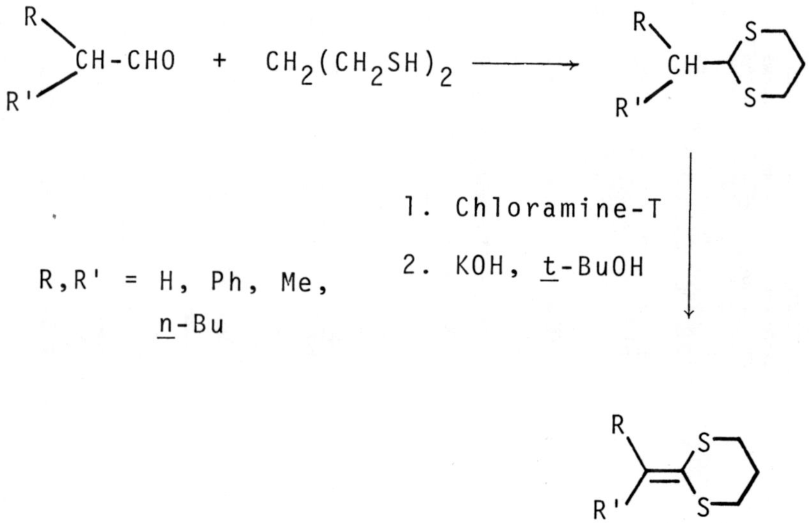

R,R' = H, Ph, Me, *n*-Bu

~60-75%

VI.A.19-15 H. Yoshida et al., Synthesis, 551 (1976).

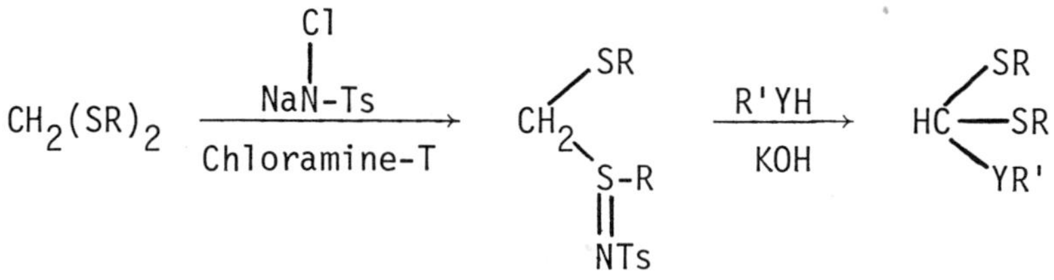

R = alkyl, aryl, cyclic   Y = O, S   ~60-80%

VI.A.19-16 V. I. Cohen, Helv. Chem. Acta, 59, 840 (1976).

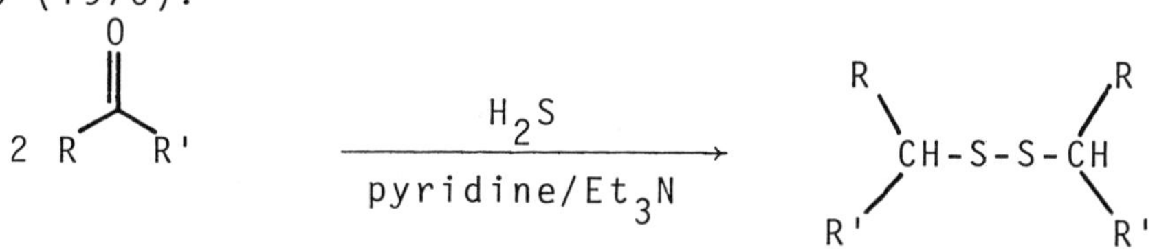

R,R' = H, alkyl, aryl, cyclic   ~60-80%

VI.A.19-17 P. Dubs and R. Stüssi, Helv. Chim. Acta, 59, 1307 (1976).

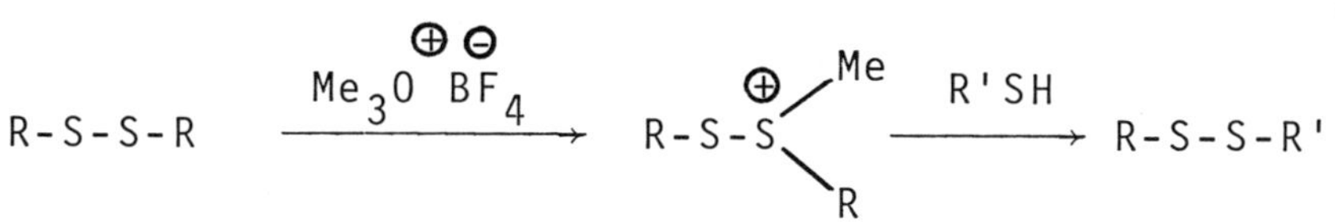

R = Me, Et, n-Pr, i-Pr   52-82%

R' = alkyl, aryl, heterocyclic

VI.A.19-18 J. R. Grunwell and D. L. Foerst, Synth. Commun., 6, 453 (1976).

$$RCOOH + R'SH \xrightarrow{DCC} R\text{-}\overset{\overset{\displaystyle O}{\|}}{C}\text{-}SR'$$

R = CH_3, Ph R' = Ar, 1°,2° alkyl 24-74%

VI.A.19-19 F. Souto-Bachiller, G. S. Bates and S. Masamune, J. Chem. Soc., Chem. Commun., 719 (1976).

A summary of the uses of I as a condensing agent in the preparation of thiol esters and amides.

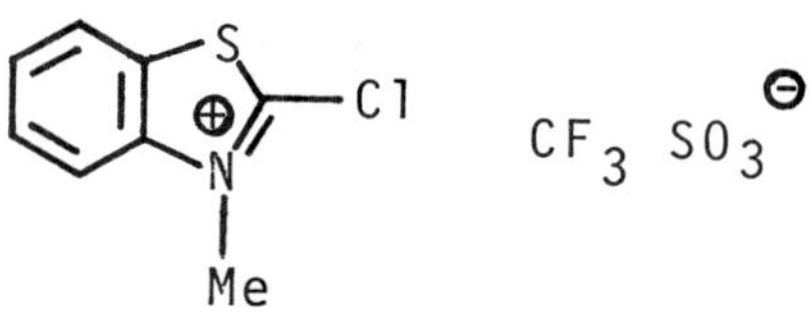

I

VI.A.19-20 F. Le Goffic, S. Sicsic, and C. Vincent, Tetrahedron Lett., 2845 (1976).

Use of a membrane-bound pyridine catalyst

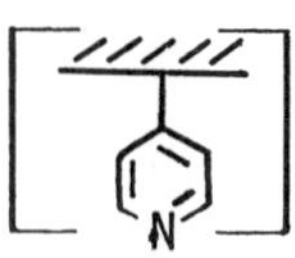

in the synthesis of the thioesters of thiol-amino acids. Most yields ~90%.

VI.A.19-21 W. H. Baarschers, Can. J. Chem., **54**, 3056 (1976).

$$Ar\text{-}SO_2\text{-}OR \xrightarrow{R'Li} Ar\text{-}SO_2R'$$

R = Ph, n-alkyl — 38-98%

R' = alkyl, aryl, heterocyclic

VI.A.19-22 J. D. Coyle, Chem. Soc. Rev., **4**, 523

Review: "Photochemistry of Organic Sulfur Compounds"

VI.B.1-1 H.J.J. Loozen, W.M.M. Robben, and H. M. Buck, Rec. Trav. Chim. Pays-Bas, **95**, 248 (1976).

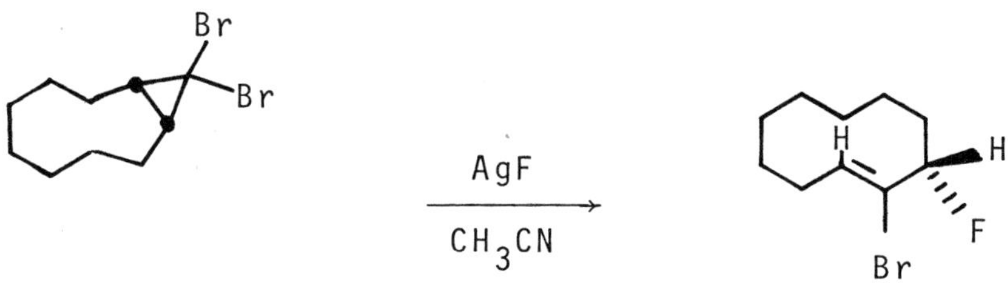

~30%

VI.B.1-2 H.J.J. Loozen et al., J. Org. Chem., 41, 384 (1976).

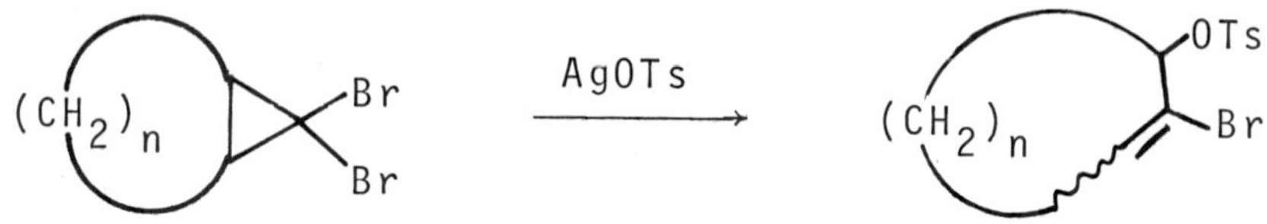

n = 5, 6, 7 81-93%

VI.B.1-3 H. H. Wasserman, M. J. Hearn, B. Haveaux, and M. Thyes, J. Org. Chem., 41, 153 (1976).

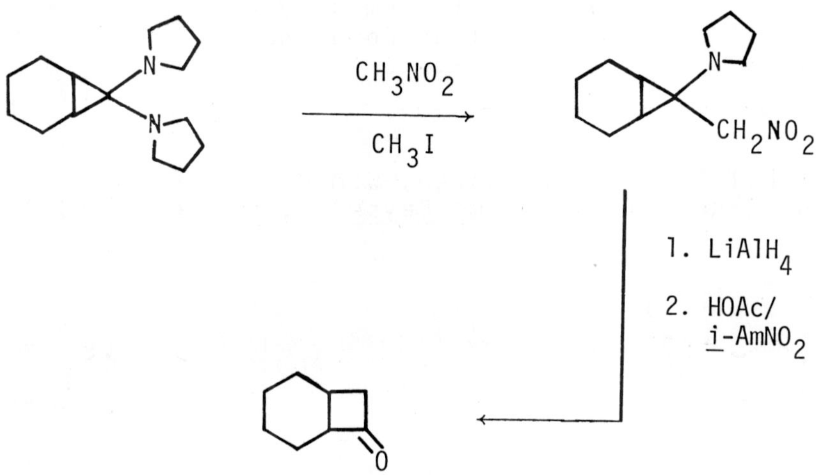

~18% overall

VI.B.1-4 K. Hafner, H. Diehl, and H. U. Süss, Angew. Chem. Int. Ed., 15, 104 (1976).

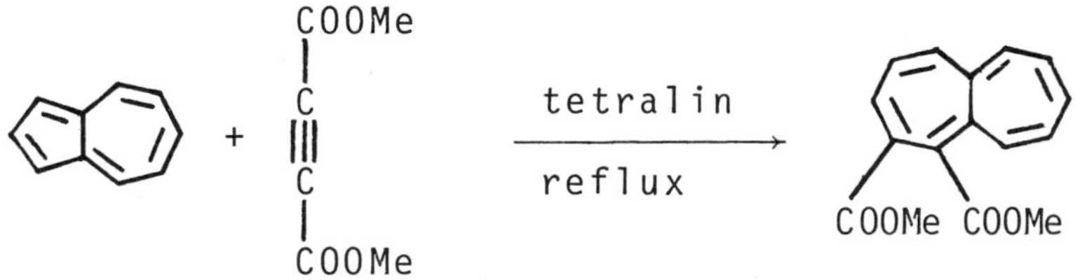

15%

VI.B.1-5 M. Sato, J. Tsunetsugu, and S. Ebine, Bull. Chem. Soc. Japan, 49, 2230 (1976).

A study of the following reaction:

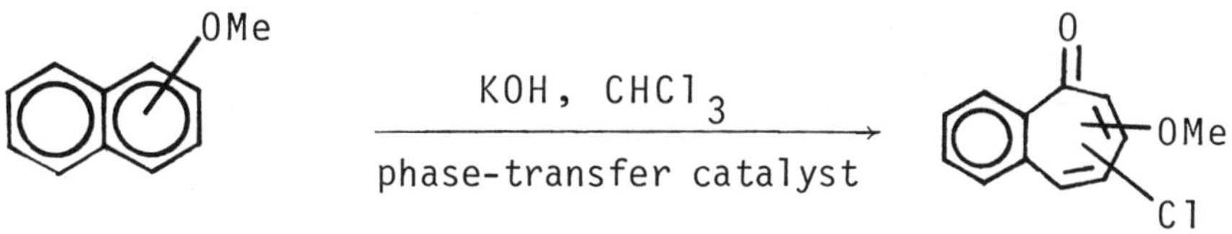

Various isomers, widely varying yields.

VI.B.1-6 S. Kwon et al., Synthesis, 249 (1976).

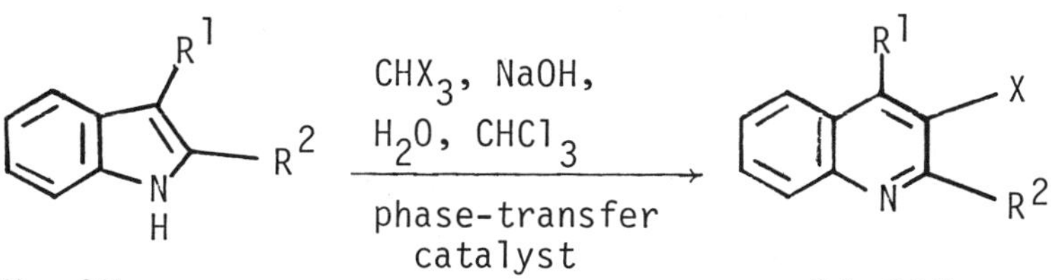

R^1 = H, CH_3

R^2 = H, CH_3, Ph

24-68%

VI.B.1-7 M. B. Hasiak, Comptes Rendus (C), 282, 1003 (1976).

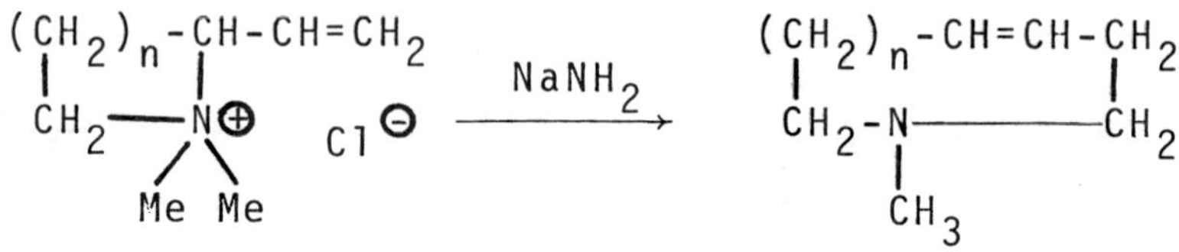

94% (n=3)

35% (n=2)

VI.B.1-8 Y. Sato, H. Kojima, and H. Shirai, J. Org. Chem., 41, 195 (1976).

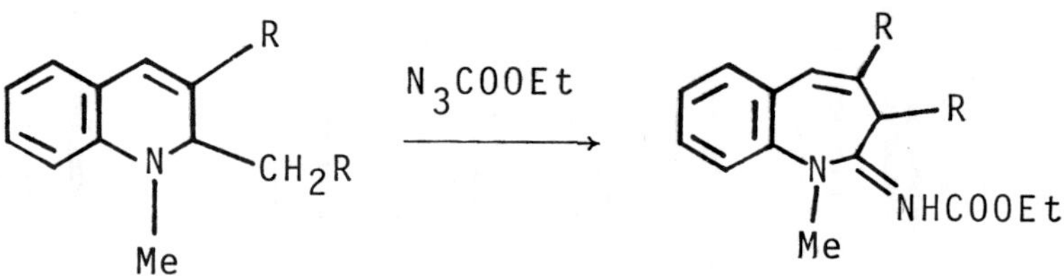

R:	Me	Et	i-Pr
% yield:	67	83	30

VI.B.2-1 G. Ortar and A. Romeo, J. Chem. Soc., Perkin I, 111 (1976).

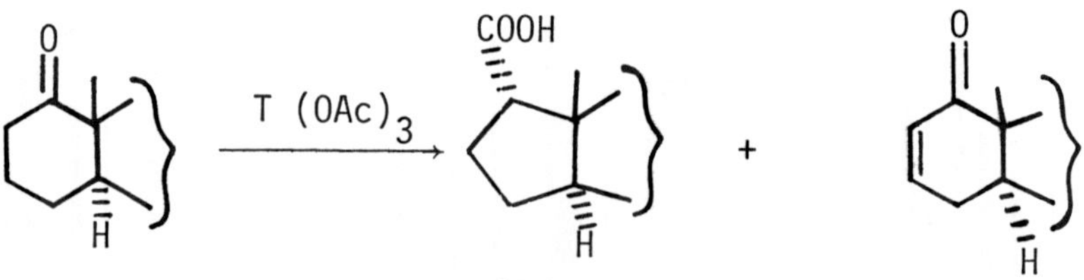

27% 28%

VI.B.2-2 H. Neunhoeffer and V. Boehnisch, Liebigs Ann. Chem., 153 (1976).

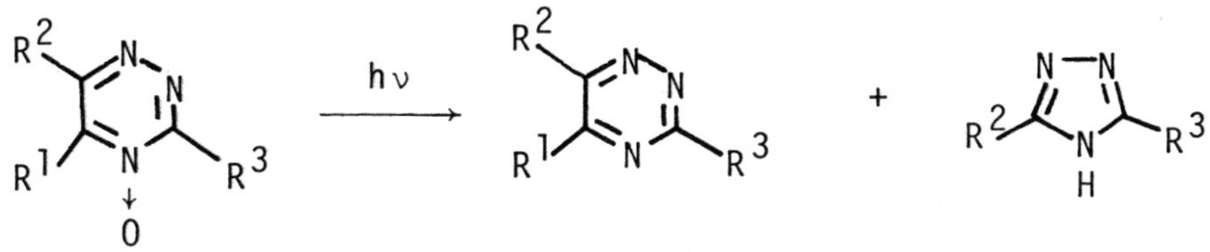

R's = H, Me, Ph 48-82% total yield

VI.C.1-1 S. F. Martin, J. Org. Chem., 41, 3337 (1976).

A novel spiroannelation sequence:

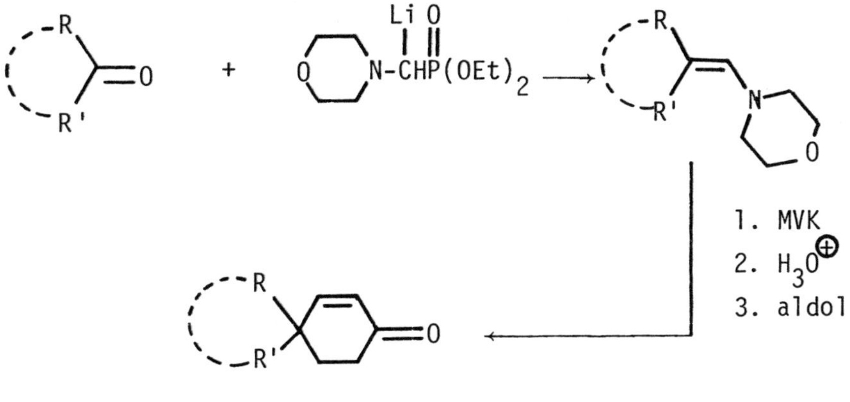

30-36%

VI.C.1-2   V. V. Kane, Synth. Commun., 6, 237 (1976).

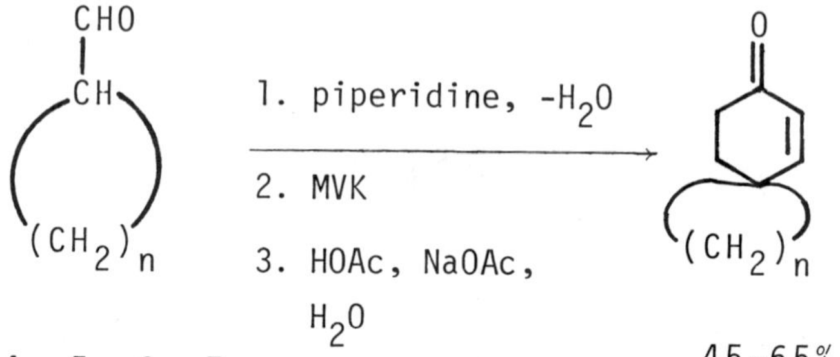

n = 4, 5, 6, 7

45-65%

VI.C.1-3   R. R. Wroble and D. S. Watt, J. Org. Chem., 41, 2939 (1976).

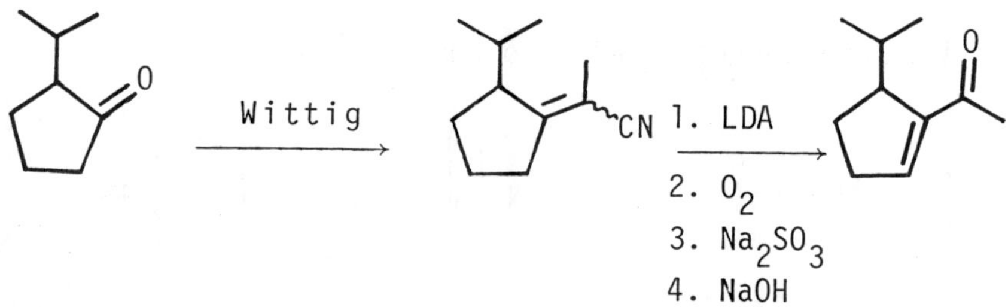

10 examples, yields generally ~70%

VI.C.1-4 J. E. Baldwin, O. W. Lever, Jr., and N. R. Tzodikov, J. Org. Chem., 41, 2312 (1976).

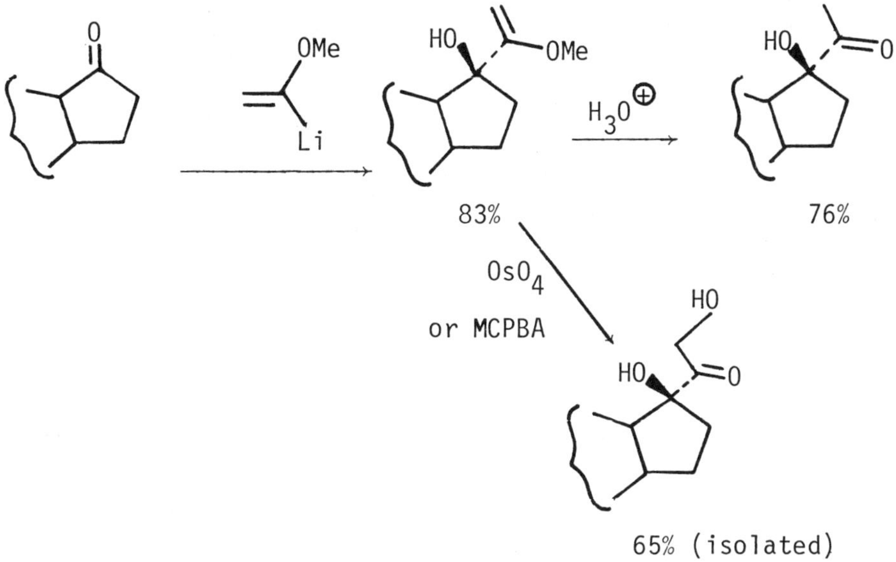

VI.C.1-5 B. M. Trost and D. E. Keeley, J. Am. Chem. Soc., 98, 248 (1976).

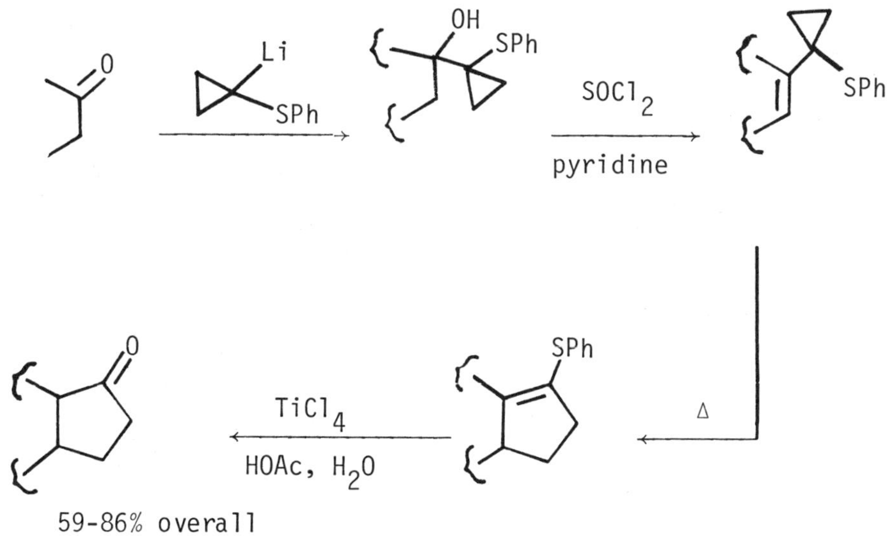

VI.C.1-6 T. Shono, I. Nishiguchi, and M. Nitta, Chem. Lett., 1319 (1976).

An alkylative carbonyl-transposition:

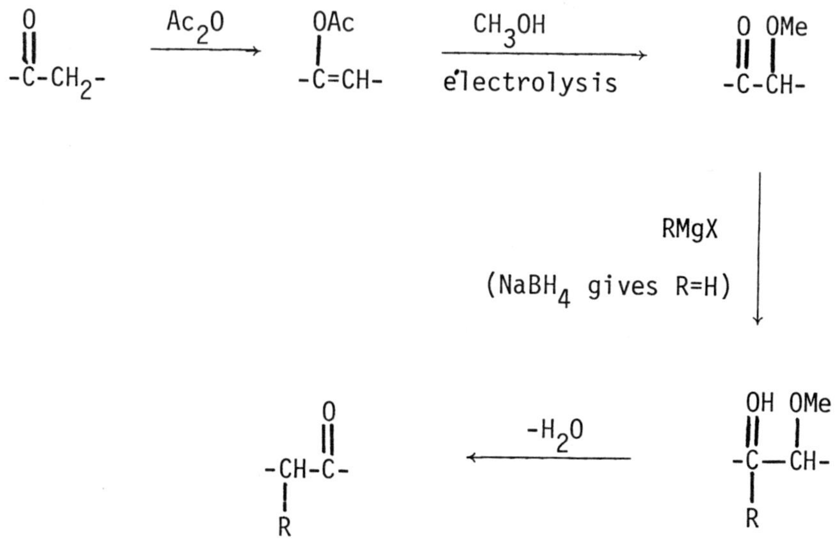

VI.C.1-7 G. Stork and G. A. Kraus, J. Am. Chem. Soc., 98, 2351 (1976).

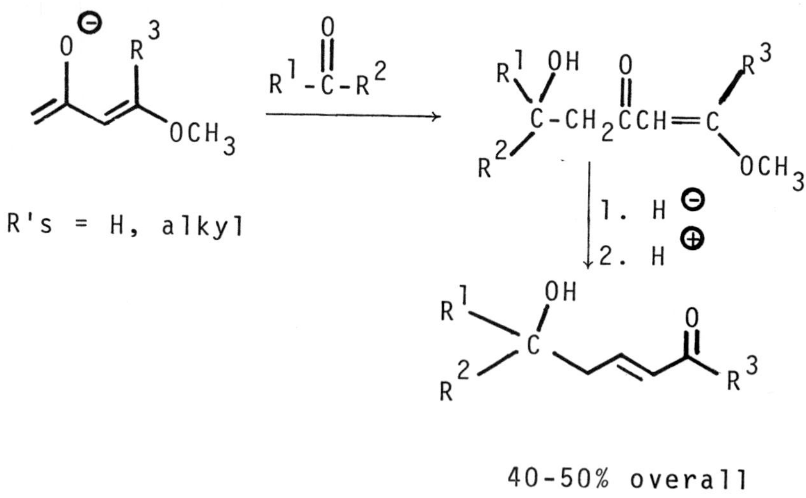

VI.C.1-8 B. M. Trost and K. Hiroi, J. Am. Chem. Soc., 98, 4313 (1976).

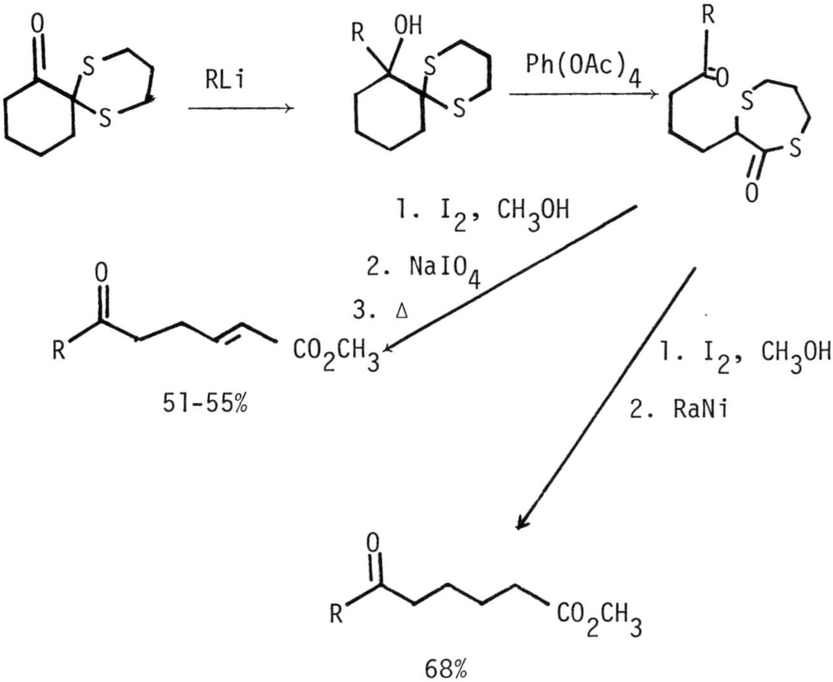

VI.C.1-9 V. Jäger and H. Grund, Angew. Chem. Int. Ed., 15, 50 (1976).

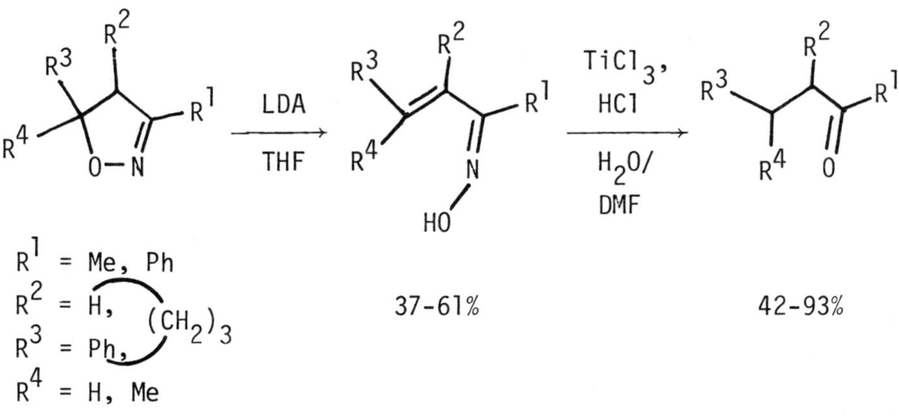

VI.C.1-10 P. Blatcher, J. I. Grayson, and S. Warren, J. Chem. Soc., Chem. Commun., 547 (1976).

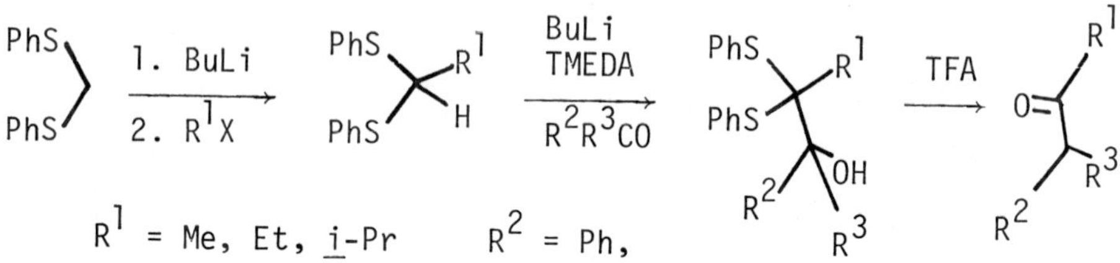

R^1 = Me, Et, i-Pr R^2 = Ph, 1° alkyl

R^3 = H, $-(CH_2)_5-$ Overall yields ~40%

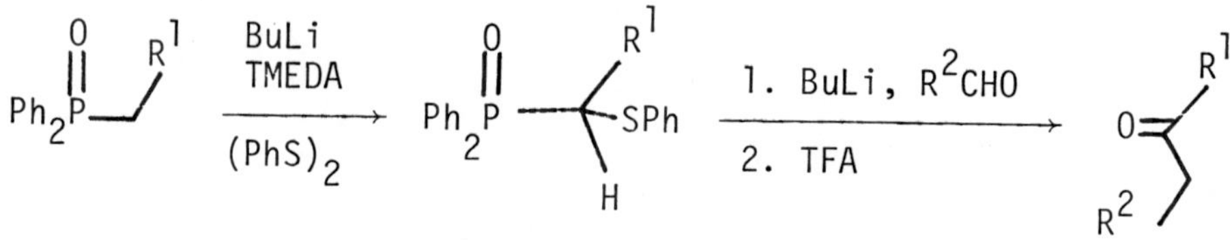

R^1 = Et, Bz, i-Pr R^2 = aryl, hexyl 53-69%

VI.C.2-1 A. Yasuda, H. Yamamoto, and H. Nozaki, Tetrahedron Lett., 2621 (1976).

A 1,3-transposition reaction of allylic alcohols.

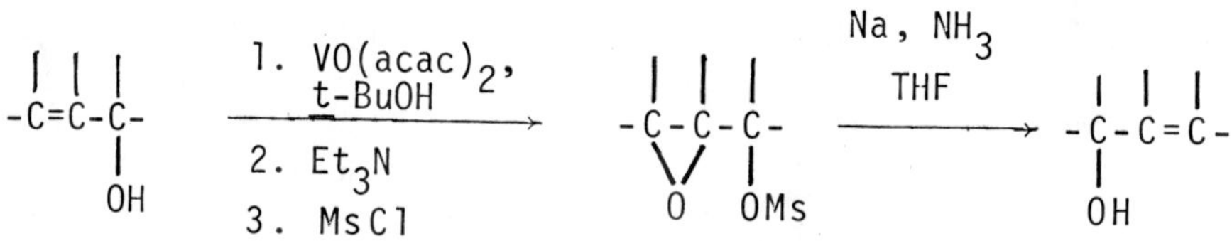

Yields typically 70%.

VI.C.2-2 S. S. Hall, C.-K. Sha, and F. Jordan, J. Org. Chem., 41, 1494 (1976).

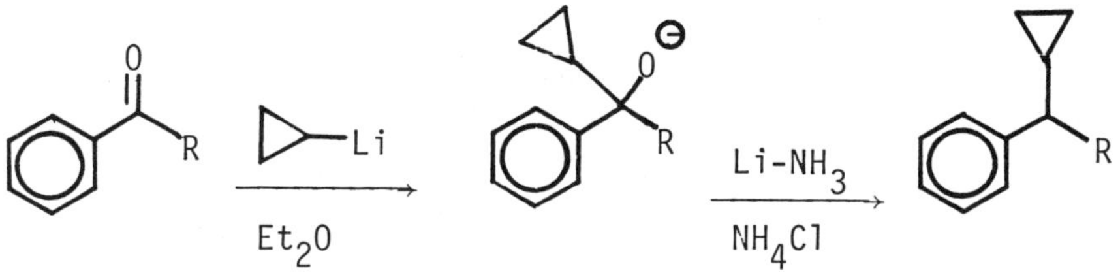

51-98%

VI.C.2-3 B. M. Trost, T. N. Salzmann, and K. Hiroi, J. Am. Chem. Soc., 98, 4887 (1976).

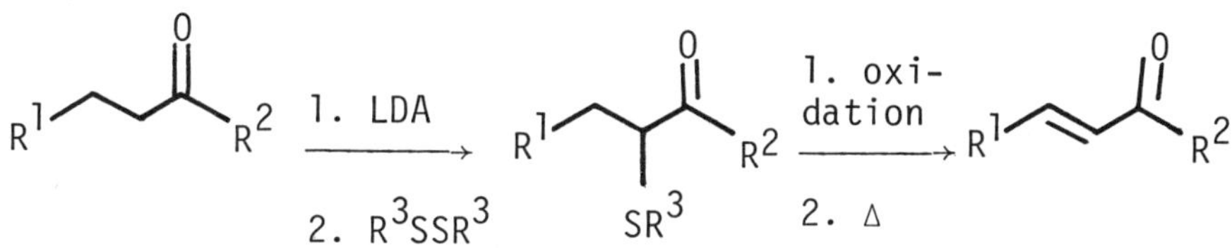

R^2 = alkyl, aryl, alkoxy

R^3 = Me, Ph

overall yields generally 60-80%

VI.C.2-4 L. A. Paquette, W. P. Melega, and J. D. Kramer, Tetrahedron Lett., 4033 (1976).

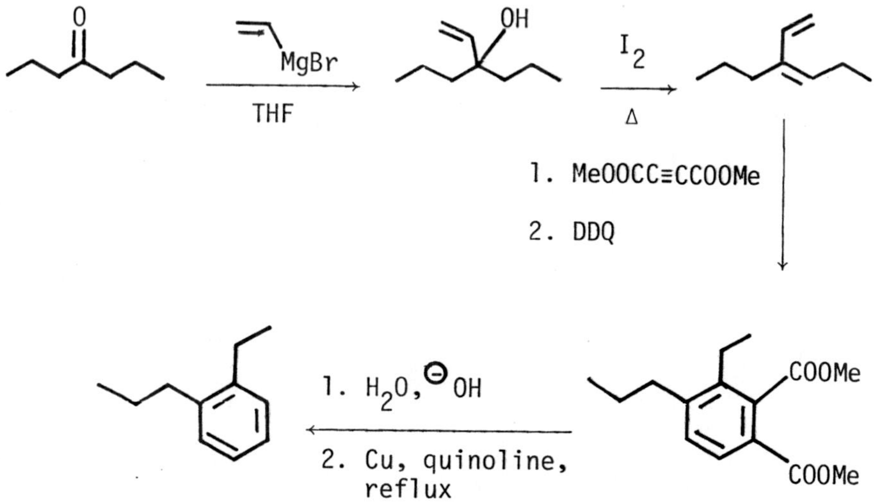

20% overall

VI.C.2-5 J. Rémion, W. Dumont, and A. Krief, Tetrahedron Lett., 1385 (1976).

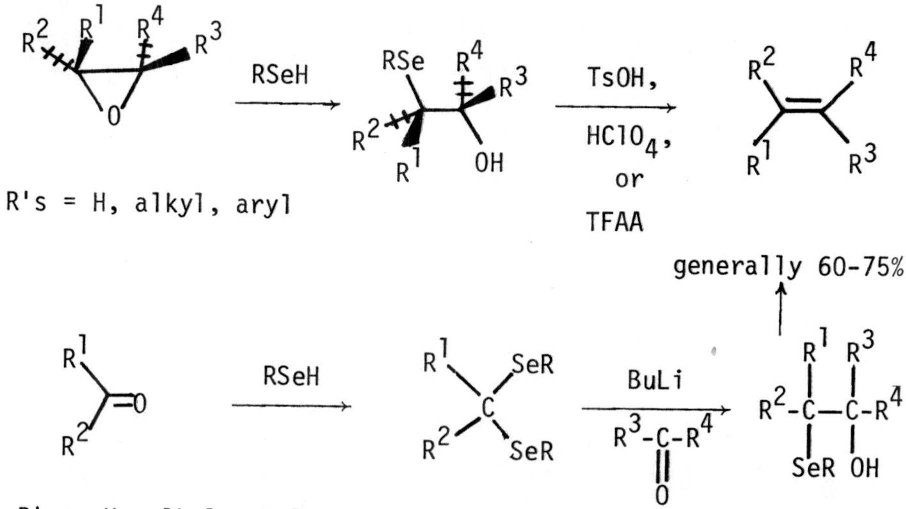

R's = H, alkyl, aryl

VII.1 A. G. Schultz and M. H. Berger, J. Org. Chem., 41, 585 (1976).

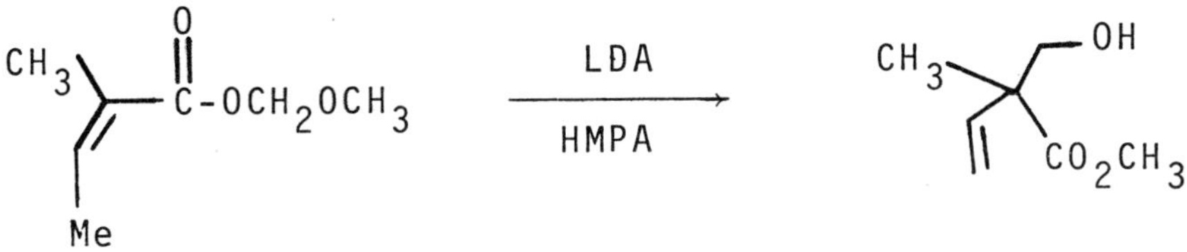

61%

VII.2 W. Oppolzer, T. Sarkar, and K. K. Mahalanabis, Helv. Chim. Acta, 59, 2012 (1976).

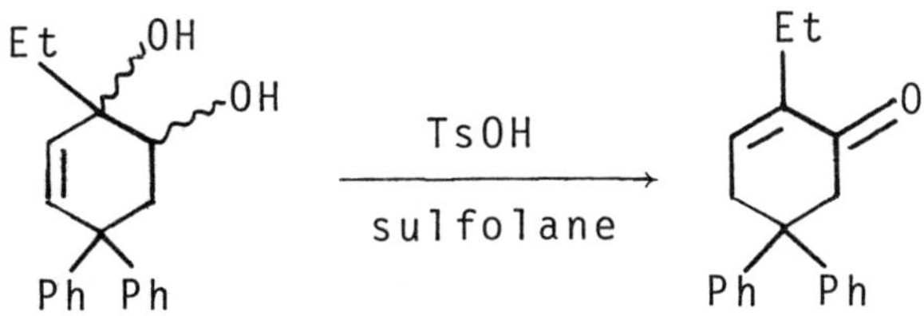

92%

VII.3 C. W. Jefford and A. F. Boschung, Helv. Chim. Acta, 59, 962 (1976).

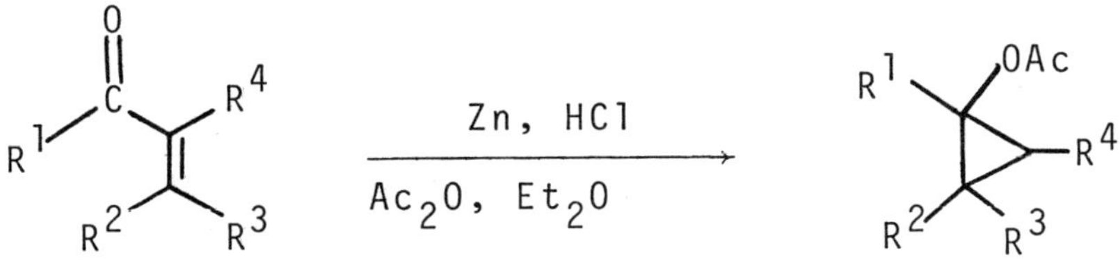

R's = alkyl, H

~40-50%

VII.4 R.L.N. Harris, J. L. Huppatz, and J. N. Phillips, Angew. Chem. Int. Ed., 15, 498 (1976).

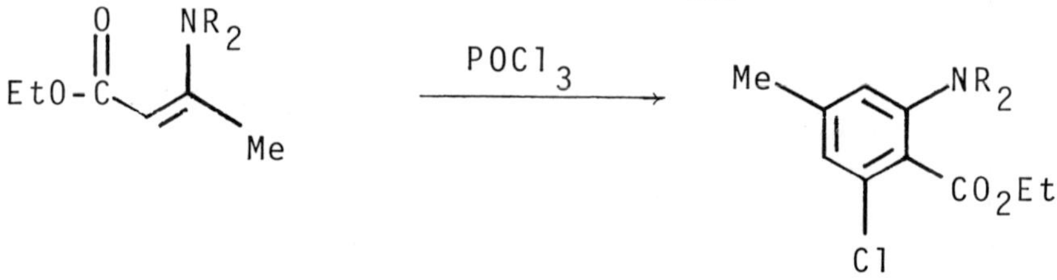

R_2 = Me_2, Et_2, $-(CH_2)_{4,5}$

78-89%

VII.5 N. Dennis, A. R. Katritzky, and Y. Takeuchi, Angew. Chem. Int. Ed., 15, 1 (1976).

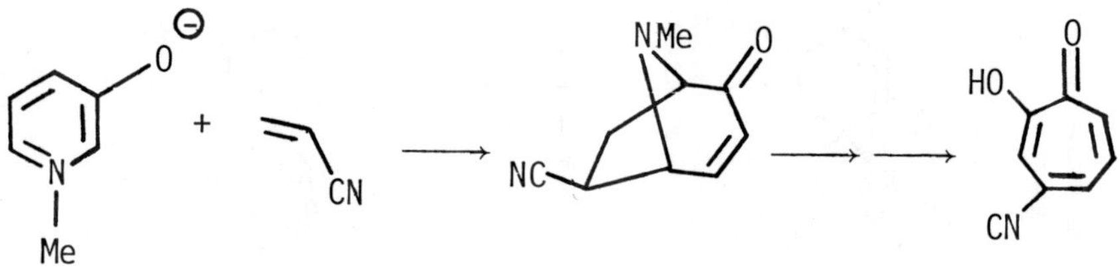

VII.6 B. Ganem, Tetrahedron Lett., 1951 (1976).

Synthesis of hydroxamic acids from oximinoesters:

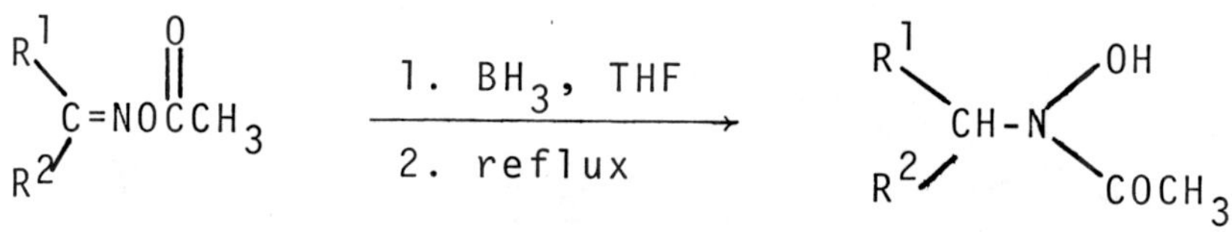

R^1 = H, alkyl

25-60%

R^2 = alkyl, aryl

VII.7 M. W. Gittos *et al.*, *J. Chem. Soc. Perkin I*, 141 (1976).

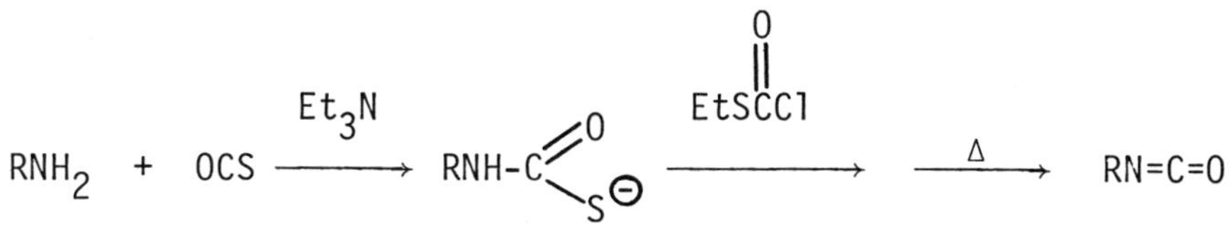

R = alkyl, aryl-substituted alkyl 58-96%

VIII.1 J. B. Hendrickson, Fortachritte der Chem. Forschung, 62, 49 (1976).

Review: "A General Protocol for Systematic Synthesis Design"

VIII.2 M. Bersohn and A. Esack, Chem. Rev., 76, 269 (1976).

Review: "Computers and Organic Synthesis"

VIII.3 T. Kametani and K. Fukumoto, Accounts Chem. Res., 9, 319 (1976).

Review: "Total Synthesis of Natural Products by Retero Mass Spectral Synthesis"

VIII.4 G. W. Gokel and H. D. Durst, Synthesis, 168 (1976).

Review: "Principles and Synthetic Applications in Crown Ether Chemistry"

VIII.5 J. I. Crowley and H. Rapoport, Accounts Chem. Res., 9, 135 (1976).

Review: "Solid Phase Organic Synthesis: Novelty or Fundamental Concept?"

VIII.6 A. Patchornik and M. A. Kraus, *Pure and Appl. Chem.*, **46**, 183 (1976).

Review: "Recent Advances in the Use of Polymers as Chemical Reagents"

VIII.7 B. M. Mikhailov, *Russ. Chem. Rev.*, **45**, 557 (1976).

Review: "Methods of Synthesis and Properties of Allylboranes"

VIII.8 H. C. Brown, *Pure and Appl. Chem.*, **47**, 49 (1976).

Review: "Organoboranes -- The Modern Miracle"

VIII.9 J. Weill-Raynal, *Synthesis*, 633 (1976).

Review: "Formation of Carbon-Carbon Bonds by Using Organoboranes"

VIII.10 G. Ohloff, *Pure and Appl. Chem.*, **43**, 481 (1975).

Review: "Singlet Oxygen: A Reagent in Organic Synthesis"

VIII.11 D. H. Kim, J. Het. Chem., 13, 179 (1976).

Review: "Acetic Anhydride as a Synthetic Reagent"

VIII.12 J. J. Renard and H. I. Bolker, Chem. Rev., 76, 487 (1976).

Review: "The Chemistry of Chlorine Monoxide (Dichlorine Monoxide)"

VIII.13 J. K. Rasmussen and A. Hassner, Chem. Rev., 76, 389 (1976).

Review: "Recent Developments in the Synthetic Uses of Chlorosulfonyl Isocyanate"

VIII.14 W. Kirmse, Angew. Chem. Int. Ed., 15, 251 (1976).

Review: "Nitrogen as a Leaving Group: Aliphatic Diazonium Ions"

VIII.15 T. Mukaiyama, Angew. Chem. Int. Ed., 15, 94 (1976).

Review: "Oxidation-Reduction Condensation"

VIII.16 K. A. Korinek, Chem. and Ind., 931 (1976).

Review: "Electro-Organic Oxidation and Reduction"

VIII.17 F. Minisci, Fortschritte der Chem. Forschung, 62, 1 (1976).

Review: "Recent Aspects of Homolytic Aromatic Substitutions"

VIII.18 J. A. Zoltewicz, Fortschritte der Chem. Forschung, 59, 33 (1975).

Review: "New Directions in Aromatic Nucleophilic Substitution"

VIII.19 R. A. Abramovitch and I. Shinkai, Accounts Chem. Res., 9, 192 (1976).

Review: "Aromatic Substitution via New Rearrangements of Heteroaromatic N-Oxides"

VIII.20 J. d'Angelo, Tetrahedron, 32, 2979 (1976).

Review: "Ketone Enolates: Regiospecific Preparation and Synthetic Uses"

VIII.21 R. E. Gawley, Synthesis, 777 (1976).

Review: "The Robinson Annelation and Related Reactions"

VIII.22 H. Stetter, Angew. Chem. Int. Ed., 15, 639 (1976).

Review: "Catalyzed Additions of Aldehydes to Activated Double Bonds -- A New Synthetic Approach"

VIII.23 W. S. Johnson, Angew. Chem. Int. Ed., 15, 9 (1976).

Review: "Biomimetic Polyene Cyclizations"

VIII.24 A. P. Krapcho, Synthesis, 425 (1976).

Review: "Synthesis of Carbocyclic Spiro Compounds via Rearrangement Routes"

VIII.25 J. F. Liebman and A. Greenberg, Chem. Rev., 76, 311 (1976).

Review: "A Survey of Strained Organic Molecules"

VIII.26 A. Padwa, Angew. Chem. Int. Ed., 15, 123 (1976).

Review: "Intramolecular 1,3-Dipolar Cycloaddition Reactions"

VIII.27 C. W. Spangler, Chem. Rev., 76, 187 (1976).

Review: "Thermal [1,j] Sigmatropic Rearrangements"

VIII.28 E. Campaigne and S. W. Schneller, Synthesis, 705 (1976).

Review: "Cyclization of Ylidenemalononitriles"

VIII.29 T. Wagner-Jauregg, Synthesis, 349 (1976).

Review: "Reactions of Azines and Imines with Dienophiles"

VIII.30 F. J. McQuillin, Chem. and Ind., 941 (1976).

Review: "Homogeneous Catalysis in the Functionalisation of Alkenes and Alkanes"

VIII.31 I. Ernest, Angew. Chem. Int. Ed., 15, 207 (1976).

Review: "A Synthesis of Prostaglandins: Strategy and Reality"

VIII.32 N. Cohen, Accounts Chem. Res., 9, 412 (1976).

Review: "Asymmetric Induction in 19-Norsteroid Total Synthesis"

VIII.33 F. Kienzle, Pure and Appl. Chem., 47, 183 (1976).

Review: "The Technical Syntheses of Carotenoids"

VIII.34 B.C.L. Weedon, Pure and Appl. Chem., 47, 161 (1976).

Review: "Synthesis of Carotenoids and Related Polyenes"

VIII.35 H. Pommer and A. Nürrenbach, Pure and Appl. Chem., 43, 527 (1975).

Review: "Industrial Synthesis of Terpene Compounds"

VIII.36 A. J. Floyd, S. F. Dyke, and S. E. Ward, Chem. Rev., 76, 509 (1976).

Review: "The Synthesis of Phenanthrenes"

VIII.37 N. V. Averina and N. S. Zefirov, Russ. Chem. Rev., 45, 544 (1976).

Review: "Advances in the Synthesis of Heteroadamantanes"

VIII.38 P. Vollhardt, Fortschritte der Chem. Forschung, 59, 113 (1975).

Review: "Cyclobutadienoids"

VIII.39 W.-H. Kunau, Angew. Chem. Int. Ed., 15, 61 (1976).

Review: "Chemistry and Biochemistry of Unsaturated Fatty Acids"

VIII.40 H. G. Aurich and W. Weiss, Fortschritte der Chem. Forschung, 59, 65 (1975).

Review: "Formation and Reactions of Aminyloxides" (nitroxides)

VIII.41 A. P. Kozikowski and H. F. Wetter, Synthesis, 561 (1976).

Review: "Transition Metals in Organic Synthesis"

VIII.42 J. Schwartz and J. A. Labinger, Angew. Chem. Int. Ed., 15, 333 (1976).

Review: "Hydrozirconation: A New Transition Metal Reagent for Organic Synthesis"

VIII.43 R. D. Rieke, Fortschritte der Chem. Forschung, 59, 1 (1975).

Review: "Use of Activated Metals in Organic and Organometallic Synthesis"

VIII.44 P. Beak, Accounts Chem. Res., 9, 230 (1976).

Review: "Silver-Assisted Reactions of Chloroformates: A New Route to Reactive Carbocations"

VIII.45 T. J. Marks, Accounts Chem. Res., 9, 223 (1976).

Review: "Actinide Organometallic Chemistry"

VIII.46 K. C. Bishop III, Chem. Rev., **76**, 461 (1976).

Review: "Transition Metal Catalyzed Rearrangements of Small Ring Organic Molecules"

VIII.47 I. P. Beletskaya, G. A. Artamkina, and O. A. Reutov, Russ. Chem. Rev., **45**, 330 (1976).

Review: "The Interaction of Organometallic Derivatives with Inorganic Halides"

VIII.48 P. J. Smith, Chem. and Ind., 1025 (1976).

Review: "Organic Synthesis via Organotin Intermediates -- Some Recent Developments"

VIII.49 A. McKillop, Pure and Appl. Chem., **43**, 463 (1975).

Review: "Applications of Thallium(III)-nitrate (TTN) to Organic Synthesis"

AUTHOR INDEX

AUTHOR INDEX

A
B 7
C 8
D 9
E 0
F 1
G 2
H 3
I 4
J 5